Nanomaterial-Plant Interactions

Plant Exposure to Engineered Nanoparticles

Uptake, Transformation, Molecular and Physiological Responses

Edited by

Cyren M. Rico
Department of Chemistry and Biochemistry,
Missouri State University,
Springfield, MO, United States

ELSEVIER

ACADEMIC PRESS
An imprint of Elsevier

Academic Press is an imprint of Elsevier
125 London Wall, London EC2Y 5AS, United Kingdom
525 B Street, Suite 1650, San Diego, CA 92101, United States
50 Hampshire Street, 5th Floor, Cambridge, MA 02139, United States
The Boulevard, Langford Lane, Kidlington, Oxford OX5 1GB, United Kingdom

ISBN: 978-0-323-85032-2

Publisher: Nikki Levy
Acquisitions Editor: Nancy J. Maragioglio
Editorial Project Manager: Andrae Akeh
Production Project Manager: Punithavathy Govindaradjane
Cover Designer: Mark Rogers

Typeset by MPS Limited, Chennai, India

Plant Exposure to Engineered Nanoparticles

Uptake, Transformation, Molecular and Physiological Responses

Contents

List of contributors

Oluwasegun Michael Abolade Department of Chemistry and Biochemistry, Missouri State University, Springfield, MO, United States

Adewole T. Adetunji Department of Agriculture, Cape Peninsula University of Technology, Wellington, South Africa

Charles O. Adetunji Applied Microbiology, Biotechnology and Nanotechnology Laboratory, Department of Microbiology, Edo State University Uzairue, Iyamho, Auchi, Nigeria

Hamaad Raza Ahmad Institute of Soil and Environmental Sciences, University of Agriculture, Faisalabad, Punjab, Pakistan

Ojuolape F. Akande Institute of Ecology and Environmental Studies, Obafemi Awolowo University, Ile-Ife, Nigeria

Christian P. Andersen Consulting Plant Physiologist, Corvallis, OR, United States

Muhammad Ashar Ayub Institute of Soil and Environmental Sciences, University of Agriculture, Faisalabad, Punjab, Pakistan; Indian River Research and Education Centre, Institute of Food and Agricultural Sciences, University of Florida, Gainesville, FL, United States; Institute of Agro-Indsutry and Environment, The Islamia University of Bahawalpur, Punjab, Pakistan

Jesus Cantu Department of Chemistry and Biochemistry, The University of Texas at El Paso, El Paso, TX, United States

Keni Cota-Ruiz MSU DOE-Plant Research Laboratory, Michigan State University, East Lansing, MI, United States

Wanqin Dai Key Laboratory for Biological Effects of Nanomaterials and Nanosafety, Institute of High Energy Physics, Chinese Academy of Sciences, Beijing, China; School of Nuclear Science and Technology, University of the Chinese Academy of Sciences, Beijing, China

Chaonan Dong Key Laboratory for Biological Effects of Nanomaterials and Nanosafety, Institute of High Energy Physics, Chinese Academy of Sciences, Beijing, China

Paul O. Fatoba Environmental Biology Unit, Department of Plant Biology, University of Ilorin, Ilorin, Nigeria

Guido Fellet Department of Agrifood, Environmental and Animal Science (DI4A), University of Udine, Udine, Italy

Kenneth Flores Department of Chemistry and Biochemistry, The University of Texas at El Paso, El Paso, TX, United States

Jorge Gardea-Torresdey Department of Chemistry and Biochemistry, The University of Texas at El Paso, El Paso, TX, United States; Environmental Science and Engineering Ph.D. Program, The University of Texas at El Paso, El Paso, TX, United States; UC Center for Environmental Implications of Nanotechnology (UC CEIN), The University of Texas at El Paso, El Paso, TX, United States; NSF-ERC Nanotechnology-Enabled Water Treatment Center (NEWT), Rice University, Houston, TX, United States

Jie Hong College of Environment, Zhejiang University of Technology, Hangzhou, Zhejiang, China

Siying Jia College of Environment, Zhejiang University of Technology, Hangzhou, Zhejiang, China

Naum J. Kirwa Department of Chemistry and Biochemistry, Missouri State University, Springfield, MO, United States

Daniel Lizzi Department of Agrifood, Environmental and Animal Science (DI4A), University of Udine, Udine, Italy

Xingmao Ma Department of Civil and Environmental Engineering, Texas A&M University, College Station, TX, United States

Luca Marchiol Department of Agrifood, Environmental and Animal Science (DI4A), University of Udine, Udine, Italy

Marta Marmiroli Department of Chemistry, Life Sciences and Environmental Sustainability, University of Parma, Parco Area delle Scienze 11/A, Parma, Italy

Polycarp C. Ofoegbu Department of Chemistry and Biochemistry, Missouri State University, Springfield, MO, United States

Clement O. Ogunkunle Environmental Biology Unit, Department of Plant Biology, University of Ilorin, Ilorin, Nigeria; Department of Plant Biology, Osun State University, Osogbo, Nigeria

Luca Pagano Department of Chemistry, Life Sciences and Environmental Sustainability, University of Parma, Parco Area delle Scienze 11/A, Parma, Italy

Laura Pilotto Department of Agrifood, Environmental and Animal Science (DI4A), University of Udine, Udine, Italy

Cyren M. Rico Department of Chemistry and Biochemistry, Missouri State University, Springfield, MO, United States

Lorenzo Rossi Indian River Research and Education Centre, Institute of Food and Agricultural Sciences, University of Florida, Gainesville, FL, United States

Wajid Umar Institute of Environmental Science, Department of Agrochemistry, Hungarian University of Agriculture and Life Sciences, Gödöllő, Hungary

Muhammad Zia ur Rehman Institute of Soil and Environmental Sciences, University of Agriculture, Faisalabad, Punjab, Pakistan

Carolina Valdes Department of Chemistry and Biochemistry, The University of Texas at El Paso, El Paso, TX, United States

Dane C. Wagner Department of Chemistry and Biochemistry, Missouri State University, Springfield, MO, United States

Xiaoxuan Wang Department of Civil and Environmental Engineering, Texas A&M University, College Station, TX, United States

Ejaz Ahmad Waraich Faculty of Agriculture, Department of Agronomy, University of Agriculture, Faisalabad, Punjab, Pakistan

Ye Yuqing Department of Chemistry and Biochemistry, The University of Texas at El Paso, El Paso, TX, United States

Peng Zhang Key Laboratory for Biological Effects of Nanomaterials and Nanosafety, Institute of High Energy Physics, Chinese Academy of Sciences, Beijing, China; School of Geography, Earth and Environmental Science, University of Birmingham, Birmingham, United Kingdom

Zhiyong Zhang Key Laboratory for Biological Effects of Nanomaterials and Nanosafety, Institute of High Energy Physics, Chinese Academy of Sciences, Beijing, China; School of Nuclear Science and Technology, University of the Chinese Academy of Sciences, Beijing, China

Chapter 1

Full life cycle exposure of plants to nanomaterials: impact on productivity

Muhammad Ashar Ayub[1,2,3], Wajid Umar[4], Muhammad Zia ur Rehman[1], Lorenzo Rossi[2], Hamaad Raza Ahmad[1] and Ejaz Ahmad Waraich[5]

[1]*Institute of Soil and Environmental Sciences, University of Agriculture, Faisalabad, Punjab, Pakistan,* [2]*Indian River Research and Education Centre, Institute of Food and Agricultural Sciences, University of Florida, Gainesville, FL, United States,* [3]*Institute of Agro-Indsutry and Environment, The Islamia University of Bahawalpur, Punjab, Pakistan,* [4]*Institute of Environmental Science, Department of Agrochemistry, Hungarian University of Agriculture and Life Sciences, Gödöllő, Hungary,* [5]*Faculty of Agriculture, Department of Agronomy, University of Agriculture, Faisalabad, Punjab, Pakistan*

Chapter Outline

1.1 Introduction

Nanotechnology is not a novel aspect and has been studied widely for the past 4 decades; however, numerous aspects of this technology are being explored lately and used in interdisciplinary sciences. The nanoparticle (NP)

Plant Exposure to Engineered Nanoparticles. DOI: https://doi.org/10.1016/B978-0-323-85032-2.00001-4

is defined as a particle with at least 1 dimension of less than 100 nm (Fig. 1.1).

The NPs can be classified into zero-, one-, or two-dimensional particles depending on their nanoscale chemistry and can differ from their bulk counterparts in mobility, reactivity, and energy absorption (Murthy, 2007). Depending on their chemical nature, size or shape, NPs can be classified into various types as organic and organic carbon NPs, carbon-based inorganic NPs, metal/metal oxide-based, metalloids, and ceramic NPs. A wide range of NP synthesis methods have been reported in the literature depending on raw material availability and use (Rane et al., 2018). Among the widely used methods, bottom-up and top-down methods are prominent to acquire the needed characteristics and properties of NPs in literature (Fig. 1.2). Various kinds of NPs have variable applications and characteristics which are controlled by preparation methods, raw material, and experimental conditions (Sun and Zeng, 2002). The surface characteristics and shapes of NPs are important criteria for selection and application in various

Definitions of Nanoparticles

(A) Nanotechnology is the understanding and control of matter at dimensions between approximately 1 and 100 nanometers, where unique phenomena enable novel applications. Encompassing nanoscale science, engineering and technology, nanotechnology involves imaging, measuring, modeling and manipulating matter at this length scale. A nanometer is one-billionth of a meter. A sheet of paper is about 100,000 nanometer thick; a single gold atom is about a third of a nanometer in diameter. Dimensions between approximately 1 and 100 nm are known as nanoscale. Unusual physical, chemical and biological properties can emerge in materials at nanoscale. These properties may differ in important ways from properties of bulk materials and single atom or molecule.

(B) Engineered nanomaterial: Any material that is deliberately created such that it is composed of discrete functional parts, either internally or at the surface, many of which will have one or more dimensions of the order of 100 nm or less.

(C) SCENIHR (2009) proposed the use of a limit of specific surface area (above 60 m^2/g) to regard a particulate substance 'nanomaterial'.

FIGURE 1.1 Literature available definitions of nanoparticles. Definition is made from information compiled by Kreyling et al. (2010) with permission. (A) The US-National Nanotechnology Initiative Strategic Plan (2007), (B) EC Scientific Committee on Emerging and Newly Identified Health Risks (SCENIHR) (2007), (C) SCENIHR (2009).

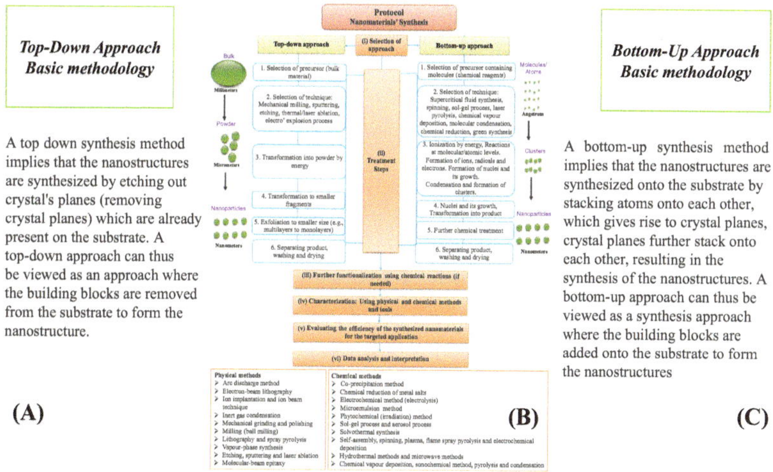

FIGURE 1.2 (A and C) Answer given by *Deon Neveling* on research gate available at https://www.researchgate.net/post/What-is-the-difference-between-top-down-and-bottom-up-methods-for-creating-nano-structures; (B) protocol of nanomaterial synthesis as given by Saleh (2021) used with permission.

studies (Fu et al., 2018), while chemical properties and effect on living things are other important factors of NPs selection and application in various fields (Khan et al., 2019; Ziarani et al., 2019).

Agriculture is under the influence of various biotic and abiotic factors affecting its sustainability (Adger et al., 2009), and plants adopt various escape and tolerance strategies to mitigate stress on cost of energy. The use of modern agricultural technologies is a promising approach to tackle this issue (Clark and Tilman, 2017), and nanotechnology can play a vital role in it (Servin and White, 2016; Servin et al., 2017). Significant research on the use of NPs for the integrated management of agricultural issues has been conducted (Rajput et al., 2018a, 2018b) as well as the potential toxicity of NPs has been widely studied (Du et al., 2007, 2019). Depending on the types of NPs and crops investigated, a wide range of positive results have been reported in various crops (Amingad et al., 2017; Mozafari et al., 2018; Pinedo-Guerrero et al., 2017; Singh et al., 2018). Over the past decade (2010–20), nanotechnology has experienced a linear increase in US$ market value and has increased from US$15.7 billion in 2010 to US$75.8 billion in 2020 making it a fast-growing market and its impact on agroecosystem must be understood (https://www.statista.com/statistics/1073886/global-market-value-nanotechnology/).

From NPs preparation to their chemistry, application, and modification, significant developments have been made in the past few decades making their application in agriculture an interesting subject of study. This chapter deals with the long-term exposure of crop plants to NPs, and the screening and identification of beneficial as well as detrimental effects.

1.2 Classification, synthesis, and characterization of nanoparticles

1.2.1 Classification

The classification of NPs can be made on the basis of their dimensions, morphology, structure, and size of at least one dimension less than 100 nm (Gleiter, 2000). Based on physical shape and dimension, first-class of NPs are zero-dimensional NPs (0D NPs) including spherical NPs, cubes, nanorods, and hollow spheres with all dimensions less than 100 nm. The second class of NPs is one-dimensional NP consisting of NPs with two dimensions of nanoscale while one dimension bigger than nanoscale and includes polymeric/ceramic/metallic NPs, nanotubes, nanorods, nanowires, and fibers. The third class is 2D including multilayered, crystalline or amorphous, thin films and nanoplates consisting of just one dimension less than 100 nm. The last class is 3D NPs with more than two dimensions more than 100 nm in size and are of various types like fibers, carbon nanotubes, pillars, polycrystals, and honeycomb-shaped NPs (Asghari et al., 2016; Aversa et al., 2018; Shiau et al., 2018).

Regarding the morphological nature of NPs, classes are developed based on the flatness and spherical character aspect ratio. According to uniformity in the morphology of NPs, isometric and inhomogeneous or dispersed are major classes of NPs. Nanorods, nano-zigzags/hooks/stars/cubes/helices, and nanoplates are various NPs that can be classified based on morphology (Saleh, 2020). According to the physical, chemical, and magnetic properties of NPs, some other classes are also developed like magnetic NPs, carbonaceous NPs, metallic NPs, branched dendrimers, and nanocomposites. Fig. 1.3 represents the methods of synthesis and classification of NPs. Based on the nature of the material, there are different types of NPs, that is, organic and inorganic NPs, ceramic NPs, and biological NPs. Liposomes, ferritin, dendrimers, and others come under the category of organic NPs. These are nontoxic and ideal for drug delivery. The term polymeric NPs is also used for organic NPs (Ijaz et al., 2020). The NPs comprising of inorganic molecules are called inorganic NPs and include metal, metal oxide, metalloid and composite NPs. Their fate in environment, effect on plants and human largely depends on their chemical nature, chemistry, and surface properties. Ceramic NPs are nonmetallic inorganic NPs prepared with successive cooling and heating and can be amorphous, hollow, and dense in nature (Sigmund et al., 2006). The NPs of biological origin are also in range of 1–100 nm and can be viruses and lipoproteins (Ijaz et al., 2020).

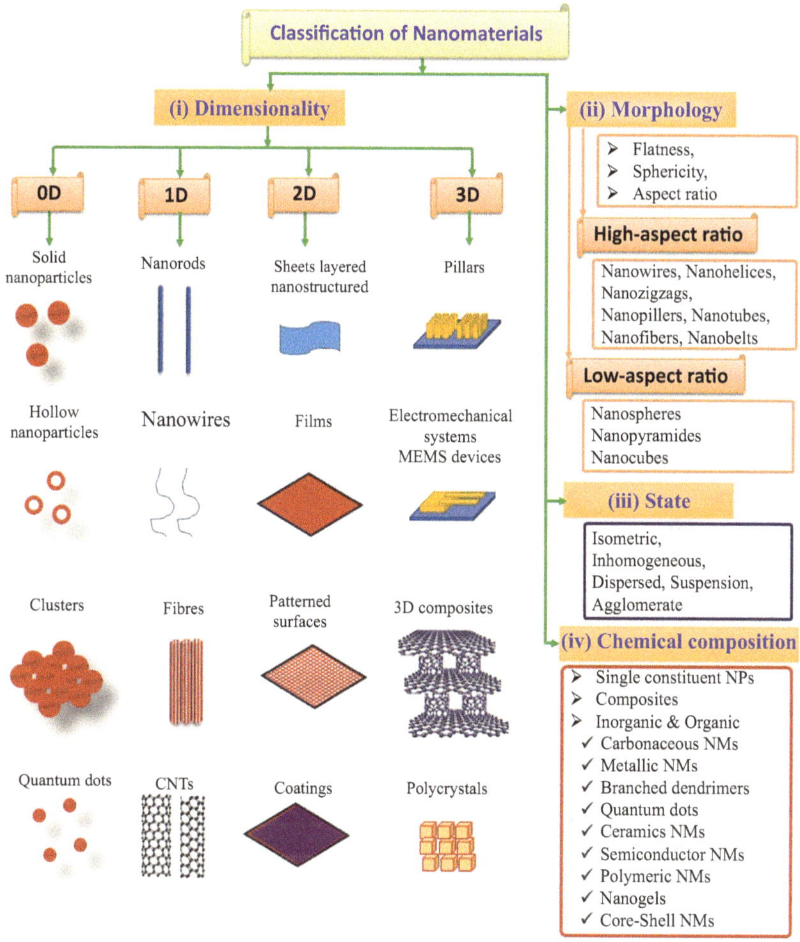

FIGURE 1.3 Classification of nanoparticles based on morphology and dimensionality acquired from Saleh (2020) with permission and license from Elsevier.

1.2.2 Synthesis

NPs can be synthesized by the following two main approaches: (1) Bottom-up approach (2) Top-down approach. These approaches include different methods such as chemical, biological, and physical. The bottom-up approach is also known as the chemical reduction of the materials. It is a simple, easy, and efficient approach to synthesize NPs of desired properties. For Top-down approach, the bulk materials are divided or reduced in size by physical forces. A brief description of the physical, chemical, and biological methods is given in Tables 1.1–1.3.

TABLE 1.1 Physical approach and methods for nanoparticles preparation.

Method	Description	Example	References
Thermal decomposition	• It is an endothermic process • Heat is responsible for the decomposition • Metal decomposition at specific temperatures resulted in the synthesis of NPs	Gd_2O_3-NPs, Fe_2O_3-NPs, MnO_2-NPs	Ahab et al. (2016), Unni et al. (2017), de la Torre and Bennewitz (2020)
Mechanical/ ball milling	• Simple and cheap method • Nanoparticles are produced by a reduction mechanism in this process • The kinetic energy of the grinder is transferred to grinding material • It is an industrial process to synthesize different metal alloys	ZnO-NPs, CuO-NPs, TiO_2-NPs, Ag-NPs	Salah et al. (2011), Yang and Chen (2017), Yadav et al. (2015), Khayati and Janghorban (2013)
Laser ablation	• Different NPs can be synthesized in solution from different solvents • Irradiation of submerged metals by laser ablation in solution leads to the condensation of plasma for the synthesis of NPs • It is an unconventional method and does not require any stabilizing agent	ZnO-NPs, CuO-NPs, TiO_2-NPs, Au-NPs; Ag-NPs	Al-Dahash et al. (2018), Abdulateef et al. (2016), Singh et al. (2016), Khumaeni et al. (2017), Valverde-Alva et al. (2015)
Sputtering	• A layer of NPs is deposited with the help of an ejection particle • Annealing is very important in this process • The deposition and properties of NPs depend on temperature, substrate, and time of annealing	ZnO-NPs, CuO-NPs, TiO_2-NPs, Au-NPs; Fe-NPs	Das et al. (2016), Jaiswal et al. (2015), Rashid et al. (2015), Simionescu et al. (2019), Hu et al. (2013), Xing et al. (2016)

TABLE 1.2 Chemical approach and methods for nanoparticles preparation.

Method	Description	Example	References
Chemical vapor deposition	• A thin film of reactant gas is deposited on the substrate material • Reaction occurred between the reactant gas and substrate surface • The NPs in the form of a deposited layer on the substrate surface are separated • Nanoparticles produced via this method have complex surface chemistry • Can release toxic gas as a by-product	TiO_2-NPs, Si-NPs; graphene NPs	Lee et al. (2011), Lee et al. (2016), Wei et al. (2009)
Sol–gel method	• A simple and efficient method that involves condensation and hydrolysis processes. A metal precursor and precipitating agent are used • Different techniques such as centrifugation and filtration are employed to separate the solid phase from the solution	TiO_2-NPs, ZnO-NPs, SiO_2-NPs, alumina NPs	Kumar and Rao (2014), Khan et al. (2016), Mohamad et al. (2019), Dubey et al. (2015)
Spinning	• Spinning disk reactors are used in this method • There is an option to control the physical parameters of the spinning disk • The reaction chamber is filled with nitrogen gas to avoid any chemical reactions • All the required reaction materials are pumped inside the chamber • The properties of NPs in this method depend on the speed of the spinning disk, temperature, surface area of the disk, and the ratio of precursor/liquid	TiO_2-NPs, ZnO-NPs, Magnetic NPs, Starch NPs	Smith et al. (2006), Mohammadi et al. (2014), Stoller and Ochando-Pulido (2020), Sana et al. (2019)
Pyrolysis	• The process is carried out in a high-temperature furnace wherein burning of the precursor is carried out using flame, laser, or plasma • The precursor is used in the form of vapor or liquid • After evaporation, the solid NPs are recovered from the furnace • It is the most commonly used method to synthesize NPs at an industrial scale	ZnO-NPs, Si@C NPs, Ceramic NPs	D'Amato et al. (2013), Sourice et al. (2015), Ghaffarian et al. (2011)

TABLE 1.3 Biological approach and methods for nanoparticles preparation.

Method	Description	Example	References
Plant extract	• Various plant extracts can reduce the metal particle size and can be used for the synthesis of NPs • Different biological molecules such as carbohydrates, proteins, and coenzymes can reduce the metal particle size • Plant parts such as roots, leaves, fruits, and stems have been used for the synthesis of NPs	ZnO-NPs, Fe-NPs, Au-NPs, TiO_2-NPs, Cu-NPs, Ag-NPs	Umar et al. (2019), Rao et al. (2013), Raghunandan et al. (2010), Bussamara et al. (2013), Sharma et al. (2018), Prathna et al. (2011)
Bacteria	• Bacteria are used for the synthesis of NPs • Enzymes released by bacteria can efficiently reduce the metal particle size • Different bacterial species have been used for this purpose	Fe_2O_3-NPs, Ag-NPs, TiO_2-NPs, Au-NPs	Philipse and Maas (2002), Sunkar and Nachiyar (2012), Wadhwani et al. (2016), Du et al. (2007)
Fungi	• Fungi act as a biological agent for the synthesis of NPs because of the presence of intracellular enzymes • It is a more efficient method compared to bacteria because fungi can produce higher amounts of NPs • Nanoparticles of well-defined morphology have been synthesized	Au-NPs, Ag-NPs, TiO_2-NPs, ZnO-NPs, Pt-NPs	Chauhan et al. (2011), Xue et al. (2016), Raliya et al. (2015a,2015b), Mashrai et al. (2017), Govender et al. (2009)
Yeast	• Yeasts are unicellular microorganisms, and several species of yeast have been used for the synthesis of NPs	Au-NPs, Ag-NPs, Zr-NPs	Tian et al. (2010), Mourato et al. (2011)

1.2.3 Characterization

NPs occupy various application domains that are relevant to our lives. According to the type of application, it is essential to characterize NPs for their chemical, physical, and other properties. Different instruments that are used to obtain these properties are listed in Table 1.4. The significant physical and chemical properties of NPs include their composition, size, and shape, and surface area as well as distribution, stability, surface chemistry, roughness, and topography. The following are the major techniques used for characterization of NPs (Fig. 1.4).

XRD: X-ray diffraction technique is used for the qualitative and quantitative determination of the atomic structure of materials. It can also determine the size of crystal NPs and the crystal structure of NPs. It is a nondestructive method. Materials of different degrees of crystallinity (e.g., noncrystalline, amorphous, and crystalline) can be analyzed by this technique.

SEM-EDS/X: Energy dispersive spectroscopy is the technique used to determine the elemental composition of materials. It is used in combination with the SEM instrument. The surface of the material is targeted by electrons of high energy to excite the surface atoms and then each atom of the material develops a specific spectrum which is then detected by a detector.

SEM: In this instrument, electron is used instead of light to form an image of the object. The beam of electrons scans the surface of the objects. It is used to determine the size, morphology, distribution, and shape of the NPs.

TEM: Transmission electron microscopy is used for the detailed analysis of the size, shape, and morphology of NPs, which is not possible with SEM. In this technique, the beam of electrons passed through the object that can develop images of much higher resolution as compared to SEM.

FTIR: Fourier transform infrared spectroscopy is used to analyze the bonding (inorganic or organic) in the particles at the molecular level. Wavelengths from 700 nm to 1 mm are used to develop spectra. Based on the bonding nature in the material, a portion of wavelengths is reflected while some portion is transmitted.

DLS: Dynamic light scattering is the technique used to determine the size distribution of small particles in solution. It can also be used to analyze complex fluids. Particles of different sizes in the solution diffract the directed light and the instrument generates a specific image.

SP-ICP-MS: Single particle-ICP-MS is the advanced form of ICP-MS. In this technique the size and mass of every single particle of the material are determined directly.

TABLE 1.4 General techniques used in the characterization of nanoparticles.

Nanoparticle	Characterization technique	References
ZnO-NPs	• XRD, SEM-EDS, SEM, FTIR, TEM, XPS	Geetha et al. (2016)
CuO-NPs	• SEM, TEM, XRD, DLS, EDS	Mali et al. (2019), Wongpisutpaisan et al. (2011)
Fe-NPs	• SEM, TEM, XRD, HRTEM	Arenas-Alatorre et al. (2019), Prodan et al. (2013)
Ag-NPs	• UV-VIS, XRD, EDS, SEM, TEM, FTIR, FESEM	Rautela et al. (2019), Hamouda et al. (2019)
Au-NPs	• SEM, TEM, FTIR, DLS, EDX, UV-VIS	Dzimitrowicz et al. (2019), Correa et al. (2016)
CeO_2-NPs	• SEM, XRD, TEM/HRTEM, UV-VIS	Ravishankar et al. (2015), Kumar et al. (2013)
TiO_2-NPs	• SP-ICP-MS, XRD, FTIR, FESEM, XRPD	Wojcieszek et al. (2020), Bessi et al. (2017), Lusvardi et al. (2017)

SEM, Scanning electron microscope; *HRTEM*, high-resolution-transmission electron microscopy; *TEM*, transmission electron microscope; *XRD*, X-ray diffraction; *FTIR*, Fourier Transform Infrared; *DLS*, dynamic light scattering; *EDX*, energy dispersive X-ray analysis; *XPS*, X-ray photoelectron spectroscopy; *SP-ICP-MS*, single particle inductively coupled mass spectroscopy.

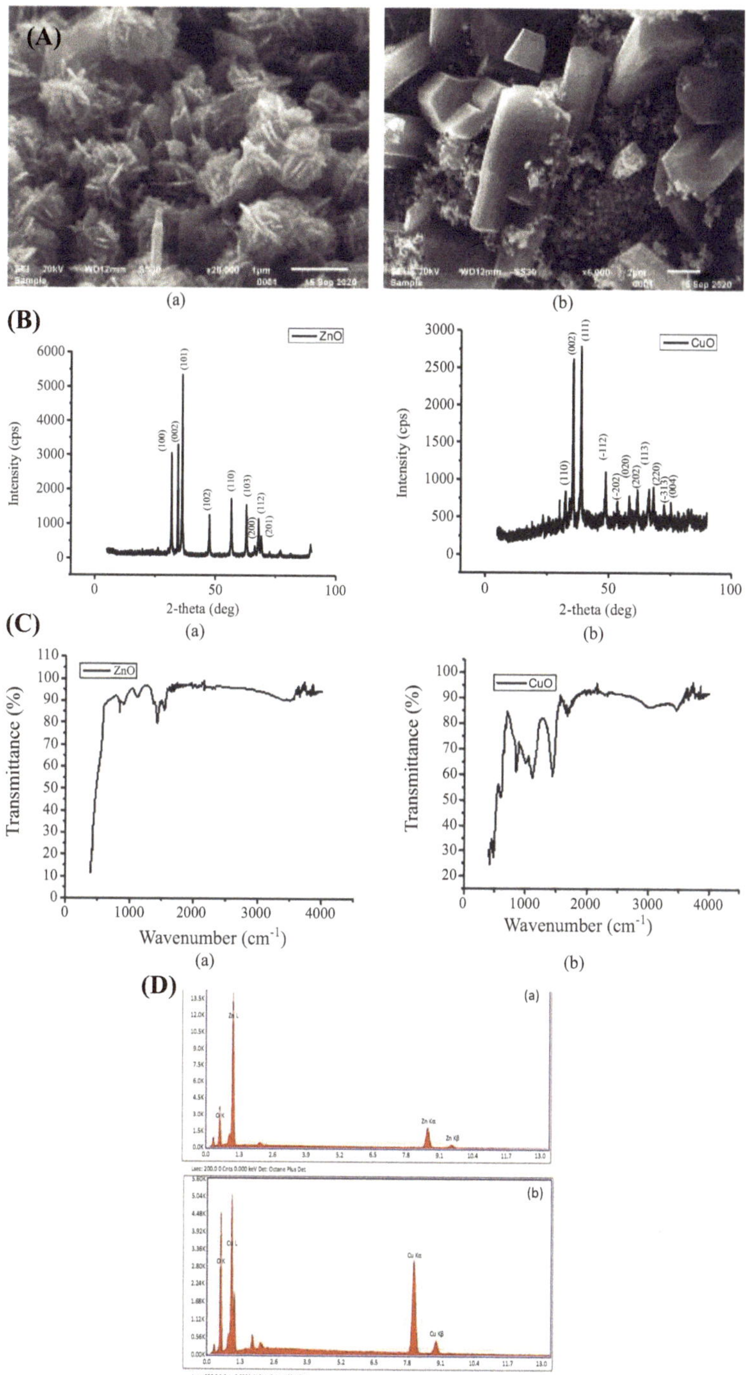

FIGURE 1.4 Characterization of ZnO and CuO-NPs prepared with green synthesis (*Ficus benghalensis* leaf extract) reported by Yadav et al. (2022). Acquired with permission to reuse from Elsevier. (A) SEM images of ZnO-NPs (a), CuO-NPs (b); (B) X-ray Diffraction ZnO-NPs (a), CuO-NPs (b); (C) FT-IR spectra ZnO-NPs (a), CuO-NPs (b); (D) Energy dispersive spectroscopy (EDS) ZnO-NPs (a), CuO-NPs (b).

1.3 Exposure of nanoparticles to plants

1.3.1 Modes of exposure of nanoparticles to plants

Exposure of NPs to plants is inevitable as plants provide multiple entry points and large surfaces for NPs entrance (Dietz and Herth, 2011). On the release of NPs to terrestrial environment, 28% of total NPs end up in the soil and terrestrial plant species absorb a significant amount of these NPs (Keller et al., 2013). The main entry points of NPs in plants are roots, leaf pores, and other aboveground parts (Rajput et al., 2018a,b). Once NPs enter the root, the Casparian strips can sometimes act as the final trafficking point for NPs translocation which can hinder some NPs like CuO (Peng et al., 2015). The NPs making their entry into plant tissue via air dispersal can enter through stomata and make their way into plant intercellular space (Pullagurala et al., 2018). The metal NPs have different fate in plants as different metal ions have variable chemistry in different plants. The metal-tolerant plants can stop the uptake of metallic NPs into their upper plant tissues via restricting them in their root endodermis or storing them in their root cortex cells while metal hyperaccumulators can accumulate more in upper plant tissues (Manceau et al., 2008). Though net accumulation of these NPs in plant tissues can be monitored, exact mechanisms of NPs translocation, accumulation and toxicity in plants are not well understood. Some researchers have reported that NPs can cause several morphological, physiological, genetic, or epigenetic changes in plants disturbing their normal nutritional homeostasis.

The tiny size of NPs facilitates the entrance, translocation, and penetration into plant tissue via absorption from roots and translocation through the vascular system to aboveground parts (Rico et al., 2011). Some NPs adhere to plant roots and are reluctant to absorption but most of them either get taken up from root or aboveground tissues (Verma et al., 2019). Literature suggests that plants can take up NPs in their original nanoparticulate form and translocate them into plant tissues where they get assimilated into cell walls, cellular organelles, vacuoles, lipid envelops, and others (Ahmed et al., 2018a,b; Rajput et al., 2018a,b; Verma et al., 2018). In other cases, dissolution of NPs can also occur in soil releasing respective metal/nonmetal ions in soil and plant can uptake these ions or NPs up taken by plants get ionized inside plant cell/tissue (Ahmed et al., 2018a,b; Keller et al., 2017).

1.3.2 Mechanisms of nanoparticles uptake, translocation, and intracellular assimilation

Literature has proven uptake of NPs in plants via active transport mechanism (Wang et al., 2012) wherein NPs first adhere to plant roots and are then transported into root via apoplastic pathway (Anjum et al., 2015). On cell exposure, NPs can pierce the cell wall and membrane of the epidermal cell and translocate to stele via a symplastic pathway from where they can make their way into the different parts of the plants (Tripathi et al., 2016). A passive pathway (endodermal apoplast) is also adopted (Judy et al., 2016). The xylem is the main channel for the aboveground translocation of NPs in plants (Aslani et al., 2014; Held et al., 2015). The transpiration pull in the xylem, responsible for water uptake is the main driving force of NPs uplifting as reported in various crop plants (Aslani et al., 2014; Thuesombat et al., 2014). To cope with the upsurge of NPs, plant cells also adopt endocytic pathways (Tonigold and Mailänder, 2016) which can be either dependent and independent of clathrin (Miralles et al., 2012). In the clathrin-dependent pathway, sheathed or folded cell membrane is developed which transforms into clathrin-coated vesicles (Ng et al., 2015). In the clathrin-independent pathway, the three major processes of phagocytosis, caveolae, and fluid-phase endocytosis are involved (Monti et al., 2015), and these processes helped in the assimilation of Au NPs in *Nicotiana tabacum* (Onelli et al., 2008). Si NPs also get internalized via endocytosis pathways (Walker et al., 2016). Other major classes of engineered NPs such as multiwalled carbon nanotubes get assimilated in plants via endosomal pathways as cell wall is semipermeable to some materials (Majzoub et al., 2016; Pauly and Keegstra, 2016). Another mode of entry of engineered NPs into intercellular space is via parenchymatous spaces aiding in the diffusion of materials to cotyledons (Ma et al., 2015) while in the seed coat, aquaporins serve this purpose (Yulianti et al., 2016). In top-down mode of transfer, NPs enter via leaves by two pathways: penetration through stomatal and entrapment inside cracks of the cuticle. Entrapment is followed by translocation and redistribution of these NPs in stem and root cells via phloem channels (Hong et al., 2014; Raliya et al., 2015a,b).

1.3.3 Methods of intra/intertissue detection

Once NPs are taken up by plants, its accumulation in various plant tissues should be detected to check their intertissue fate. If metallic NPs are of concern and their respective metal ion/ contents need to be determined, then destructive sampling is performed and samples are run on atomic absorption spectroscopy or inductively coupled plasma mass spectrometry. A combination of various techniques (i.e., spectroscopy, diffraction, and imaging) can be more effective in the detection and measurement of intact NPs. For imaging, scanning electron microscopy (SEM) can be used for the visual localization of NPs with higher density. The SEM is often equipped with an energy disruptive X-ray analyzer, which helps in the quantitative determination of metals/elements present in plant samples and in the understanding of the chemistry of NPs. If NPs absorbed in plants are very small (out of the resolution range of SEM), then high-resolution-transmission electron microscopy (HRTEM) is used along with selective area electron diffraction for NPs phase contribution and distribution quantification and imaging. There are some microscopes equipped with an electron energy loss spectra camera which can detect oxidation and transition state (in 3D) of an element in a nanoscale. The NPs detection methods (based on imaging) can sometimes get limited results, and special preparation of samples should be done for effective NPs detection.

1.4 Effects of nanoparticles on plant growth and productivity

As explained earlier, NPs can be of various types, and their effects on plants largely depend on chemical nature, particle size, and surface properties of NPs. The widespread application of NPs in agroecosystem and their fate in soil-plant continuum is an interesting and well-studied phenomenon (Anjum et al., 2016). The uptake and accumulation of engineered NPs can also make their way into human food chain very easily and various kinds of concerns are associated with it (Deng et al., 2014). Owing to their various point and nonpoint sources, characterization of NPs in plants and their input into human food chain is very complex to monitor (Rajput et al., 2018a,b). Basic positive and negative effects of NPs on plants are given in Fig. 1.5.

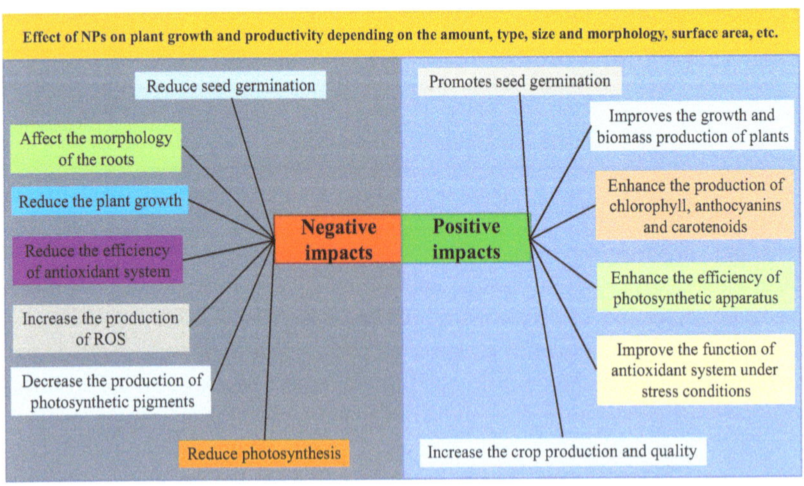

FIGURE 1.5 General positive and negative effects of nanoparticles on plants.

1.4.1 Macronutrient and beneficial-element nanoparticles

Macronutrient nano-fertilizers are the nano-based materials of primary plant nutrients (nitrogen, phosphorus, and potassium) used as fertilizer for plants to increase nutrient use efficiency, growth, and productivity. Conventional fertilizers have low uptake efficiencies and require larger application amounts (Raliya et al., 2017). Owing to depleting supply of macro plant nutrients and huge losses during plant cultivation, nano-sized formulation of fertilizers is a promising approach being explored worldwide. Nano-fertilizers of macronutrients have the potential to substitute traditionally used commercial fertilizers due to long persistence, easy and eco-friendly application, and higher efficiencies (depending on chemistry and surface properties). A wide range of N, P, K, Ca, S, and Mg nano-fertilizers have been prepared and tested and recommended for potential utilization on a commercial level. Table 1.5 shows some relevant studies regarding the preparation and application of nano-formulations of macronutrients.

Similarly, some elements, though not necessary for plant growth, have some reported beneficial effects on the plant such as titanium and silicon. Several works have been conducted on their potential benefits and toxicity in crop plants. Few relevant studies have been summarized in Table 1.5.

1.4.2 Micronutrient nano-fertilizers

1.4.2.1 Zinc oxide nanoparticles

Among these, the share of ZnO-NPs is significant, and its global production ranges between 440–33,400 tons, making it the third most-produced and used metal NP (Bondarenko et al., 2013; Connolly et al., 2016; Peng et al., 2017). In terms of plant absorption, the NPs are 15–20 times more likely to be absorbed than their counterpart bulk particles (Srivastav et al., 2016). ZnO-NPs are expected to be present at around 3.1–31 $\mu g\ kg^{-1}$ in soil and 76–760 $\mu g\ L^{-1}$ in water (Boxall et al., 2007; Ghosh et al., 2016). Widespread use of ZnO-NPs in the field of photooxidation and photocatalyst has a huge impact on its concentration in the environment (Sirelkhatim et al., 2015). Soil is a heterogeneous material and ZnO-NPs contents in the soil can be available to plant depending on soil properties. Soil physicochemical and biological properties control the net fate of ZnO-NPs in soil. Due to its small size and high surface area, the ZnO-NPs can adhere to soil colloids and usually have less mobility at various ionic strengths (Zhao et al., 2012). The sorption of ZnO-NPs in plants is usually high and it is influenced by soil pH as ZnO-NPs toxicity was reported to be more in acidic soils rather than alkaline or neutral soils (Shen et al., 2015). A work conducted by Miglietta et al. (2015) has investigated the potential toxicity of ZnO-NPs, ZnO bulk, and ionic Zn to *Lepidium sativum* and applied ZnO-NPs via dry mixing in soil. Results show that dry mixing of ZnO-NPs showed significant

TABLE 1.5 Effect of macronutrient nanoparticles and beneficial-element nanoparticles on plants.

Plant name, scientific name, and exposure period	Nanoparticle dose, characteristics, and crop growing conditions	Effect on plants	References
Macronutrients nanoparticles (potential beneficial effects and toxicity)			
Maize (Zea mays)	Nano-zeolite-based urea (nanozeourea)	• Results from greenhouse investigation have shown that application of prepared nano-fertilizer has resulted in a 28% increase of N contents in grains of plants	Manikandan and Subramanian (2016)
Italian ryegrass (*Lolium multiflorum*)	Clinoptilolite-NH_4	• Another greenhouse investigation using prepared nano-N fertilizer has suggested that a net increase in N use efficiency can be observed with prepared NPs application	Zareabyaneh and Bayatvarkeshi (2015)
Sweet almond (*Prunus dulcis*)	Nano-urea modified with hydroxyapatite	• A field investigation of nano-urea modified with HA has proved overall great efficacy of nano-fertilizer to provide good quality crop due to slow and longer-lasting release of nutrients	Badran and Savin (2018)
Potato (*Solanum tuberosum*)	NPK nano-fertilizers	• High economical yield, starch contents, and NPK use efficiency was observed in field investigation proving the efficacy of prepared nano-fertilizer	Abd El-Azeim et al. (2020)

Pearl Millet (*Pennisetum americanum* L.) cv. HHB 67	Bio-inspired Nitrogen fertilizer prepared from ammonium nitrate	• The optimum foliar dose was found to be of 80 ppm concentration and has shown a significant effect on crop growth	Thomas et al. (2016)
Soil N release kinetic study	Urea-modified hydroxyapatite nanoparticles	• Prepared nano-fertilizer was found to be very effective in the slow release of urea/N in soil for a long time	Kottegoda et al. (2011)
	Urea-hydroxyapatite nanohybrids	• Prepared hybrid nanoparticles were found to retain 40% N and have significant potential of slow N release and potentially can be used as an effective N fertilizer	Kottegoda et al. (2017)
Ladies' finger (*Abelmoschus esculentus*)	Hybrid nano fertilizer (urea-modified hydroxyapatite)	• With reduced dose of prepared nano-fertilizer more beneficial effects was evident suggesting its suitable applicability for use as a N and P fertilizer. Moreover, the physical characteristics of prepared nanoparticles were very ideal to be used as a slow-release fertilizer	Tarafder et al. (2020)
	Urea-silica nanohybrids for slow and effective release of Nitrogen	• Prepared nanoparticles have 36% urea w/w and urea holding capacity of 86% making them a suitable slow-release alternative N fertilizer	de Silva et al. (2020)
Soybean (glycine max), sorghum (*Sorghum bicolor*), pea (*Pisum sativum*), and bok choy-Chinese cabbage (Brassica rapa subsp. chinensis)	Hydroxyapatite nanoparticles	• Concentration-dependent positive and negative effects were observed. Positive effects were due to N and P fertilizer supply while the negative effect was reported due to excess Ca availability	Maghsoodi et al. (2020)

(*Continued*)

TABLE 1.5 (Continued)

Plant name, scientific name, and exposure period	Nanoparticle dose, characteristics, and crop growing conditions	Effect on plants	References
Tomato (*Solanum lycopersicum L.*)	Hydroxyapatite nanoparticles	• Nano-HA prepared in this study has shown no toxicity to model plants and has shown potential to be used as and P provider and carrier	Marchiol et al. (2019)
Mung bean (*Vigna radiata*) seedlings	Hydroxyapatite nanoparticles (biotoxicity evaluation)	• Prepared nanoparticles were tested for potential toxicity at 5 mg mL^{-1} culture dose. Toxic amount of Ca can be responsible for this effect	Jiang et al. (2014)
Bok choy-Chinese cabbage (*Brassica rapa subsp. chinensis*)	Nano-hydroxyapatite for effective Cd toxicity amelioration	• Prepared nanoparticles were applied to reverse the adverse effect of Cd toxicity via nutrition provision and specific Cd adsorption. Results have shown that n-HA is significantly beneficial in this regard	Li and Huang (2014)
Rice (*Oryza sativa*)	Calcium phosphate nanoparticles	• Calcium is an important plant nutrient and its application via nano form has shown a net positive effect on crop growth and development along with improved physiological attributes	Upadhyaya et al. (2017)
Maize (*Zea mays*)	Calcium phosphate synergistic effect with fungi	• Calcium nanoparticles were found to be beneficial with further increment observed when applied synergistically with fungi	Rane et al. (2015)

Cucumber (*Cucumis sativus*)	Urea functionalized calcium phosphate nanoparticles	• In vivo studies on cucumber show that prepared NPs were as effective as conventional N fertilizer in terms of plant growth and development suggesting it as a more appropriate form of N and Ca/P fertilizer	Carmona et al. (2021)
Maize (*Zea mays*)	Hydroxyapatite nanoparticles functionalized with humic substances	• Compared with commercial fertilizers, a significant increase in crop growth, abiotic stress tolerance, rhizosphere chemistry of biota, and the nutrient acquisition was observed which were attributed to better nutrient release and humic substance production due to treatments. The huge compatibility of this fertilizer, slow-release, and high economical value make it suitable for industrial use as an effective alternative fertilizer	Yoon et al. (2020)
Sunflower (*Helianthus annuus*)	Nano-HA, P fertilizer	• Prepared nano-HA are suitable for effective use as P fertilizer	Xiong (2019)
Sunflower (*Helianthus annuus*)	Nano-HA, P fertilizer	• P-deficient soils (a Ultisol and a Vertisol) were added with prepared nano-HA fertilizers more suitable for acidic soils	Xiong et al. (2018)
Alfalfa (*Medicago sativa*)	Potassium sulfate nanoparticles	• The application of NPs has shown a net increase in crop vigor, growth, and Na/K ratio with reduced electrolyte leakage and increase in proline and catalase contents	El-Sharkawy et al. (2017)
Water spinach (*Ipomoea aquatic*)	Potassium-coated zeolite nanoparticle	• Better uptake of K was observed compared to counterpart commercial fertilizers and soil physicochemical properties and available P and K contents were improved	Rajonee et al. (2017)

(Continued)

TABLE 1.5 (Continued)

Plant name, scientific name, and exposure period	Nanoparticle dose, characteristics, and crop growing conditions	Effect on plants	References
Potato (*Solanum tuberosum*)	NPK nano-fertilizers	• Higher K uptake was observed with more production and starch contents in potato tubers	Abd El-Azeim et al. (2020)
Rice (*Oryza sativa* L.)	Calcium phosphate NPs	• With the application of prepared nanoparticles, higher root length, shoot fresh, dry weight, and biomass was observed at lower levels of application	Upadhyaya et al. (2017)
Apple (*Malus domestica* L.)	Nano-Ca	• Overall fruit quality was improved compared to $CaCl_2$-treated plants	Ranjbar et al. (2018)
Pea (*Vigna unguiculate*)	Mg-NPs	• Foliar application of Mg-NPs caused an overall improvement in growth, physiological stability, and shoot Mg contents in plants	Delfani et al. (2014)
Tomato (*Solanum lycopersicum*)	Sulfur nanoparticles (SNPs)	• High-quality tomatoes were obtained with foliar application of prepared nanoparticles	Salem et al. (2016)
	Surface-modified nano-zeolite as a source of slow-release Sulfur	• Prepared nano-zeolite has shown a significant potential of slow release of SO_4 ion making it suitable to be used as nano-S fertilizer	Thirunavukkarasu and Subramanian (2014a,b)

Beneficial nutrients nanoparticles (potential beneficial effects and toxicity)			
Brassica napus	Nano-titanium (TiO_2)	• A net promotion of seed germination and seedling growth was observed with the application of Ti nanoparticles	Mahmoodzadeh et al. (2013)
Wheat (*Triticum aestivum*)	Nano-Titanium (TiO_2) (Field Experiment)	• Up to certain concentration of nanoparticles, a net increase in plant growth and biomass production was observed	Rafique et al. (2018)
	Nano-titanium (TiO_2) (lab experiment)	• A net enhancement in abscisic and jasmonic acids was observed suggesting its role in triggering phytohormones production in wheat	Jiang et al. (2017)
Spinach (*Spinacia oleracea*)	Nano anatase-TiO_2	• A net enhancement in enzymes involved in N-assimilation was observed	Yang et al. (2006)
Borad beans (*Vicia faba*)	TiO_2-NPs	• The applied NPs were found to be assimilated in root tissues and antioxidants (peroxidase-POD) contents were found to be increased with higher application	Castiglione et al. (2016)
Moldavian Balm (*Moldavian dragonhead*)	TiO-NPs	• The application of Ti-NPs can reverse the adverse effects of salinity in tested plants	Gohari et al. (2020)
Wheat (*Triticum aestivum L.*)	TiO-NPs application in wheat to control fungal disease	• Exogenous application of TiO-NPs has helped in the control of fungal diseases in wheat	Satti et al. (2021)

(Continued)

TABLE 1.5 (Continued)

Plant name, scientific name, and exposure period	Nanoparticle dose, characteristics, and crop growing conditions	Effect on plants	References
Maize (*Zea mays*)	Nano-SiO_2	• In hydroponic and field experiments it was observed that the application of Si-NPs has enhanced proteins, chlorophyll contents, and phenolic products with more silica accumulation in plant leaves	Suriyaprabha et al. (2012a,b)
Tomato (*Lycopersicum esculentum*)	Si-NPs	• A net increase in seed germination, vigor, germination index, and weight was observed	Siddiqui et al. (2014)
Fenugreek (*Trigonella foenum*)	Si-NPs	• Under a hydroponic investigation, the increased uptake of Si-NPs in plants helped in cell wall lignification and the formation of stress-responsive enzymes	Nazaralian et al. (2017)
Wheat (*Triticum aestivum*)	Si-NPs	• Under a field investigation, Si-NPs can help in the reversal of UVB radiation effects	Tripathi et al. (2017a,b)
Wheat (*Triticum aestivum*)	Si-NPs	• A lower concentration of Si-NPs can be beneficial but higher concentrations are hazardous for plant growth	Karimi and Mohsenzadeh (2016)

bioavailability of ZnO-NPs proving it to be an acceptable approach. Similarly, Waalewijn-Kool et al. (2012) has reported that ZnO-NPs toxicity is not dependent on size and significant bioavailability can be observed at any form it is available.

The use of ZnO-NPs for better agriculture is well reported regarding disinfection of seeds (Babajani et al., 2019), enhanced nutrition and Zn accumulation (Umar et al., 2020; Munir et al., 2018), regulation of plants metabolism (Hussain et al., 2018; Vafaee Moghadam et al., 2021) and stress mitigation such as cadmium in wheat (Hussain et al., 2018, 2021), maize (Itroutwar et al., 2020; Rizwan et al., 2019a,b), and barley (Plaksenkova et al., 2020). Sometimes, higher concentrations of ZnO-NPs can be toxic to plants as well resulting in cytotoxicity and loss in productivity (Al-Salama 2010; Watson et al., 2015; Zhang et al., 2015). The molecular response of plants against ZnO-NPs under various conditions is still not fully understood (Iranbakhsh et al., 2021). The effect of ZnO-NPs on the plant is a widely studied phenomenon and well reported. Literature has shown net positive as well as negative effects of ZnO-NPs in plants as described in Table 1.6.

1.4.2.2 Copper oxide nanoparticles

The use of copper NPs in commercial products such as agrochemicals, semiconductors, catalysts, sensors, paint, and antimicrobial products has led to a surge in their production and release into terrestrial and aquatic environments (Keller et al., 2017). The direct use of these NPs in nano-fertilizers and pesticides has also become a source of input in agricultural soil (Kah, 2015; Lazareva and Keller, 2014). The antimicrobial activities of CuO-NPs have raised much concern regarding soil microbial community disturbance.

Though Cu is an essential plant nutrient, the effect of CuO-NPs on plants is not well understood. Available studies on the effect of CuO-NPs use above natural limits of CuO-NPs to check their possible toxicity (Frenk et al., 2013; Rousk et al., 2012) while the bioavailable concentration in soil is way less than what has been used in studies (Garner and Keller, 2014). The bioavailability and toxicity of CuO-NPs, like other NPs, depends on soil properties such as organic matter content, aggregate stability, pH, texture, redox potential, and soil solution chemistry making the establishment of a generalized statement regarding its effect on plants an issue (Keller et al., 2017). Cu is also an essential plant nutrient, but higher ionic contents of Cu are toxic to plants (Arguello et al., 2013). As the release of CuO-NPs is dependent on soil properties plus the Cu accumulation potential of various plants vary, a generalize exclusion regarding their effect on plants is difficult. Table 1.7 provides a detailed review of the latest literature regarding the effect of CuO-NPs (positive and negative) with a special focus on the cause of beneficial and toxic effects.

TABLE 1.6 Effect of zinc nanoparticles on plants.

Plant name (scientific name), and exposure period	Nanoparticle dose, characters, and crop growing conditions	Potential beneficial effects				Other specific responses	References	Phytotoxicity	References
		Plant germination and seedling growth	Crop growth and biomass	Grain and seed production	Biochemical attributes				
Barley (*Hordeum vulgare L.*), full maturity	• Prepared from salt solution and the average size of 32 nm applied in levels of 0, 1, 2 and 4 mg L^{-1}	• Increased germination, seedling root, and shoot elongation	• Increased root/shoot length and biomass on 14-day exposure	–	• Net cytotoxicity was observed with increasing ZnNPs doses, increased oxidative stress	–	Plaksenkova et al. (2020)	• Higher levels of ZnO-NPs application can disrupt the normal functioning of crop plants suggesting the phytotoxicity potential of ZnO-NPs	Al-Salama (2010)
Wheat (*Triticum aestivum*), seedlings	• ZnO-NPs priming @ 10 ppm	• Higher seed germination and growth	• Higher biomass production	–	• Enhanced α-amylase activity, enhanced photosynthetic activity, higher antioxidants activity		Rai-Kalal, and Jajoo (2021)		
Wheat (*Triticum aestivum*), full maturity	• Foliar Exposure of ZnO-NPs @ 0, 25, 50, 100 mg L^{-1} under cadmium and drought conditions	• Higher seedling growth	• Higher biomass production and potential reversal of Cd toxicity and drought stress	• Higher grain yield with lower Cd contents compared to control	• Decreased oxidative stress (due to water deficit and Cd toxicity)	• Decreased Cd uptake and translocation and potential reversal of toxicity compared to control	Adrees et al. (2021)		

	• ZnO-NPs were applied @ 0, 25, 50, 75, and 100 ppm as soil and foliar (sperate studies) under Cadmium stress (spiked soils)	• Cd toxicity decreased seedling growth	• Higher plant growth and biomass compared to control in both applications	• Higher grain with less Cd accumulation thus decreased risk associated with its consumption	• Decreased electrolyte leakage, higher antioxidants levels, and toxicity reversal compared to control	• Decreased Cd uptake in both applications of ZnO-NPs (soil and foliar)	Hussain et al. (2018)		
Wheat (*Triticum aestivum*), full maturity under field conditions	• Zn, Fe, and Si-NPs foliar application single as well as combined under Cd contaminated field conditions	• The least growth was observed in the control	• Higher plant growth and biomass production	• Higher grain yield (1000 grain weight and total grain yield) was observed in all NPs treatments	• A decrease in oxidative stress was observed with a net increase in antioxidants	• Zn and Fe application resulted in higher grain Zn and Fe contents respectively with decreased Cd contents accumulation	Hussain et al. (2021)		
Wheat (*Triticum aestivum*), full maturity	• ZnO-NPs, as well as $ZnSO_4$ (bulk salt), were tested for effective toxicity vs biofortification in germination and pot experiment	• A moderate dose of both sources was found to be beneficial	• Moderate doses increased the growth and biomass of plants. Root structural damage was more in bulk Zn compared to Zn NPs	• Moderate doses of Zn were found to be beneficial for crop plants	• Deregulation of antioxidants was more in bulk Zn compared to Zn NPs	• Higher Zn contents in grains were observed under NPs treatment while higher shoot Zn was observed under bulk Zn application	Du et al. (2019)		

(Continued)

TABLE 1.6 (Continued)

Plant name (scientific name), and exposure period	Nanoparticle dose, characters, and crop growing conditions	Potential beneficial effects				Other specific responses	References	Phytotoxicity	References
		Plant germination and seedling growth	Crop growth and biomass	Grain and seed production	Biochemical attributes				
Maize (*Zea mays*), *seedling*	• ZnO-NPs (@100 mg L^{-1}) were applied as a nano-priming agent in a paper roll towel method	• Improved Seed germination rate was observed compared to control	• Shoot, Root length, and biomass was also found to increase with ZnO application	–	–	• SEM detection has shown the adsorption of ZnO-NPs in the endosperm region	Itroutwar et al. (2020)	• Potential high availability of ZnO-NPs compared to ionic Zn can make its uptake and translocation more in newly germinating corn seedlings disrupting their physiology	Zhang et al. (2015)
Maize (*Zea mays*), *vegetative growth*	• ZnO-NPs were foliar applied @ 0, 50, 75, 100 mg L^{-1} alone and in combination with soil-applied biochar under Cd spiked soil conditions	–	• Shoot growth was enhanced as well as its biomass with ZnO-NPs applied alone as well as in combination with biochar	–	• The application of NPs has decreased oxidative stress due to Cd toxicity. Biochar-ZnO-NPs interaction was also significant in this regard	• Nanoparticles-biochar interaction has also proven to be significant in decreasing Cd root and shoot contents	Rizwan et al. (2019a, 2019b)		

TABLE 1.7 Effect of copper nanoparticles on growth of plants.

Plant name, (scientific name), and exposure period	Nanoparticle dose, characters, and crop growing conditions	Effect on crop plants				Other specific responses	References	Phytotoxicity	References
		Plant germination and seedling growth	Crop growth and biomass	Grain and seed production	Biochemical attributes				
Wheat (*Triticum aestivum*), seedling	• Copper oxide nanoparticles (CuO NPs) dispersions (0.01.0.1 g L^{-1})	• The lowest level of Cu NPs (0.01 g L^{-1}) has shown a positive effect on germination rate and seedling growth	• The lowest level of Cu NPs (0.01 g L^{-1}) has shown a positive effect on Shoot and root length and biomass	–	• Higher concentrations of NPs disrupted the antioxidants behavior	• Separate studies have shown at CuO NPs (0.001.0.0.1 g L^{-1}) application and toxicity on *Alternaria solani* fungi at higher rates	Zakharova et al. (2019)	• Higher doses of CuO-NPs were found to be toxic to wheat seedlings as well as to *Alternaria solani*	Zakharova et al. (2019)
Rice (*Oryza sativa* japonica "Koshihikari")	• Phytotoxicity of copper oxide nanoparticles (nCuO, 0.1–100 mg L^{-1}) was observed along with Arsenic	• Cu NPs significantly improved seedling growth	• Helped reverse overall As effects	–	• The overall life cycle of the Rice plant was shortened	• Helped in decreasing As accumulation in shoot parts	Liu et al. (2018)		
Brassica rapa ssp. Rapa, seedlings	• Seed treatment with CuO NPs to check its Phyto-availability and	• A net decrease in growth and biomass was observed	• A net decrease was observed in biomass and growth with	–	• A net decrease in chlorophyll and sugar contents was observed while	• CuO-NPs resulted in higher uptake of Cu causing specific toxicity	Chung et al. (2019)	• Seedlings experienced severe oxidative stress in response to CuO-NPs	Chung et al. (2019)

(*Continued*)

TABLE 1.7 (Continued)

Plant name, (scientific name), and exposure period	Nanoparticle dose, characters, and crop growing conditions	Effect on crop plants				Other specific responses	References	Phytotoxicity	References
		Plant germination and seedling growth	Crop growth and biomass	Grain and seed production	Biochemical attributes				
	effect on seedling growth	depicting the overall toxicity of CuO-NPs	the toxic accumulation of Cu		upregulation of proline and anthocyanins was observed			treatment and higher ROS, H_2O_2, and MDA contents along with DNA damage	
Rice (*Oryza sativa*, var. Jyoti)	• Phytotoxicity of CuO-NPs was studied for potential phytotoxicity to rice	• Germination rate and seedling growth was disturbed	• Biomass of plants was significantly reduced showing signs of potential toxicity	–	• Photosynthetic disturbance, stomatal conductance disruption, decrease in photosystem efficiency, and elevated oxidative stress was observed	–	Da Costa, Sharma (2016)	• Potential accumulation of Cu in root and shoot resulted in an upsurge of Cu accumulation resulting in oxidative damage to plants	Da Costa, Sharma (2016)
Maize (*Zea mays*), full maturity	• Priming of CuO-NPs in plants against drought stress	–	• Increased plant biomass	• Total seed number and yield was also increased	• Enhanced leaf water contents. Increased anthocyanin, chlorophyll, and carotenoid	–	Van Nguyen et al. (2021)	• CuO-NPs (0.1, 1, and 100 mg kg^{-1} dry soil) has shown toxicity to carbon and nitrogen cycling soil microbes	Simonin et al. (2018)

Sweet potato (*Ipomoea batatas)* Beauregard-14 (B-14, low lignin) and Covington (COV, high lignin) sweet potato	• CuO-NPs, bulk copper oxide (bCuO), and $CuCl_2$ at 25–125 mg kg^{-1} were applied and tested for the potential effect	–	• Lower doses of nano-CuO improved plant growth	–	• B-14's gas exchange parameters were less effected	• The nano-Cu can be used as potential fertilizer	Bonilla-Bird et al. (2020)		
Rice (*Oryza Sativa* L.) seedlings in hydroponics	• 62.5, 125, and 250 mg L^{-1} CuO NPs application for potential toxicity evaluation	Nano-Cu has significantly decreased the growth of seedlings due to oxidative damage.	–	–	• 250 mg L^{-1} CuO NPs has shown net oxidative stress		Yang et al. (2020)		

1.4.2.3 Iron nanoparticles

Iron is an important plant nutrient; however, due to its high immobilization in soil, it becomes unavailable for plant uptake in normal conditions due to Fe^{+2} to Fe^{+3} transformation (oxidation). Being an essential plant nutrient, Fe plays vital roles in chlorophyll biosynthesis, carbon assimilation, and redox reaction (Ye et al., 2015). The nano sources of Fe (Fe-NPs) have a diverse effect on plants and have been reported to be positive as well as lethal for various crops depending on application rate, NPs characteristics, and crop type (Barrena et al., 2009; Tripathi et al., 2017a,b).

The adverse and toxic effects of Fe-NPs on plants results from oxidative damage and disruption in membrane exchange on uptake of NPs (Tatari et al., 2012). Once NPs take entry into the cell, they can interfere with the electron transport chain of chloroplast and mitochondria due to increased oxidative damage (Huth et al., 2016). The initiation of oxidative damage via production of ROS can be a stress signal for plants but its higher concentration can be toxic for cellular machinery of plant cell (Huth et al., 2016; Tatari et al., 2012). The benefits and toxicity of Fe-NPs thus depend on their bioavailable concentration in the soil ecosystem. Studies indicating the effect of Fe-NPs on plant growth are summarized in Table 1.8.

1.4.2.4 Manganese nanoparticles

Manganese is another important plant micronutrient involved in carbon assimilation, light harvest, nitrogen metabolism, and plant defense (Datnoff et al., 2007; Millaleo et al., 2010; Pradhan et al., 2013, 2014). Several works have reported the beneficial role of Mn in crop nutrition and growth and development in maize, soybean, wheat, and sugarcane (Dimkpa and Bindraban, 2016; Fageria, 2001). In contrast, the higher concentration of Mn and its oxidation state in soil, which largely depends on soil properties such as pH and redox potential, make it potentially toxic to plants (Fageria, 2001; Millaleo et al., 2010; Rengel, 2015). Many studies have reported the use of Mn and Mn-NPs for crop growth improvement as well as its toxicity in higher amounts. As reported, Mn-NPs have shown an increment in root and shoot growth of mung bean (Pradhan et al., 2013), biomass production in eggplant (Elmer and White, 2016), nontoxicity on mustard (Landa et al., 2016) and lettuce (Liu et al., 2016), which exhibited dose-dependent plant growth. It was also reported to suppress disease progression in watermelon without significant effect on yield (Elmer et al., 2018). Table 1.9 shows few available works on Mn-NPs.

TABLE 1.8 Effect of iron nanoparticles on growth of plants.

Plant name, (scientific name), and exposure period	Nanoparticle dose, characters, and crop growing conditions	Effect on crop plants				Other specific responses	References	Phytotoxicity	References
		Plant germination and seedling growth	Crop growth and biomass	Grain and seed production	Biochemical attributes				
Arabidopsis thaliana, full life cycle	• Positive (PC) and negative (NC) charged iron oxide (Fe_2O_3) nanoparticles (IONPs), 3 and 25 mg L^{-1} doses application	• A net decrease in seedling growth was observed for 25 mg L^{-1} doses compared to control in both types of Fe-NPs • Regarding pollen viability, 3 mg L^{-1} doses were found to be toxic	• No visible effect was reported proving that sometimes NPs toxicity can go undetected	• Seed yield was significantly reduced	–	• The 3 mg L^{-1} doses have not shown any sign of toxicity but higher doses (25 mg L^{-1}) is toxic for plant growth	Bombin et al. (2015)	• Fe-specific toxicity can be a cause of toxicity. Higher does was found to be more toxic	Bombin et al. (2015)
Wheat (*Triticum aestivum* L.), 5-day hydroponic investigation	• Citric acid-coated-Fe_3O_4 nanoparticles (5, 10, 15, 20 mg L^{-1})	• No effect on the germination rate of plants	• No toxicity was observed on any level	–	• Magnetic nanoparticles are rendered in roots and not translocated into upper tissue parts. As no translocation was observed into plant tissue, no oxidative stress was observed	• Electron microscopic investigation suggest that NPs entered the root through the apoplastic pathway then and also deposited in epidermal cell	Iannone et al. (2016)		

(Continued)

TABLE 1.8 (Continued)

Plant name, (scientific name), and exposure period	Nanoparticle dose, characters, and crop growing conditions	Effect on crop plants				Other specific responses	References	Phytotoxicity	References
		Plant germination and seedling growth	Crop growth and biomass	Grain and seed production	Biochemical attributes				
Peanut (*Arachis hypogaea*)	● The use of Fe-NPs for application as effective Fe fertilizer was explored with the application of Fe_2O_3-NPs and EDTA-Fe general fertilizer	● Seedling growth was significantly increased as the peanut is Fe sensitive crop.	● A net increase of biomass was observed. Higher Fe contents in the plant body were observed proving the efficacy of Fe-NPs to be used as Fe fertilizer	–	–	● Increased phytohormones content and antioxidants activity was observed in response to Fe nanoparticles application	Rui et al. (2016)		
Wheat (*Triticum aestivum* L.), 5-day hydroponic investigation	● Fe NPs (0, 5, 10, 15, and 20 mg L^{-1}) application was done to check efficacy of Fe-NPs to reverse Cd toxicity in wheat	● Seedling growth was improved	● Higher biomass was obtained with Fe supplementation and potential reversal of Cd toxicity was observed	● Higher grain production was observed with less Cd uptake than control treatment	● Cd toxicity was reversed as Fe supplementation enhanced antioxidants activity and nutrient acquisition of wheat	● Potential reversal of Cd toxicity	Rizwan et al. (2019a, 2019b)		

Sunflower (*Helianthus annuus L*), seedlings	• Fe_3O_4 nanoparticles (Fe-NPs) were applied @ 0, 50 or 500 ppm Fe-NPs under variable magnetic technique	• No significant effect of Fe-NPs and magnetic field was observed on seed germination	–	–	• Least amount of Fe was taken up by plants suggesting possible adsorption of Fe-NPs on plant roots	• Fe-NPs has tendency to get adsorbed or release Fe for plant's uptake but high amounts remain adsorbed on root surface	Kornarzyński et al. (2020)		
Yellow medick (*Medicago falcata L.*) plants under hydroponics	• Hydroponics investigation with addition of Fe_3O_4 NPs at 1, 2 and 4 mg L^{-1}	• Plant growth and root length increased with application of Fe-NPs	–	–	• Fe-NPs application has shown an incremental effect on Chlorophyll contents and miRNA expression		Kokina et al. (2020)	• Genotoxicity and reduced genome stability of miR159c. And application of FeNPs enhanced the fungal resistance in plants	Kokina et al. (2020)
Evening primrose (*Oenothera biennis L*)	• Fe_2O_3-NPs and bulk (Fe_2O_3) in interaction with citric acid were tested for effect	• Fe_2O_3) and its bulk counterpart (bα-Fe_2O_3) with and/or without citrate interaction	• Nano-Fe increased germination of seeds as well as enhanced tolerance index	• A net increase in root tolerance index was observed. Nano-Fe and citrate combination enhanced plant growth		• Low and moderate levels of Fe-NPs have beneficial effect on plants	Asadi-Kavan et al. (2020)		

TABLE 1.9 Effect of manganese nanoparticles on growth of plants.

Plant name, (Scientific name), and exposure period	Nanoparticle dose, characters, and crop growing conditions	Effect on crop plants				Other specific responses	References	Phytotoxicity	References
		Plant germination and seedling growth	Crop growth and biomass	Grain and seed production	Biochemical attributes				
Wheat (*Triticum aestivum L.*)	● Nano-Mn was compared with ionic and bulk Mn for nutritional acquisition behavior of plants	–	–	● Nano-Mn resulted in higher Mn contents on grain	–	● Foliar application of Mn is more effective in this regard	Dimkpa et al. (2018)	● Bulk Mn application significantly disrupted the nutrient acquisition in plants suggesting its toxicity	Dimkpa et al. (2018)
Mung bean (*Vigna radiata*)	● Comparative study of Mn-sulphate and Mn nanoparticles	● Overall N assimilation was improved	–	–	● Mn-NPs were found to be enhancing the net flux of N-assimilation through NIR-NiR and GS-GOGAT pathways	● Excess elemental Mn was found to be toxic to the cell organelle of tested mice (mitochondria)	Pradhan et al. (2014)	● NO toxicity was shown to mice mitochondria in vivo study suggesting its nontoxicology to cell organelles	Pradhan et al. (2014)
Vigna radiata (lab incubation study)	● Mn-NPs	● Overall plant growth was improved compared to control		–	● Phosphorylation and oxygen evolution	–	Chung et al. (2019)		
Lactuca sativa	● Manganese oxide nanoparticles (MOx NPs)	● The growth of lettuce was stimulated with the application of prepared nanoparticles		–	–	–	Da Costa and Sharma (2016)		

1.5 Summary

Nanotechnology is becoming a widely acceptable and diverse discipline of agriculture research due to immense importance of nanomaterials. A wide range of NPs is being produced to serve different purposes including nutrition provision, growth promotion, stress management, disease control, as well as soil and water pollution management.

References

Abd El-Azeim, M.M., Sherif, M.A., Hussien, M.S., Haddad, S.A., 2020. Temporal impacts of different fertilization systems on soil health under arid conditions of potato monocropping. J. Soil. Sci. Plant. Nutr. 20 (2), 322–334.

Abdulateef, S.A., MatJafri, M.Z., Omar, A.F., Ahmed, N.M., Azzez, S.A., Ibrahim, I.M., et al., 2016. Preparation of CuO nanoparticles by laser ablation in liquid. AIP Conf. Proc. 1733 (1), 020035. AIP Publishing LLC.

Adger, W.N., Dessai, S., Goulden, M., Hulme, M., Lorenzoni, I., Nelson, D.R., et al., 2009. Are there social limits to adaptation to climate change? Clim. Change 93 (3–4), 335–354.

Adrees, M., Khan, Z.S., Hafeez, M., Rizwan, M., Hussain, K., Asrar, M., et al., 2021. Foliar exposure of zinc oxide nanoparticles improved the growth of wheat (*Triticum aestivum* L.) and decreased cadmium concentration in grains under simultaneous Cd and water deficient stress. Ecotoxicol. Environ. Saf. 208, 111627.

Ahab, A., Rohman, F., Iskandar, F., Haryanto, F., Arif, I., 2016. A simple straightforward thermal decomposition synthesis of PEG-covered Gd_2O_3 (Gd_2O_3@ PEG) nanoparticles. Adv. Powder Technol. 27 (4), 1800–1805.

Ahmed, B., Khan, M.S., Musarrat, J., 2018a. Toxicity assessment of metal oxide nano-pollutants on tomato (Solanum lycopersicon): a study on growth dynamics and plant cell death. Environ. Pollut. 240, 802–816.

Ahmed, B., Khan, M.S., Saquib, Q., Al-Shaeri, M., Musarrat, J., 2018b. Interplay between engineered nanomaterials (ENMs) and edible plants: a current perspective. Phytotoxicity of Nanoparticles. Springer, Cham, pp. 63–102.

Al-Dahash, G., Mubder Khilkala, W., Abd Alwahid, S.N., 2018. Preparation and characterization of ZnO nanoparticles by laser ablation in NaOH aqueous solution. Iran. J. Chem. Chem. Eng. (IJCCE) 37 (1), 11–16.

Al-Salama, Y., 2010. Uptake and accumulation of zinc oxide nanoparticles by barley plants. Res. J. of Aleppo Univ. Agricultural Science Series No. 87.

Amingad, V., Sreenivas, K.N., Fakrudin, B., Seetharamu, G.K., Shankarappa, T.H., Venugopalan, R., 2017. Comparison of silver nanoparticles and other metal nanoparticles on postharvest attributes and bacterial load in Cut Roses var. Taj Mahal. Int. J. Pure Appl. Biosci. 5, 579–584.

Anjum, N.A., Adam, V., Kizek, R., Duarte, A.C., Pereira, E., Iqbal, M., et al., 2015. Nanoscale copper in the soil–plant system–toxicity and underlying potential mechanisms. Environ. Res. 138, 306–325.

Anjum, N.A., Rodrigo, M.A.M., Moulick, A., Heger, Z., Kopel, P., Zítka, O., et al., 2016. Transport phenomena of nanoparticles in plants and animals/humans. Environ. Res. 151, 233–243.

Arenas-Alatorre, J., Lukas, O., Rodríguez-Gómez, A., Reyes, R.H., Tapia-del León, C., 2019. Synthesis and characterization of iron oxide nanoparticles grown via a non-conventional chemical method using an external magnetic field. Mater. Lett. 242, 13–16.

Arguello, J.M., Raimunda, D., Padilla-Benavides, T., 2013. Mechanisms of copper homeostasis in bacteria. Front. Cell. Infect. Microbiol. 3, 73. Available from: https://doi.org/10.3389/fcimb.2013.00073.

Asadi-Kavan, Z., Khavari-Nejad, R.A., Iranbakhsh, A., Najafi, F., 2020. Cooperative effects of iron oxide nanoparticle (α-Fe_2O_3) and citrate on germination and oxidative system of evening primrose (*Oenothera biennis* L.). J. Plant. Interact. 15 (1), 166–179.

Asghari, F., Jahanshiri, Z., Imani, M., Shams-Ghahfarokhi, M., Razzaghi-Abyaneh, M., 2016. Antifungal nanomaterials: synthesis, properties, and applications. Nanobiomaterials in Antimicrobial Therapy. William Andrew Publishing, pp. 343–383.

Aslani, F., Bagheri, S., Muhd Julkapli, N., Juraimi, A.S., Hashemi, F.S.G., Baghdadi, A., 2014. Effects of engineered nanomaterials on plants growth: an overview. Sci. World J. 2014.

Aversa, R., Modarres, M.H., Cozzini, S., Ciancio, R., Chiusole, A., 2018. The first annotated set of scanning electron microscopy images for nanoscience. Sci. data 5, 180172.

Babajani, A., Iranbakhsh, Ardebili, Z.O., Eslami, B., 2019. Differential growth, nutrition, physiology, and gene expression in Melissa officinalis mediated by zinc oxide and elemental selenium nanoparticles. Env. Sci. Poll. Res. 26, 24430–24444.

Badran, A., Savin, I., 2018. Effect of nano-fertilizer on seed germination and first stages of bitter almond seedlings' growth under saline conditions. BioNanoScience 8 (3), 742–751.

Barrena, R., Casals, E., Colon, J., Font, X., Sanchez, A., Puntes, V., 2009. Evaluation of the ecotoxicity of model nanoparticles. Chemosphere 75, 850–857. CrossRef CAS.

Bessi, A., Boudine, B., Boudaren, C., 2017. Fabrication and characterization of TiO_2 nanoparticles. In: International Symposium on Materials and Sustainable Development. Springer, Cham, pp. 296–300.

Bombin, S., LeFebvre, M., Sherwood, J., Xu, Y., Bao, Y., Ramonell, K.M., 2015. Developmental and reproductive effects of iron oxide nanoparticles in Arabidopsis thaliana. Int. J. Mol. Sci. 16 (10), 24174–24193.

Bondarenko, O., Juganson, K., Ivask, A., Kasemets, K., Mortimer, M., Kahru, A., 2013. Toxicity of Ag, CuO and ZnO nanoparticles to selected environmentally relevant test organisms and mammalian cells in vitro: a critical review. Arch Toxicol 87 (7), 1181–1200.

Bonilla-Bird, N.J., Ye, Y., Akter, T., Valdes-Bracamontes, C., Darrouzet-Nardi, A.J., Saupe, G. B., et al., 2020. Effect of copper oxide nanoparticles on two varieties of sweetpotato plants. Plant. Physiol. Biochem. 154, 277–286.

Boxall, A.B., Chaudhry, Q., Sinclair, C., Jones, A., Aitken, R., Jefferson, B., et al., 2007. Current and Future Predicted Environmental Exposure to Engineered Nanoparticles. Central Science Laboratory, Department of the Environment and Rural Affairs, London, UK, p. 89.

Bussamara, R., Melo, W.W., Scholten, J.D., Migowski, P., Marin, G., Zapata, M.J., et al., 2013. Controlled synthesis of Mn_3O_4 nanoparticles in ionic liquids. Dalton Trans. 42 (40), 14473–14479.

Carmona, F.J., Dal Sasso, G., Ramírez-Rodríguez, G.B., Pii, Y., Delgado-López, J.M., Guagliardi, A., et al., 2021. Urea-functionalized amorphous calcium phosphate nanofertilizers: optimizing the synthetic strategy towards environmental sustainability and manufacturing costs. Sci. Rep. 11 (1), 1–14.

Castiglione, M.R., Giorgetti, L., Becarelli, S., Siracusa, G., Lorenzi, R., Di Gregorio, S., 2016. Polycyclic aromatic hydrocarbon-contaminated soils: bioaugmentation of autochthonous bacteria and toxicological assessment of the bioremediation process by means of Vicia faba L. Environ. Sci. Pollut. Res. 23 (8), 7930–7941.

Chauhan, A., Zubair, S., Tufail, S., Sherwani, A., Sajid, M., Raman, S.C., et al., 2011. Fungus-mediated biological synthesis of gold nanoparticles: potential in detection of liver cancer. Int. J. Nanomed. 6, 2305.

Chung, I.M., Rekha, K., Venkidasamy, B., Thiruvengadam, M., 2019. Effect of copper oxide nanoparticles on the physiology, bioactive molecules, and transcriptional changes in Brassica rapa ssp. rapa seedlings. Water, Air, Soil. Pollut. 230 (2), 48.

Clark, M., Tilman, D., 2017. Comparative analysis of environmental impacts of agricultural production systems, agricultural input efficiency, and food choice. Environ. Res. Lett. 12 (6), 064016.

Connolly, M., Fernández, M., Conde, E., Torrent, F., Navas, J.M., Fernández-Cruz, M.L., 2016. Tissue distribution of zinc and subtle oxidative stress effects after dietary administration of ZnO nanoparticles to rainbow trout. Sci Total Environ 551, 334–343.

Correa, S.N., Naranjo, A.M., Herrera, A.P., 2016. Biosynthesis and characterization of gold nanoparticles using extracts of *Tamarindus indica* L leaves. J. Phys.: Conf. Ser. 687 (1), 012082. IOP Publishing.

Da Costa, M.V.J., Sharma, P.K., 2016. Effect of copper oxide nanoparticles on growth, morphology, photosynthesis, and antioxidant response in *Oryza sativa*. Photosynthetica 54 (1), 110–119.

D'Amato, R., Falconieri, M., Gagliardi, S., Popovici, E., Serra, E., Terranova, G., et al., 2013. Synthesis of ceramic nanoparticles by laser pyrolysis: from research to applications. J. Anal. Appl. Pyrolysis 104, 461–469.

Das, A., Kushwaha, A., Bansal, N.R., Suresh, V., Dinda, S., Chattopadhyay, S., et al., 2016. Copper oxide nano-particles film on glass by using sputter and chemical bath deposition. Adv. Mater. Lett. (7(8)), 600-603

Datnoff, L.E., Elmer, W.H., Huber, D.M., 2007. Mineral Nutrition and Plant Disease. American Phytopathological Society, Paul, MI, USA.

de la Torre, C.M., Bennewitz, M.F., 2020. Manganese oxide nanoparticle synthesis by thermal decomposition of manganese (II) acetylacetonate. JoVE (J. Vis. Exp.) (160), e61572.

de Silva, M., Siriwardena, D.P., Sandaruwan, C., Priyadarshana, G., Karunaratne, V., Kottegoda, N., 2020. Urea-silica nanohybrids with potential applications for slow and precise release of nitrogen. Mater. Lett. 272, 127839.

Delfani, M., Baradarn Firouzabadi, M., Farrokhi, N., Makarian, H., 2014. Some physiological responses of black-eyed pea to iron and magnesium nanofertilizers. Commun. soil. Sci. Plant. Anal. 45 (4), 530–540.

Deng, Y.Q., White, J.C., Xing, B.S., 2014. Interactions between engineered nanomaterials and agricultural crops: implications for food safety. J Zhejiang Univ Sci A 15 (8), 552–572.

Dietz, K.J., Herth, S., 2011. Plant nanotoxicology. Trends Plant. Sci. 16 (11), 582–589.

Dimkpa, C.O., Bindraban, P.S., 2016. Micronutrients fortification for efficient agronomic production. Agron. Sustain. Dev. 36, 1–26 [CrossRef].

Dimkpa, C.O., Singh, U., Adisa, I.O., Bindraban, P.S., Elmer, W.H., Gardea-Torresdey, J.L., et al., 2018. Effects of manganese nanoparticle exposure on nutrient acquisition in wheat (*Triticum aestivum* L.). Agronomy 8 (9), 158.

Du, L., Jiang, H., Liu, X., Wang, E., 2007. Biosynthesis of gold nanoparticles assisted by Escherichia coli DH5α and its application on direct electrochemistry of hemoglobin. Electrochem. Commun. 9 (5), 1165–1170.

Du, W., Yang, J., Peng, Q., Liang, X., Mao, H., 2019. Comparison study of zinc nanoparticles and zinc sulphate on wheat growth: from toxicity and zinc biofortification. Chemosphere 227, 109–116.

Dubey, R.S., Rajesh, Y.B.R.D., More, M.A., 2015. Synthesis and characterization of SiO_2 nanoparticles via sol-gel method for industrial applications. Mater. Today: Proc. 2 (4–5), 3575–3579.

Dzimitrowicz, A., Jamroz, P., Sergiel, I., Kozlecki, T., Pohl, P., 2019. Preparation and characterization of gold nanoparticles prepared with aqueous extracts of Lamiaceae plants and the effect of follow-up treatment with atmospheric pressure glow microdischarge. Arab. J. Chem. 12 (8), 4118–4130.

EC Scientific Committee on Emerging and Newly Identified Health Risks (SCENIHR) 2007. The existing and proposed definitions relating to products of nanotechnologies; Subsection 3.3.3.1, p. 11. <https://ec.europa.eu/health/ph_risk/committees/04_scenihr/docs/scenihr_o_012.pdf> (accessed 17.09.21.).

Elmer, W., White, J.C., 2016. The use of metallic oxide nanoparticles to enhance growth of tomatoes and eggplants in disease infested soil or soilless medium. Environ. Sci. Nano 3, 1072–1079.

Elmer, W., De La Torre-Roche, R., Pagano, L., Majumdar, S., Zuverza-Mena, N., Dimkpa, C., et al., 2018. Effect of metalloid and metallic oxide nanoparticles on Fusarium wilt of watermelon. Plant. Dis. 102, 1394–1401.

El-Sharkawy, M.S., El-Beshsbeshy, T.R., Mahmoud, E.K., Abdelkader, N.I., Al-Shal, R.M., Missaoui, A.M., 2017. Response of alfalfa under salt stress to the application of potassium sulfate nanoparticles. Am. J. Plant. Sci. 8 (08), 1751.

Fageria, N.K., 2001. Adequate and toxic levels of copper and manganese in upland rice, common bean, corn, soybean, and wheat grown on an oxisol. Commun. Soil. Sci. Plant. Anal. 32, 1659–1676 [CrossRef].

Frenk, S., Ben-Moshe, T., Dror, I., Berkowitz, B., Minz, D., 2013. Effect of metal oxide nanoparticles on microbial community structure and function in two different soil types. PLoS One 8, e84441. Available from: https://doi.org/10.1371/journal.pone.0084441.

Fu, X., Cai, J., Zhang, X., Li, W.D., Ge, H., Hu, Y., 2018. Top-down fabrication of shape-controlled, monodisperse nanoparticles for biomedical applications. Adv. Drug. Deliv. Rev. 132, 169–187.

Garner, K.L., Keller, A.A., 2014. Emerging patterns for engineered nanomaterials in the environment: a review of fate and toxicity studies. J. Nanopart. Res. 16, 2503. Available from: https://doi.org/10.1007/s11051-014-2503-2.

Geetha, M.S., Nagabhushana, H., Shivananjaiah, H.N., 2016. Green mediated synthesis and characterization of ZnO nanoparticles using Euphorbia Jatropa latex as reducing agent. J. Sci.: Adv. Mater. Dev. 1 (3), 301–310.

Ghaffarian, H.R., Saiedi, M., Sayyadnejad, M.A., Rashidi, A.M., 2011. Synthesis of ZnO nanoparticles by spray pyrolysis method. Iran. J. Chem. Chem. Eng., 30 (1), 1–6.

Ghosh, M., Sinha, S., Jothiramajayam, M., Jana, A., Nag, A., Mukherjee, A., 2016. Cytogenotoxicity and oxidative stress induced by zinc oxide nanoparticle in human lymphocyte cells in vitro and Swiss albino male mice in vivo. Food Chem. Toxicol. 97, 286–296.

Gleiter, H., 2000. Nanostructured materials: basic concepts and microstructure. Acta Mater. 48 (1), 1–29.

Gohari, G., Mohammadi, A., Akbari, A., Panahirad, S., Dadpour, M.R., Fotopoulos, V., et al., 2020. Titanium dioxide nanoparticles (TiO_2 NPs) promote growth and ameliorate salinity stress effects on essential oil profile and biochemical attributes of Dracocephalum moldavica. Sci. Rep. 10 (1), 1–14.

Govender, Y., Riddin, T., Gericke, M., Whiteley, C.G., 2009. Bioreduction of platinum salts into nanoparticles: a mechanistic perspective. Biotechnol. Lett. 31 (1), 95–100.

Hamouda, R.A., Hussein, M.H., Abo-Elmagd, R.A., Bawazir, S.S., 2019. Synthesis and biological characterization of silver nanoparticles derived from the cyanobacterium Oscillatoria limnetica. Sci. Rep. 9 (1), 1–17.

Held, M.A., Jiang, N., Basu, D., Showalter, A.M., Faik, A., 2015. Plant cell wall polysaccharides: structure and biosynthesis. Polysaccharides: Bioactivity Biotechnol 3–54.

Hong, G., Zou, Y., Antaris, A.L., Diao, S., Wu, D., Cheng, K., et al., 2014. Ultrafast fluorescence imaging in vivo with conjugated polymer fluorophores in the second near-infrared window. Nat. Commun. 5 (1), 1–9.

Hu, X.L., Takai, O., Saito, N., 2013. Synthesis of gold nanoparticles by solution plasma sputtering in various solvents. J. Phys.: Conf. Ser. 417 (1), 012030. IOP Publishing.

Hussain, A., Ali, S., Rizwan, M., ur Rehman, M.Z., Javed, M.R., Imran, M., et al., 2018. Zinc oxide nanoparticles alter the wheat physiological response and reduce the cadmium uptake by plants. Environ. Pollut. 242, 1518–1526.

Hussain, A., Rizwan, M., Ali, S., ur Rehman, M.Z., Qayyum, M.F., Nawaz, R., et al., 2021. Combined use of different nanoparticles effectively decreased cadmium (Cd) concentration in grains of wheat grown in a field contaminated with Cd. Ecotoxicol. Environ. Saf. 215, 112139.

Huth, C., Mertz-Henning, L.M., Lopes, S.J., Tabaldi, L.A., Rossato, L.V., Krzyzanowski, F.C., et al., 2016. Susceptibility to weathering damage and oxidative stress on soybean seeds with different lignin contents in the seed coat. J. Seed Sci. 38 (4), 296–304.

Iannone, M.F., Groppa, M.D., de Sousa, M.E., van Raap, M.B.F., Benavides, M.P., 2016. Impact of magnetite iron oxide nanoparticles on wheat (*Triticum aestivum* L.) development: evaluation of oxidative damage. Environ. Exp. Botany 131, 77–88.

Ijaz, I., Gilani, E., Nazir, A., Bukhari, A., 2020. Detail review on chemical, physical and green synthesis, classification, characterizations and applications of nanoparticles. Green. Chem. Lett. Rev. 13 (3), 223–245.

Iranbakhsh, A., Oraghi Ardebili, Z., Oraghi Ardebili, N., 2021. Chapter 3 in Plant responses to nanomaterials: recent interventions, and physiological and biochemical responses. In: Singh, V.P., Singh, S., Prasad, S.M., Chauhan, D.K., Tripathi, D.K. (Eds.), Synthesis and Characterization of Zinc Oxide Nanoparticles and Their Impact on Plants. Springer International Publishing.

Itroutwar, P.D., Kasivelu, G., Raguraman, V., Malaichamy, K., Sevathapandian, S.K., 2020. Effects of biogenic zinc oxide nanoparticles on seed germination and seedling vigor of maize (Zea mays). Biocatalysis Agric. Biotechnol. 29, 101778.

Jaiswal, J., Chauhan, S., Chandra, R., 2015. Influence of sputtering parameters on structural, optical and thermal properties of copper nanoparticles synthesized by dc magnetron sputtering. Int. J. Sci. Technol. Manage 4 (01), 678–688.

Jiang, H., Liu, J.K., Wang, J.D., Lu, Y., Zhang, M., Yang, X.H., et al., 2014. The biotoxicity of hydroxyapatite nanoparticles to the plant growth. J. Hazard. Mater. 270, 71–81.

Jiang, F., Shen, Y., Ma, C., Zhang, X., Cao, W., Rui, Y., 2017. Effects of TiO_2 nanoparticles on wheat (*Triticum aestivum* L.) seedlings cultivated under super-elevated and normal CO_2 conditions. PLoS One 12 (5), e0178088.

Judy, J.D., Kirby, J.K., McLaughlin, M.J., Cavagnaro, T., Bertsch, P.M., 2016. Gold nanomaterial uptake from soil is not increased by arbuscular mycorrhizal colonization of Solanum Lycopersicum (Tomato). Nanomaterials 6 (4), 68.

Kah, M., 2015. Nanopesticides and nanofertilizers: emerging contaminants or opportunities for risk mitigation? Front. Chem. 3, 64. Available from: https://doi.org/10.3389/fchem.2015.00064.

Karimi, J., Mohsenzadeh, S., 2016. Effects of silicon oxide nanoparticles on growth and physiology of wheat seedlings. Russian J. Plant. Physiol. 63 (1), 119–123.

Keller, A.A., McFerran, S., Lazareva, A., Suh, S., 2013. Global life cycle releases of engineered nanomaterials. J. Nanopart. Res. 15 (6), 1692.

Keller, A.A., Adeleye, A.S., Conway, J.R., Garner, K.L., Zhao, L., Cherr, G.N., et al., 2017. Comparative environmental fate and toxicity of copper nanomaterials. NanoImpact 7, 28–40. Available from: https://doi.org/10.1016/j.impact.2017.05.003.

Khan, M.F., Ansari, A.H., Hameedullah, M., Ahmad, E., Husain, F.M., Zia, Q., et al., 2016. Sol-gel synthesis of thorn-like ZnO nanoparticles endorsing mechanical stirring effect and their antimicrobial activities: potential role as nano-antibiotics. Sci. Rep. 6 (1), 1–12.

Khan, I., Saeed, K., Khan, I., 2019. Nanoparticles: properties, applications and toxicities. Arab. J. Chem. 12 (7), 908–931.

Khayati, G.R., Janghorban, K., 2013. Preparation of nanostructure silver powders by mechanical decomposing and mechanochemical reduction of silver oxide. Trans. Nonferrous Met. Soc. China 23 (5), 1520–1524.

Khumaeni, A., Budi, W.S., Sutanto, H., 2017. Synthesis and characterization of high-purity gold nanoparticles by laser ablation method using low-energy Nd: YAG laser 1064nm. J. Phys.: Conf. Ser. 909 (1), 012037. IOP Publishing.

Kokina, I., Plaksenkova, I., Jermaļonoka, M., Petrova, A., 2020. Impact of iron oxide nanoparticles on yellow medick (*Medicago falcata* L.) plants. J. Plant. Interact. 15 (1), 1–7.

Kornarzyński, K., Sujak, A., Czernel, G., Wiącek, D., 2020. Effect of Fe_3O_4 nanoparticles on germination of seeds and concentration of elements in *Helianthus annuus* L. under constant magnetic field. Sci. Rep. 10 (1), 1–10.

Kottegoda, N., Munaweera, I., Madusanka, N., Karunaratne, V., 2011. A green slow-release fertilizer composition based on urea-modified hydroxyapatite nanoparticles encapsulated wood. Curr sci 73–78.

Kottegoda, N., Sandaruwan, C., Priyadarshana, G., Siriwardhana, A., Rathnayake, U.A., Berugoda Arachchige, D.M., et al., 2017. Urea-hydroxyapatite nanohybrids for slow release of nitrogen. ACS Nano 11 (2), 1214–1221.

Kreyling, W.G., Semmler-Behnke, M., Chaudhry, Q., 2010. A complementary definition of nanomaterial. Nano Today 5 (3), 165–168.

Kumar, S.G., Rao, K.K., 2014. Polymorphic phase transition among the titania crystal structures using a solution-based approach: from precursor chemistry to nucleation process. Nanoscale 6 (20), 11574–11632.

Kumar, E., Selvarajan, P., Muthuraj, D., 2013. Synthesis and characterization of CeO_2 nanocrystals by solvothermal route. Mater. Res. 16 (2), 269–276.

Landa, P., Cyrusova, T., Jerabkova, J., Drabek, O., Vanek, T., Podlipna, R., 2016. Effect of mtal oxides on plant germination: phytotoxicity of nanoparticles, bulk materials and metal ions. Water Soil. Air Pollut. 227, 448.

Lazareva, A., Keller, A.A., 2014. Estimating potential life cycle releases of engineered nanomaterials from wastewater treatment plants. ACS Sustain. Chem. Eng. 2, 1656–1665. Available from: https://doi.org/10.1021/sc500121w.

Lee, H., Song, M.Y., Jurng, J., Park, Y.K., 2011. The synthesis and coating process of TiO_2 nanoparticles using CVD process. Powder Technol. 214 (1), 64–68.

Lee, S.H., Jung, J.S., Lee, S.S., Lee, S.B., Hwang, N.M., 2016. Low temperature deposition of polycrystalline silicon thin films on a flexible polymer substrate by hot wire chemical vapor deposition. J. Cryst. Growth 453, 151–157.

Li, Z., Huang, J., 2014. Effects of nanoparticle hydroxyapatite on growth and antioxidant system in pakchoi (Brassica chinensis L.) from cadmium-contaminated soil. J. Nanomater. 2014.

Liu, R., Zhang, H., Lal, R., 2016. Effects of stabilized nanoparticles of copper, zinc, manganese, and iron oxides in low concentrations on lettuce (Lactuca sativa) seed germination: nanotoxicants or nanonutrients? Water Air Soil. Pollut. 227, 42.

Liu, J., Simms, M., Song, S., King, R.S., Cobb, G.P., 2018. Physiological effects of copper oxide nanoparticles and arsenic on the growth and life cycle of rice (Oryza sativa japonica 'Koshihikari'). Environ. Sci. Technol. 52 (23), 13728–13737.

Lusvardi, G., Barani, C., Giubertoni, F., Paganelli, G., 2017. Synthesis and characterization of TiO_2 nanoparticles for the reduction of water pollutants. Materials 10 (10), 1208.

Ma, C., White, J.C., Dhankher, O.P., Xing, B., 2015. Metal-based nanotoxicity and detoxification pathways in higher plants. Environ. Sci. Technol. 49 (12), 7109–7122.

Maghsoodi, M.R., Ghodszad, L., Lajayer, B.A., 2020. Dilemma of hydroxyapatite nanoparticles as phosphorus fertilizer: potentials, challenges and effects on plants. Environ. Technol. Innov. 100869.

Mahmoodzadeh, H., Nabavi, M. Kashefi, H., 2013. Effect of nanoscale titanium dioxide particles on the germination and growth of canola (*Brassica napus*). J. Ornam. Hortic. 3 (1), 25–32.

Majzoub, R.N., Wonder, E., Ewert, K.K., Kotamraju, V.R., Teesalu, T., Safinya, C.R., 2016. Rab11 and lysotracker markers reveal correlation between endosomal pathways and transfection efficiency of surface-functionalized cationic liposome–DNA nanoparticles. J. Phys. Chem. B 120 (26), 6439–6453.

Mali, S.C., Raj, S., Trivedi, R., 2019. Biosynthesis of copper oxide nanoparticles using Enicostemma axillare (Lam.) leaf extract. Biochem. Biophys. Rep. 20, 100699.

Manceau, A., Nagy, K.L., Marcus, M.A., Lanson, M., Geoffroy, N., Jacquet, T., et al., 2008. Formation of metallic copper nanoparticles at the soil − root interface. Environ. Sci. Technol. 42 (5), 1766–1772.

Manikandan, A., Subramanian, K.S., 2016. Evaluation of zeolite based nitrogen nano-fertilizers on maize growth, yield and quality on inceptisols and alfisols. Int J Plant Soil Sci 9 (4), 1–9.

Marchiol, L., Filippi, A., Adamiano, A., Degli Esposti, L., Iafisco, M., Mattiello, A., et al., 2019. Influence of hydroxyapatite nanoparticles on germination and plant metabolism of tomato (Solanum lycopersicum L.): preliminary evidence. Agronomy 9 (4), 161.

Mashrai, A., Khanam, H., Aljawfi, R.N., 2017. Biological synthesis of ZnO nanoparticles using C. albicans and studying their catalytic performance in the synthesis of steroidal pyrazolines. Arab. J. Chem. 10, S1530–S1536.

Miglietta, M.L., Rametta, G., Manzo, S., Salluzzo, A., Rimauro, J., Di Francia, G., 2015. Methodological issues about techniques for the spiking of standard OECD soil with nanoparticles: evidence of different behaviours. J. Nanopart. Res. 17 (7), 1–12.

Millaleo, R., Reyes-Díaz, M., Ivanov, A.G., Mora, M.L., Alberdi, M., 2010. Manganese as essential and toxic element for plants: transport, accumulation and resistance mechanisms. J. Soil. Sci. Plant. Nutr. 10, 470–481 [CrossRef].

Millaleo, R., Reyes-Díaz, M., Ivanov, A.G., Mora, M.L., Alberdi, M., 2010. Manganese as essential and toxic element for plants: transport, accumulation and resistance mechanisms. J Soil Sci Plant Nutr 10 (4), 470–481.

Miralles, P., Johnson, E., Church, T.L., Harris, A.T., 2012. Multiwalled carbon nanotubes in alfalfa and wheat: toxicology and uptake. J. R. Soc. Interface 9 (77), 3514–3527.

Mohamad, S.N.S., Mahmed, N., Halin, D.S.C., Razak, K.A., Norizan, M.N., Mohamad, I.S., 2019. Synthesis of alumina nanoparticles by sol-gel method and their applications in the removal of copper ions (Cu^{2+}) from the solution. IOP Conf. Ser.: Mater. Sci. Eng. 701 (1), 012034. IOP Publishing.

Mohammadi, S., Harvey, A., Boodhoo, K.V., 2014. Synthesis of TiO_2 nanoparticles in a spinning disc reactor. Chem. Eng. J. 258, 171–184.

Monti, D.M., Guarnieri, D., Napolitano, G., Piccoli, R., Netti, P., Fusco, S., et al., 2015. Biocompatibility, uptake and endocytosis pathways of polystyrene nanoparticles in primary human renal epithelial cells. J. Biotechnol. 193, 3–10.

Mourato, A., Gadanho, M., Lino, A.R., Tenreiro, R., 2011. Biosynthesis of crystalline silver and gold nanoparticles by extremophilic yeasts. Bioinorganic Chem. Appl. 2011.

Mozafari, A.A., Havas, F., Ghaderi, N., 2018. Application of iron nanoparticles and salicylic acid in in vitro culture of strawberries (Fragaria × ananassa Duch.) to cope with drought stress. Plant. Cell Tissue Organ. Cult. (PCTOC) 132 (3), 511–523.

Munir, T., Rizwan, M., Kashif, M., Shahzad, A., Ali, S., Amin, N., et al., 2018. Effect of zinc oxide nanoparticles on the growth and Zn uptake in wheat (Triticumaestivum L.) by seed priming method. Dig. J. Nanomater. Biostruct. (DJNB) 13 (1).

Murthy, S.K., 2007. Nanoparticles in modern medicine: state of the art and future challenges. Int. J. Nanomed. 2 (2), 129.

Nazaralian, S., Majd, A., Irian, S., Najafi, F., Ghahremaninejad, F., Landberg, T., et al., 2017. Comparison of silicon nanoparticles and silicate treatments in fenugreek. Plant. Physiol. Biochem. 115, 25–33.

Ng, C.T., Tang, F.M.A., Li, J.J.E., Ong, C., Yung, L.L.Y., Bay, B.H., 2015. Clathrin-mediated endocytosis of gold nanoparticles in vitro. Anat. Rec. 298 (2), 418–427.

Onelli, E., Prescianotto-Baschong, C., Caccianiga, M., Moscatelli, A., 2008. Clathrin-dependent and independent endocytic pathways in tobacco protoplasts revealed by labelling with charged nanogold. J. Exp. Bot. 59 (11), 3051–3068.

Pauly, M., Keegstra, K., 2016. Biosynthesis of the plant cell wall matrix polysaccharide xyloglucan. Annu. Rev. Plant. Biol. 67, 235–259.

Peng, C., Duan, D., Xu, C., Chen, Y., Sun, L., Zhang, H., et al., 2015. Translocation and biotransformation of CuO nanoparticles in rice (Oryza sativa L.) plants. Environ. Pollut. 197, 99–107.

Peng, Y.H., Tsai, Y.C., Hsiung, C.E., Lin, Y.H., Shih, Y.H., 2017. Influence of water chemistry on the environmental behaviors of commercial ZnO nanoparticles in various water and wastewater samples. J Hazard Mater 322, 348–356.

Philipse, A.P., Maas, D., 2002. Magnetic colloids from magnetotactic bacteria: chain formation and colloidal stability. Langmuir 18 (25), 9977–9984.

Pinedo-Guerrero, Z.H., Hernández-Fuentes, A.D., Ortega-Ortiz, H., Benavides-Mendoza, A., Cadenas-Pliego, G., 2017. Cu nanoparticles in hydrogels of chitosan-PVA affects the characteristics of post-harvest and bioactive compounds of jalapeño pepper. Molecules 22 (6), 926.

Plaksenkova, I., Kokina, I., Petrova, A., Jermaļonoka, M., Gerbreders, V., Krasovska, M., 2020. The impact of zinc oxide nanoparticles on cytotoxicity, genotoxicity, and miRNA expression in barley (Hordeum vulgare L.) seedlings. Sci. World J. 2020.

Pradhan, S., Patra, P., Das, S., Chandra, S., Mitra, S., Dey, K.K., et al., 2013. Photochemical modulation of biosafe manganese nanoparticles on Vigna radiata: a detailed molecular, biochemical, and biophysical study. Environ. Sci. Technol. 47 (22), 13122–13131.

Pradhan, S., Patra, P., Mitra, S., Dey, K.K., Jain, S., Sarkar, S., et al., 2014. Manganese nanoparticles: impact on non-nodulated plant as a potent enhancer in nitrogen metabolism and toxicity study both in vivo and in vitro. J. Agric. Food Chem. 62 (35), 8777–8785.

Prathna, T.C., Chandrasekaran, N., Raichur, A.M., Mukherjee, A., 2011. Biomimetic synthesis of silver nanoparticles by Citrus limon (lemon) aqueous extract and theoretical prediction of particle size. Colloids Surf. B: Biointerfaces 82 (1), 152–159.

Prodan, A.M., Iconaru, S.L., Ciobanu, C.S., Chifiriuc, M.C., Stoicea, M., Predoi, D., 2013. Iron oxide magnetic nanoparticles: characterization and toxicity evaluation by in vitro and in vivo assays. J. Nanomater. 2013.

Pullagurala, V.L.R., Rawat, S., Adisa, I.O., Hernandez-Viezcas, J.A., Peralta-Videa, J.R., Gardea-Torresdey, J.L., 2018. Plant uptake and translocation of contaminants of emerging concern in soil. Sci. Total. Environ. 636, 1585–1596.

Rafique, R., Zahra, Z., Virk, N., Shahid, M., Pinelli, E., Park, T.J., et al., 2018. Dose-dependent physiological responses of *Triticum aestivum* L. to soil applied TiO_2 nanoparticles: alterations in chlorophyll content, H_2O_2 production, and genotoxicity. Agric. Ecosyst. Environ. 255, 95–101.

Raghunandan, D., Bedre, M.D., Basavaraja, S., Sawle, B., Manjunath, S.Y., Venkataraman, A., 2010. Rapid biosynthesis of irregular shaped gold nanoparticles from macerated aqueous extracellular dried clove buds (*Syzygium aromaticum*) solution. Colloids Surf. B: Biointerfaces 79 (1), 235–240.

Rai-Kalal, P., Jajoo, A., 2021. Priming with zinc oxide nanoparticles improve germination and photosynthetic performance in wheat. Plant. Physiol. Biochem. 160, 341–351.

Rajonee, A.A., Zaman, S., Huq, S.M.I., 2017. Preparation, characterization and evaluation of efficacy of phosphorus and potassium incorporated nano fertilizer. Adv. Nanopart. 6 (02), 62.

Rajput, V.D., Minkina, T., Suskova, S., Mandzhieva, S., Tsitsuashvili, V., Chapligin, V., et al., 2018a. Effects of copper nanoparticles (CuO NPs) on crop plants: a mini review. BioNanoScience 8 (1), 36–42.

Rajput, V.D., Minkina, T.M., Behal, A., Sushkova, S.N., Mandzhieva, S., Singh, R., et al., 2018b. Effects of zinc-oxide nanoparticles on soil, plants, animals and soil organisms: a review. Environ. Nanotechnol. Monit. Manag. 9, 76–84.

Raliya, R., Biswas, P., Tarafdar, J.C., 2015a. TiO_2 nanoparticle biosynthesis and its physiological effect on mung bean (Vigna radiata L.). Biotechnol. Rep. 5, 22–26.

Raliya, R., Nair, R., Chavalmane, S., Wang, W.N., Biswas, P., 2015b. Mechanistic evaluation of translocation and physiological impact of titanium dioxide and zinc oxide nanoparticles on the tomato (Solanum lycopersicum L.) plant. Metallomics 7 (12), 1584–1594.

Raliya, R., Saharan, V., Dimkpa, C., Biswas, P., 2017. Nanofertilizer for precision and sustainable agriculture: current state and future perspectives. J Agric Food Chem 66 (26), 6487–6503.

Rane, M., Bawskar, M., Rathod, D., Nagaonkar, D., Rai, M., 2015. Influence of calcium phosphate nanoparticles, Piriformospora indica and Glomus mosseae on growth of Zea mays. Adv. Nat. Sci.: Nanosci. Nanotechnol. 6 (4), 045014.

Rane, A.V., Kanny, K., Abitha, V.K., Thomas, S., 2018. Methods for synthesis of nanoparticles and fabrication of nanocomposites. Synthesis of Inorganic Nanomaterials. Woodhead Publishing, pp. 121–139.

Ranjbar, S., Rahemi, M., Ramezanian, A., 2018. Comparison of nano-calcium and calcium chloride spray on postharvest quality and cell wall enzymes activity in apple cv. Red Delicious. Sci. Hortic. 240, 57–64.

Rao, A., Bankar, A., Kumar, A.R., Gosavi, S., Zinjarde, S., 2013. Removal of hexavalent chromium ions by Yarrowia lipolytica cells modified with phyto-inspired $Fe0/Fe_3O_4$ nanoparticles. J. Contaminant Hydrol. 146, 63–73.

Rashid, J., Barakat, M.A., Salah, N., Habib, S.S., 2015. ZnO-nanoparticles thin films synthesized by RF sputtering for photocatalytic degradation of 2-chlorophenol in synthetic wastewater. J. Ind. Eng. Chem. 23, 134–139.

Rautela, A., Rani, J., Das, M.D., 2019. Green synthesis of silver nanoparticles from Tectona grandis seeds extract: characterization and mechanism of antimicrobial action on different microorganisms. J. Anal. Sci. Technol. 10 (1), 1–10.

Ravishankar, T.N., Ramakrishnappa, T., Nagaraju, G., Rajanaika, H., 2015. Synthesis and characterization of CeO_2 nanoparticles via solution combustion method for photocatalytic and antibacterial activity studies. Chem.Open 4 (2), 146–154.

Rengel, Z., 2015. Availability of Mn, Zn and Fe in the rhizosphere. J. Soil. Sci. Plant. Nutr. 15, 397–409.

Rico, C.M., Majumdar, S., Duarte-Gardea, M., Peralta-Videa, J.R., Gardea-Torresdey, J.L., 2011. Interaction of nanoparticles with edible plants and their possible implications in the food chain. J. Agric. Food Chem. 59 (8), 3485–3498.

Rizwan, M., Ali, S., Ali, B., Adrees, M., Arshad, M., Hussain, A., et al., 2019a. Zinc and iron oxide nanoparticles improved the plant growth and reduced the oxidative stress and cadmium concentration in wheat. Chemosphere 214, 269–277.

Rizwan, M., Ali, S., ur Rehman, M.Z., Adrees, M., Arshad, M., Qayyum, M.F., et al., 2019b. Alleviation of cadmium accumulation in maize (Zea mays L.) by foliar spray of zinc oxide nanoparticles and biochar to contaminated soil. Environ. Pollut. 248, 358–367.

Rousk, J., Ackermann, K., Curling, S.F., Jones, D.L., 2012. Comparative toxicity of nanoparticulate CuO and ZnO to soil bacterial communities. PLoS One 7, e34197. Available from: https://doi.org/10.1371/journal.pone.0034197.

Rui, M., Ma, C., Hao, Y., Guo, J., Rui, Y., Tang, X., et al., 2016. Iron oxide nanoparticles as a potential iron fertilizer for peanut (Arachis hypogaea). Front. plant. Sci. 7, 815.

Salah, N., Habib, S.S., Khan, Z.H., Memic, A., Azam, A., Alarfaj, E., et al., 2011. High-energy ball milling technique for ZnO nanoparticles as antibacterial material. Int. J. Nanomed. 6, 863.

Saleh, T.A., 2020. Nanomaterials: classification, properties, and environmental toxicities. Environ. Technol. Innov. 101067.

Saleh, T.A., 2021. Protocols for synthesis of nanomaterials, polymers, and green materials as adsorbents for water treatment technologies. Environ. Technol. Innov. 101821.

Salem, N.M., Albanna, L.S., Awwad, A.M., 2016. Green synthesis of sulfur nanoparticles using Punica granatum peels and the effects on the growth of tomato by foliar spray applications. Environ. Nanotechnol. Monit. Manag. 6, 83–87.

Sana, S., Boodhoo, K., Zivkovic, V., 2019. Production of starch nanoparticles through solvent-antisolvent precipitation in a spinning disc reactor. Green. Process. Synth. 8 (1), 507–515.

Satti, S.H., Raja, N.I., Javed, B., Akram, A., Mashwani, Z.-u-.R, Ahmad, M.S., et al., 2021. Titanium dioxide nanoparticles elicited agro-morphological and physicochemical modifications in wheat plants to control Bipolaris sorokiniana. PLoS One 16 (2), e0246880. Available from: https://doi.org/10.1371/journal.pone.0246880.

Servin, A.D., White, J.C., 2016. Nanotechnology in agriculture: next steps for understanding engineered nanoparticle exposure and risk. NanoImpact 1, 9–12.

Servin, A.D., Pagano, L., Castillo-Michel, H., De la Torre-Roche, R., Hawthorne, J., Hernandez-Viezcas, J.A., et al., 2017. Weathering in soil increases nanoparticle CuO bioaccumulation within a terrestrial food chain. Nanotoxicology 11 (1), 98–111.

Sharma, P., Pant, S., Poonia, P., Kumari, S., Dave, V., Sharma, S., 2018. Green synthesis of colloidal copper nanoparticles capped with Tinospora cordifolia and its application in catalytic degradation in textile dye: an ecologically sound approach. J. Inorg. Organomet. Polym. Mater. 28 (6), 2463–2472.

Shen, Z., Chen, Z., Hou, Z., Li, T., Lu, X., 2015. Ecotoxicological effect of zinc oxide nanoparticles on soil microorganisms. Front. Environ. Sci. Eng. 9 (5), 912–918.

Shiau, B.W., Lin, C.H., Liao, Y.Y., Lee, Y.R., Liu, S.H., Ding, W.C., Lee, J.R., 2018. The characteristics and mechanisms of Au nanoparticles processed by functional centrifugal procedures. J Phys Chem Solids 116, 161–167.

Siddiqui, M.H., Al-Whaibi, M.H., Faisal, M., Al Sahli, A.A., 2014. Nano-silicon dioxide mitigates the adverse effects of salt stress on Cucurbita pepo L. Environ. Toxicol. Chem. 33 (11), 2429–2437.

Sigmund, W., Yuh, J., Park, H., Maneeratana, V., Pyrgiotakis, G., Daga, A., et al., 2006. Processing and structure relationships in electrospinning of ceramic fiber systems. J. Am. Ceram. Soc. 89 (2), 395–407.

Simionescu, O.G., Romaniţan, C., Tutunaru, O., Ion, V., Buiu, O., Avram, A., 2019. RF magnetron sputtering deposition of TiO_2 thin films in a small continuous oxygen flow rate. Coatings 9 (7), 442.

Simonin, M., Cantarel, A.A., Crouzet, A., Gervaix, J., Martins, J.M., Richaume, A., 2018. Negative effects of copper oxide nanoparticles on carbon and nitrogen cycle microbial activities in contrasting agricultural soils and in presence of plants. Front. Microbiol. 9, 3102.

Singh, A., Vihinen, J., Frankberg, E., Hyvärinen, L., Honkanen, M., Levänen, E., 2016. Pulsed laser ablation-induced green synthesis of TiO_2 nanoparticles and application of novel small angle X-ray scattering technique for nanoparticle size and size distribution analysis. Nanoscale Res. Lett. 11 (1), 1–9.

Singh, A., Prasad, S.M., Singh, S., 2018. Impact of nano ZnO on metabolic attributes and fluorescence kinetics of rice seedlings. Environ. Nanotechnol. Monit. Manag. 9, 42–49.

Sirelkhatim, A., Mahmud, S., Seeni, A., Kaus, N.H.M., Ann, L.C., Bakhori, S.K.M., et al., 2015. Review on zinc oxide nanoparticles: antibacterial activity and toxicity mechanism. Nano-micro Lett. 7 (3), 219–242.

SCENIHR, 2009. Risk assessment of products of nanotechnologies. <http://ec.europa.eu/health/phrisk/committees/04 scenihr/docs/scenihro023.pdf> (accessed 21/04/2022.).

Smith, N., Raston, C.L., Saunders, M., & Woodward, R., 2006. Synthesis of magnetic nanoparticles using spinning disc processing. In: Technical Proceedings of the 2006 NSTI Nanotechnology Conference and Trade Show. Nano Science and Technology Institute, 343–346.

Sourice, J., Quinsac, A., Leconte, Y., Sublemontier, O., Porcher, W., Haon, C., et al., 2015. One-step synthesis of Si@ C nanoparticles by laser pyrolysis: high-capacity anode material for lithium-ion batteries. ACS Appl. Mater. Interfaces 7 (12), 6637–6644.

Srivastav, A.K., Kumar, M., Ansari, N.G., Jain, A.K., Shankar, J., Arjaria, N., et al., 2016. A comprehensive toxicity study of zinc oxide nanoparticles vs their bulk in Wistar rats: toxicity study of zinc oxide nanoparticles. Hum. Exp. Toxicol. 35 (12), 1286–1304.

Stoller, M., Ochando-Pulido, J.M., 2020. ZnO nano-particles production intensification by means of a spinning disk reactor. Nanomaterials 10 (7), 1321.

Sun, S., Zeng, H., 2002. Size-controlled synthesis of magnetite nanoparticles. J. Am. Chem. Soc. 124 (28), 8204–8205.

Sunkar, S., Nachiyar, C.V., 2012. Biogenesis of antibacterial silver nanoparticles using the endophytic bacterium Bacillus cereus isolated from Garcinia xanthochymus. Asian Pac. J. Tropical Biomed. 2 (12), 953–959.

Suriyaprabha, R., Karunakaran, G., Yuvakkumar, R., Rajendran, V., Kannan, N., 2012a. Silica nanoparticles for increased silica availability in maize (Zea mays. L) seeds under hydroponic conditions. Curr. Nanosci. 8 (6), 902–908.

Suriyaprabha, R., Karunakaran, G., Yuvakkumar, R., Prabu, P., Rajendran, V., Kannan, N., 2012b. Growth and physiological responses of maize (Zea mays L.) to porous silica nanoparticles in soil. J. Nanopart. Res. 14 (12), 1294.

Tarafder, C., Daizy, M., Alam, M.M., Ali, M.R., Islam, M.J., Islam, R., et al., 2020. Formulation of a hybrid nanofertilizer for slow and sustainable release of micronutrients. ACS Omega 5 (37), 23960–23966.

Tatari, M., Ghazvini, R.F., Etemadi, N., Ahadi, A.M., Mousavi, A., 2012. Analysis of antioxidant enzymes activity, lipid peroxidation and proline content of agropyron desertorum under drought stress. South. West. J. Hortic. Biol. Environ. 3 (1), 9–24.

The (US-)National Nanotechnology Initiative Strategic Plan, 2007. National nanotechnology initiative strategic plan, December 2007. <https://www.nano.gov/2007-Strategic-Plan> (accessed 17.09.21.).

Thirunavukkarasu, M., Subramanian, K.S., 2014a. Surface modified nano-zeolite used as carrier for slow release of sulphur. J. Appl. Nat. Sci. 6 (1), 19–26.

Thirunavukkarasu, M., Subramanian, K.S., 2014b. Surface modified nano-zeolite based sulphur fertilizer on growth and biochemical parameters of groundnut. Trends Biosci. 7 (7), 565–568.

Thomas, E., Rathore, I., Tarafdar, J.C., 2016. Bio-inspired synthesis of nitrogen nanoparticles and its application on pearl millet (Pennisetum americanum L) cv. HHB 67. J. Bionanosci. 10 (4), 300–306.

Thuesombat, P., Hannongbua, S., Akasit, S., Chadchawan, S., 2014. Effect of silver nanoparticles on rice (Oryza sativa L. cv. KDML 105) seed germination and seedling growth. Ecotoxicol. Environ. Saf. 104, 302–309.

Tian, X., He, W., Cui, J., Zhang, X., Zhou, W., Yan, S., et al., 2010. Mesoporous zirconium phosphate from yeast biotemplate. J. Colloid Interface Sci. 343 (1), 344–349.

Tonigold, M., Mailänder, V., 2016. Endocytosis and intracellular processing of nanoparticles in dendritic cells: routes to effective immunonanomedicines. Nanomedicine 11, 20.

Tripathi, D.K., Singh, S., Singh, V.P., Prasad, S.M., Chauhan, D.K., Dubey, N.K., 2016. Silicon nanoparticles more efficiently alleviate arsenate toxicity than silicon in maize cultiver and hybrid differing in arsenate tolerance. Front. Environ. Sci. 4, 46.

Tripathi, D.K., Shweta, Singh, S., Singh, S., Pandey, R., Singh, V.P., et al., 2017a. An overview on manufactured nanoparticles in plants: uptake, translocation, accumulation and phytotoxicity. Plant. Physiol. Biochem. 110, 2–12. CrossRef CAS.

Tripathi, D.K., Singh, S., Singh, V.P., Prasad, S.M., Dubey, N.K., Chauhan, D.K., 2017b. Silicon nanoparticles more effectively alleviated UV-B stress than silicon in wheat (*Triticum aestivum*) seedlings. Plant. Physiol. Biochem. 110, 70–81.

Umar, H., Kavaz, D., Rizaner, N., 2019. Biosynthesis of zinc oxide nanoparticles using Albizia lebbeck stem bark, and evaluation of its antimicrobial, antioxidant, and cytotoxic activities on human breast cancer cell lines. Int. J. Nanomed. 14, 87.

Umar, W., Hameed, M.K., Aziz, T., Maqsood, M.A., Bilal, H.M., Rasheed, N., 2020. Synthesis, characterization and application of ZnO nanoparticles for improved growth and Zn biofortification in maize. Arch. Agron. Soil. Sci. 1–13.

Unni, M., Uhl, A.M., Savliwala, S., Savitzky, B.H., Dhavalikar, R., Garraud, N., et al., 2017. Thermal decomposition synthesis of iron oxide nanoparticles with diminished magnetic dead layer by controlled addition of oxygen. ACS Nano 11 (2), 2284–2303.

Upadhyaya, H., Begum, L., Dey, B., Nath, P.K., Panda, S.K., 2017. Impact of calcium phosphate nanoparticles on rice plant. J. Plant. Sci. Phytopathol. 1, 1–10.

Vafaee Moghadam, A., Iranbakhsh, A., Saadatmand, S., Ebadi, M., Ardebili, Z.O., 2021. New insights into the transcriptional, epigenetic, and physiological responses to zinc oxide nanoparticles in Datura stramonium; potential species for phytoremediation. J. Plant. Growth Regul. pmid:3364969.

Valverde-Alva, M.A., García-Fernández, T., Villagrán-Muniz, M., Sánchez-Aké, C., Castañeda-Guzmán, R., Esparza-Alegría, E., et al., 2015. Synthesis of silver nanoparticles by laser ablation in ethanol: a pulsed photoacoustic study. Appl. Surf. Sci. 355, 341–349.

Van Nguyen, D., Nguyen, H.M., Le, N.T., Nguyen, K.H., Nguyen, H.T., Le, H.M., et al., 2021. Copper nanoparticle application enhances plant growth and grain yield in maize under drought stress conditions. J. Plant. Growth Regul. 1–12.

Verma, S.K., Das, A.K., Gantait, S., Kumar, V., Gurel, E., 2019. Applications of carbon nanomaterials in the plant system: a perspective view on the pros and cons. Sci. Total. Environ. 667, 485–499.

Verma, S.K., Das, A.K., Patel, M.K., Shah, A., Kumar, V., Gantait, S., 2018. Engineered nanomaterials for plant growth and development: a perspective analysis. Sci Total Environ 630, 1413–1435.

Waalewijn-Kool, P.L., Ortiz, M.D., Van Gestel, C.A., 2012. Effect of different spiking procedures on the distribution and toxicity of ZnO nanoparticles in soil. Ecotoxicology 21 (7), 1797–1804.

Wadhwani, S.A., Shedbalkar, U.U., Singh, R., Chopade, B.A., 2016. Biogenic selenium nanoparticles: current status and future prospects. Appl. Microbiol. Biotechnol. 100 (6), 2555–2566.

Walker, W.A., Tarannum, M., Vivero-Escoto, J.L., 2016. Cellular endocytosis and trafficking of cholera toxin B-modified mesoporous silica nanoparticles. J. Mater. Chem. B 4 (7), 1254–1262.

Wang, Z., Xie, X., Zhao, J., Liu, X., Feng, W., White, J.C., et al., 2012. Xylem-and phloem-based transport of CuO nanoparticles in maize (Zea mays L.). Environ. Sci. Technol. 46 (8), 4434–4441.

Watson, J.L., Fang, T., Dimkpa, C.O., Britt, D.W., McLean, J.E., Jacobson, A., et al., 2015. The phytotoxicity of ZnO nanoparticles on wheat varies with soil properties. Biometals 28 (1), 101–112.

Wei, D., Liu, Y., Wang, Y., Zhang, H., Huang, L., Yu, G., 2009. Synthesis of N-doped graphene by chemical vapor deposition and its electrical properties. Nano Lett. 9 (5), 1752–1758.

Wojcieszek, J., Jiménez-Lamana, J., Ruzik, L., Asztemborska, M., Jarosz, M., Szpunar, J., 2020. Characterization of TiO_2 NPs in Radish (Raphanus sativus L.) by single-particle ICP-QQQ-MS. Front. Environ. Sci .

Wongpisutpaisan, N., Charoonsuk, P., Vittayakorn, N., Pecharapa, W., 2011. Sonochemical synthesis and characterization of copper oxide nanoparticles. Energy Procedia 9, 404–409.

Xing, L., Ten Brink, G.H., Chen, B., Schmidt, F.P., Haberfehlner, G., Hofer, F., et al., 2016. Synthesis and morphology of iron–iron oxide core–shell nanoparticles produced by high pressure gas condensation. Nanotechnology 27 (21), 215703.

Xiong, L., 2019. Tailoring hydroxyapatite (HA) nanoparticles as a phosphorus (P) fertiliser in soils (Ph.D. thesis). The University of Queensland, Australia.

Xiong, L., Wang, P., Hunter, M.N., Kopittke, P.M., 2018. Bioavailability and movement of hydroxyapatite nanoparticles (HA-NPs) applied as a phosphorus fertiliser in soils. Environ. Sci.: Nano 5 (12), 2888–2898.

Xue, B., He, D., Gao, S., Wang, D., Yokoyama, K., Wang, L., 2016. Biosynthesis of silver nanoparticles by the fungus Arthroderma fulvum and its antifungal activity against genera of Candida, Aspergillus and Fusarium. Int. J. Nanomed. 11, 1899.

Yadav, S., Rani, N., Saini, K., 2022. Green synthesis of ZnO and CuO NPs using Ficus benghalensis leaf extract and their comparative study for electrode materials for high performance supercapacitor application. Mater Today Proc 49, 2124–2130.

Yadav, B.C., Singh, S., Yadav, T.P., 2015. Titania prepared by ball milling: its characterization and application as liquefied petroleum gas sensor. Synth. Reactivity Inorganic Metal-Organic Nano-Metal Chem. 45 (4), 487–494.

Yang, B., Chen, D., 2017. Synthesis of CuO nanoparticles for catalytic application via ultrasound-assisted ball milling. Process. Appl. Ceram. 11 (1), 39–44.

Yang, F., Hong, F., You, W., Liu, C., Gao, F., Wu, C., et al., 2006. Influence of nano-anatase TiO_2 on the nitrogen metabolism of growing spinach. Biol. Trace Elem. Res. 110 (2), 179–190.

Yang, Z., Xiao, Y., Jiao, T., Zhang, Y., Chen, J., Gao, Y., 2020. Effects of copper oxide nanoparticles on the growth of rice (Oryza Sativa L.) seedlings and the relevant physiological responses. Int. J. Environ. Res. Public. Health 17 (4), 1260.

Ye, L., Li, L., Wang, L., Wang, S., Li, S., Du, J., et al., 2015. MPK3/MPK6 are involved in iron deficiency-induced ethylene production in Arabidopsis. Front. Plant. Sci. 6, 953.

Yoon, H.Y., Lee, J.G., Esposti, L.D., Iafisco, M., Kim, P.J., Shin, S.G., et al., 2020. Synergistic release of crop nutrients and stimulants from hydroxyapatite nanoparticles functionalized with humic substances: toward a multifunctional nanofertilizer. ACS Omega 5 (12), 6598–6610.

Yulianti, L., Bramono, K., Mardliyati, E., Freisleben, H.J., 2016. Effects of Centella asiatica ethanolic extract encapsulated in chitosan nanoparticles on proliferation activity of skin fibroblasts and keratinocytes, type I and III collagen synthesis and aquaporin 3 expression in vitro. J. Pharm. Biomed. Sci. 6 (5).

Zakharova, O., Kolesnikov, E., Shatrova, N., Gusev, A., 2019. The effects of CuO nanoparticles on wheat seeds and seedlings and *Alternaria solani* fungi: in vitro study. IOP Conf. Ser.: Earth Environ. Sci. 226 (1), 012036. IOP Publishing.

Zareabyaneh, H., Bayatvarkeshi, M., 2015. Effects of slow-release fertilizers on nitrate leaching, its distribution in soil profile, N-use efficiency, and yield in potato crop. Environ. Earth Sci. 74 (4), 3385–3393.

Zhao, L.J., Peralta-Videa, J.R., Hernandez-Viezcas, J.A., Hong, J., Gardea-Torresdey, J.L., 2012. Transport and retention behavior of ZnO nanoparticles in two natural soils: effect of surface coating and soil composition. J Nano Res 17, 229–242.

Zhang, R., Zhang, H., Tu, C., Hu, X., Li, L., Luo, Y., et al., 2015. Phytotoxicity of ZnO nanoparticles and the released Zn (II) ion to corn (Zea mays L.) and cucumber (Cucumis sativus L.) during germination. Environ. Sci. Pollut. Res. 22 (14), 11109–11117.

Ziarani, G.M., Malmir, M., Lashgari, N., Badiei, A., 2019. The role of hollow magnetic nanoparticles in drug delivery. RSC adv. 9 (43), 25094–25106.

Further reading

Manikandan, R., Sahi, S.V., Venkatachalam, P., 2015. Impact assessment of mercury accumulation and biochemical and molecular response of Mentha arvensis: a potential hyperaccumulator plant. Sci. World J. 2015.

Zhou, X.H., Huang, B.C., Zhou, T., Liu, Y.C., Shi, H.C., 2015. Aggregation behavior of engineered nanoparticles and their impact on activated sludge in wastewater treatment. Chemosphere 119, 568–576.

Chapter 2

Modifications in elemental contents of edible portions of plants exposed to engineered nanomaterials

Clement O. Ogunkunle[1,2], Ojuolape F. Akande[3], Adewole T. Adetunji[4], Charles O. Adetunji[5] and Paul O. Fatoba[1]

[1]*Environmental Biology Unit, Department of Plant Biology, University of Ilorin, Ilorin, Nigeria,* [2]*Department of Plant Biology, Osun State University, Osogbo, Nigeria,* [3]*Institute of Ecology and Environmental Studies, Obafemi Awolowo University, Ile-Ife, Nigeria,* [4]*Department of Agriculture, Cape Peninsula University of Technology, Wellington, South Africa,* [5]*Applied Microbiology, Biotechnology and Nanotechnology Laboratory, Department of Microbiology, Edo State University Uzairue, Iyamho, Auchi, Nigeria*

Chapter Outline

2.1 Introduction

Nanotechnology is a multifaceted field with its application in almost all the aspects of science and technology. It is concerned with the manufacturing of

Plant Exposure to Engineered Nanoparticles. DOI: https://doi.org/10.1016/B978-0-323-85032-2.00007-5

particles at very small dimensions usually in nanometers. The bedrock of nanotechnology is nanomaterial production, which is an extensive and evolving area of research that has gained worldwide acceptance in the last decades. Recently, a new definition of nanomaterials was issued by USEPA as particles with elementary units that are deliberately produced with at least one of the dimensions measuring roughly between 1 and 100 nanometers so as to ensure compliance on the use of nanomaterials in commercial products (Bashaw, 2012). Additionally, sizeable composite materials containing nanoscale particles have been innovatively designed and built to create new properties (Göran et al., 2010). Composite and electronics products including other materials that are nanoporous are considered nanomaterials due to the incorporation of nanostructures to alter properties. The physical and chemical properties of nanomaterials (NMs) vary from those of bulk materials. A material may show properties distinctly different from those of the bulk as a result of reduction in its size without necessarily exhibiting true nanoscale features. Therefore, in describing nanomaterials, the definition must cut across properties that are derived from pure downscaling and true nanoscale features. Nanoscale is the only attribute shared by all nanomaterials as there is no inter-relationship between size, physical, chemical properties which is valid for all the materials while nonscalable properties show unique nanoscale features over a certain size (Borm et al., 2006).

Mineral elements (nutrients) are important components of soil and are necessary for the growth and development of food crops, and decrease in their levels can result in reduced growth/development and crop quality. These essential elements comprise microelements and macroelements, depending on their requirements for plant's optimal growth. The macroelements are required in larger amounts by plants for their optimal growth and are present in plant tissues in amounts ranging from 0.2% to 4.0% dw, while the microelements are required in lesser amounts (with concentration range of 5–200 ppm in tissues) (Pandey, 2015). Six (6) macroelements (Ca, K, P, N, S, Mg) and eight (8) microelements (Cu, Mo, Zn, Mn, Bo, Cl, Ni, Fe) have long been identified as essential for the normal growth and development of higher plants (Marschner and Marschner, 2012; Pandey, 2015). The essential roles of the mineral elements (nutrients) are briefly described in Table 2.1 (Miller, 2014).

Studies have shown that NPs can accumulate on roots of plants when exposure is via the soil, and the presence of NPs on root surfaces can possibly block the passage of aquaporin proteins and metal transporters that are involved in the movement of mineral elements into plants (Antisari et al., 2015; Yue et al., 2017). This process can alter or affect the mineral nutrition and alter the level of mineral nutrients of food crops. For instance, Yang et al. (2018) demonstrated that Ag-NPs can reduce nutrient level of elements such as Cu, Fe, and Zn in wheat grains. Zinc, which is an important component in the synthesis of Auxin and metalloenzyme activator decreased by 32.7% and 60% in 200 and 2000 mg Ag-NPs/kg-treated grains (Yang et al., 2018). Similarly, Cu which is

TABLE 2.1 Overview of functions of mineral elements in higher plants.

Elements		Available form(s)	Biological function(s) in plants
Macro	Calcium	Ca^{2+}	• Constituent of cell wall, enzyme activator, controls cell permeability and cell wall configuration and involved in cell signaling
	Potassium	K^{+}	• Involved in enzyme activation, osmotic balance, protein metabolism, opening and closing of stomata, and plant cell turgor
	Phosphorus	$H_2PO_4^{-}$, $HPO4^{2-}$	• Involved in photosynthesis and energy supply (e.g., ATP), constituent of nucleic acids, phospholipids in membranes
	Magnesium	Mg^{2+}	• Central element of chlorophyll and activator of some enzymes
	Nitrogen	NO_3^{-}, NO_4^{+}	• Constituent of chlorophylls, nucleic acid and proteins, and some coenzymes
	Sulfur	SO_4^{2-}	• Component of some amino acids (cysteine and methionine) and coenzymes
Micro	Copper	Cu^{+}, Cu^{2+}	• Involved in photosynthesis and mitochondrial respiration, carbon and nitrogen metabolism, oxidative stress protection and cell wall synthesis and activator of some enzymes like Cu-Zn SOD
	Zinc	Zn^{2+}	• Plays a role in enzyme activation, and involved in synthesis of hormones (auxins)
	Manganese	Mn^{2+}	• Involved in chlorophyll synthesis, activation of some enzymes (Mn-SOD), and participates in PS I reaction as water-splitting complex
	Boron	$H_2BO_3^{-}$	• Involved in protein synthesis, transport of sugars, respiration and the metabolism of plant hormones (indole acetic acid), RNA and carbohydrate, cell wall synthesis and lignification, and cell wall structure maintenance by cross-linking of polysaccharides
	Nickel	Ni^{2+}	• Metal cofactor of urease (enzyme) and involved in N metabolism

(*Continued*)

TABLE 2.1 (Continued)

Elements	Available form(s)	Biological function(s) in plants
Iron	Fe^{2+}, Fe^{3+}	• Involved in chlorophyll synthesis, oxido-reduction reaction, respiration, and part of structural components of cytochromes and ferredoxin
Chlorine	Cl^{-}	• Involved in photosynthesis and stomatal regulation (e.g., onion), stimulation of proton pumping ATPase
Molybdenum	MoO_4^{2-}	• Cofactor of enzymes involved in nitrogen metabolism and involved directly in redox reactions
Cobalt	Co^{2+}	• Involved in stem growth, expanding of leaf disks and nitrogen fixation in legumes

Source: Miller, A.J., (2014). Plant Mineral Nutrition. John Wiley & Sons, Chichester. https://doi.org/10.1002/9780470015902.a0023717 and Pandey, R., 2015. Mineral nutrition in plants. In: Bahadur, et al. (Eds.), Plant Biology and Biotechnology: Plant Diversity, Organization, Functions and Improvement. Springer, India.

a structural element for several metal proteins was markedly reduced by ~16% and ~49%, respectively in 200 and 2000 mg Ag-NPs/kg-treated grains. Most studies and reviews on impact of NPs on crop plants have mainly focused on the general physiological and biochemical responses such as germination (length of radicle and plumule), transport and accumulation of NPs, effect on biochemical components but limited data/information are available on the impact of NPs on crop quality, viz-a-viz changes in concentrations of mineral elements. This impact of NPs, either positive or negative, on the nutrient composition of food crops are yet to be fully studied and understood. And in fact, a negative effect of NPs in crop nutrition portends a great consequence on human and animal health through dietary intake of food crops with imbalanced nutrients. Therefore this chapter focuses on (1) use of nanoparticles (NPs) in crop production, (2) changes in mineral element contents of crop plants, (3) mechanisms of change of mineral elements, and (4) implications of changes to human health and generational seed vitality.

2.2 Applications of engineered nanoparticles in agriculture/crop production

Agriculture sector is facing many challenges such as environmental change, unreasonable resources use, and excessive use of chemical fertilizer and pesticides (Raliya et al., 2018). In the current era, nanotechnology is developing as the sixth revolutionary technology. The technology-based agricultural system along with the use of nanotechnology could be an excellent strategy to usher in the next agricultural revolution. It is now playing a crucial role in the field of agriculture by enhancing the quality and yield of crops while minimizing environment risks (Liu and Lal, 2015; Sekhon, 2014). The introduction of engineered nanomaterial technology into agriculture and crop production (known as nano-agriculture) has increased in recent times and facilitated the development of genetically modified crops and precision farming system (Sekhon, 2014). Due to their large surface area, efficient reactivity, and variable sizes, engineered NPs provide a great scope of novel applications in plant nutrition. Nano-agriculture has brought about improvement in quality, yield, and disease management of crop plants. This significant use in agriculture is connected to the unique characteristics of NPs such as large area-to-volume ratio and high holding capacity and size which allow them to be considered as "smart delivery systems" (Solanki et al., 2015). Direct applications of engineered NPs in agriculture generally include delivery of various pesticides, fertilizers, and plant disease and genetic improvement tools (Adhikari et al., 2010). These various applications provide reduced eco-harm, improved target activity and actual concentration with sustained release of active ingredients due to the intrinsic properties possessed by engineered nanomaterials (Lu et al., 2002; Raliya et al., 2015). Application of NP-based material in agri-sectors is touted as important path to sustainable developments. It is now proved to be a good resource for managing agricultural field and drug delivery to plant tissues in addition to maintaining fertility of agricultural soils. Several nanomaterial-based sensors such as electrochemical sensors, optical sensors and biosensor devises are now important instruments in nano-agriculture for detecting trace heavy metals in agricultural soils (Ion et al., 2010; Ogunkunle et al., 2021). In last decade, the use of nanomaterials emerges as alternative solution to protect crop from pathogen. Engineered NPs are reported to be of great use in agriculture for disease control and growth enhancement in plants (Patel et al., 2014). Nanofertilizers are also important component of sustainable agriculture as they are eco-friendly and has high nutrient use efficiency compared to conventional fertilizers. The beneficial roles of nanofertilizers over the conventional ones can be attributed to its controlled and targeted release of nutrients, thereby reducing nutrient input and increasing nutrient availability to crop plants as well as enhancing soil fertility and promoting crop yields (Ashfaq et al., 2017). Plant gene delivery is considered as one of the basic technologies for crop improvement and disease control. Viral gene delivery vectors, microinjection, agrobacterium mediated transformation, and microprojectile bombardment are some of the techniques employed to achieve genetic material transformation. However, these methods have low transformation efficiency. In recent years, the use of various nanomaterials like carbon nanotubes, magnetic NPs, and mesoporous silica NPs for DNA/RNA delivery in plant cells has been reported (Jat et al., 2020). A new technological approach that utilizes smart delivery technique for transport of organic and inorganic agrochemicals into plant genome is getting increasingly explored (Zarei et al., 2018). This technology exploits the capability of engineered nanomaterials to infiltrate or penetrate cell walls/cell membranes and deliver agrochemical and therapeutic agents (such as the polymer conjugates, liposomes, polymersomes, and microspheres) to impact positively the plant physiology/metabolism, thereby resolving critical agricultural issues (Mohammadi et al., 2014). Nanobiosensor is another application of engineered NPs in crop production. Biosensor is a probe that integrates a biological one with an electronic component to yield a measurable signal. It consists of three components: the biological recognition element, the transducer, and the signal processing electronics. A nanobiosensor is built to obtain, process, and analyze the data at the nanoscale level. At present, the nanotechnology-based biosensors are at the early stage of development (Fogel and Limson, 2016). Nanobiosensors open up new opportunities for research and provide tools for bio-analytical applications (Velasco, 2009; You et al., 2009). NP linked biosensors are especially attractive as they can be easily synthesized in bulk by techniques that do not require further modification approaches. Food quality and safety control represents a crucial effort not only to get a healthy food, but also to avoid waste of food. Nanobiosensors also find application in rapid, sensitive, and cost-effective detection of different targets to ensure food quality, safety, authenticity, and traceability along the food supply chain. Many smart and intelligent packaging systems involved the use of nanosensors as monitoring systems to measure physico-chemical properties (pH, moisture, temperature), analyze gas mixtures (oxygen and carbon dioxide), detect pathogens and toxins, (ethanol, lactic acid, acetic acid), and monitor freshness or decomposition (putrescine, cadaverine) (Vanderroost et al., 2014).

2.3 Changes in mineral elements of crop plant tissues under nanoparticle exposure

The uptake and transport (movement) of NPs inside plant tissues, either through apoplastic or symplastic pathway, is a complex process that is highly influenced by NP-type/traits and species of plant. And the pattern of movement is a strong indication of the parts of plant species that the NPs can reach, and also a reflection of the locations of bioaccumulation, either in the roots, stems, leaves, or seeds (Pérez-de-Luque, 2017). The apoplastic pathway involves external transport around the plasma membrane via the spaces of adjacent cell walls while the symplastic pathway presented movement across cells through the plasmodesmata. According to Sun et al. (2014), radial movement within the tissues of plants that permits NPs to access vascular tissues of roots and makes further movement (symplastic transport) to aboveground parts possible is first accredited to apoplastic pathway. However, movement into root xylem (vascular tissue) implies nanomaterials must overcome a barrier that always ends apoplastic transport called the Casparian strip (Pérez-de-Luque, 2017). This mechanism of movement actually dictates location of bioaccumulation of NPs because some can actually be stopped by the Casparian strip and get accumulated at the point of stoppage (Sun et al., 2014; Lv et al., 2015). In another way, symplastic transport may involve the use of phloem tissues (sieve tube elements) to allow the distribution of NPs toward nonphotosynthetic tissues and organs of plants (Raliya et al., 2016). Therefore the pattern of movement of NPs in plants is actually important in dictating what parts of plant such NPs may likely end up and get accumulated. For instance, Pérez-de-Luque (2017) opined that NPs that get transported via the xylem and not the phloem may likely reach the leaves via the shoot, and not move downwardly at all. On the contrary, NPs that translocate via the phloem will most importantly get applied through foliar spray and likely accumulate in organs that serve as sink such as grains, flowers, and fruits (Pérez-de-Luque, 2017).

The accumulation of NPs on surfaces of roots could indirectly or directly prevent the passage of aquaporin proteins and metal transports that are involved in the mobilization of metals into plants (Yue et al., 2017). And most of these metal ions are involved in the plant nutrition and may consequently influence the levels and regulation of plant nutrition. The impact of metal or metal oxide NPs on uptake, accumulation and storage of minerals in organs and tissues of plants are reported to be both positive and negative, and depends on both the NPs and plant species (Zuverza-Mena et al., 2016). The influence of nanomaterials on mineral elements of some crop plants is summarized in Table 2.2.

2.3.1 Metal oxide nanomaterials

2.3.1.1 Cu/CuO-NPs

Studies have shown that Cu-NPs and CuO-NPs portend both positive and detrimental effects on biochemical components of plant tissues/organs such as modifications in the macro/micronutrient contents. In the study of Deng et al. (2020), elemental contents of Mg, Ca, B, Mo, S and Mn were not altered in both the roots and leaves of *Brassica rapa* up to 600 mg CuO-NPs/soil kg treatment. But for macroelements such as K, concentration was reduced by ~40% in both roots and leaves while P was only reduced in roots at highest concentration. Iron in plants are often displaced, and always inhibited from entry into plants by Cu (Schmidt et al., 1997). So, Fe contents are always altered by CuO-NPs in pea plants. For instance, Fe contents in roots of beans were reduced upon exposure to CuO-NPs (Deng et al., 2020). The effects of CuO-NPs on the accumulation of mineral elements of green onion or scallion bulbs (*Allium fistulosum*) popularly used as seasoning have also been investigated (Wang et al., 2020a). In the root, macro- and micro-elements like Ca, Mg and Fe were increased significantly by exposure to CuO-NPs and at same time promoted their leaf contents. The authors affirmed that CuO-NPs can improve the nutritional values of scallion bulbs.

2.3.1.2 TiO_2-NPs

Titanium is a stimulant for increased nutrient absorptions in plants, but its compound in the nanoscale (TiO_2-NPs) presents controversial nutritional effects on crop plants. Some researchers reported that TiO_2-NPs promote mineral nutrients accumulation in crop plants whereas others reported otherwise. For instance, Song et al. (2019) exposed cucumber to TiO_2-NPs from seeds to maturity through soil medium at 100 mg kg^{-1} soil and found that uptake of mineral elements by roots (Si, Mg, Fe, K, and Mn) were reduced by 48.6%, 68.4%, 45.1%, 59.4% and 47.5%, respectively. The authors ascribed the reduction to the negative charge of the NPs and asserted that surface charge is a major factor that influences the interaction between TiO_2-NPs and mineral nutrients. In leaves, concentrations of K and Na were drastically reduced while that of Fe increased (Song et al., 2019). Servin et al. (2013) also reported increases in K, Mg, and Ca concentrations in tissues of TiO_2-NPs-exposed cucumber. In tomato plants cultivated on sewage sludge amended with TiO_2-NPs, Fe, B, and Na were found to be mobilized from roots into stem tissues up to the fruits in great concentrations while P increased only in leaves and stem

TABLE 2.2 Influence of nanomaterials on mineral elements of some food crops.

Nanoparticles	Exposed plant-type	Scientific name	Route of application	Tissue part(s)	Influence on mineral uptake	References
CuO-NPs	Turnip	*Brassica rapa*	Soil	Root & leaf	K↓	Deng et al. (2020)
	Cowpea	*Vigna unguiculata*	Soil	Root	Fe↓	Deng et al. (2020), Ogunkunle et al. (2017), Wang et al. (2020a)
	Green onion	*Allium fistulosum*	Soil	Leaf	Ca↑, Mg↑ & Fe↑	
Cu-NPs	Oregano	*Origanum vulgare*	Soil	Root	Ca↑ & Mg↑	Du et al. (2018)
				Shoot	Mn↓	
$CoFe_2O_4$-NPs	Tomato	*Solanum lycopersicum*	Hydroponic	Leaf	Ca↑ & Mg↑	Lopez-Moreno et al. (2016)
CeO_2 NPs-NPs	Radish	*Raphanus sativus*	Soil	Root & leaf	Ni↓	Corral-Diaz et al. (2014)
CuO-NPs	Lettuce	*Lactuca sativa*	Soil	Leaf	Mn↓, P↓, Ca↓ &	Trujillo-Reyes et al. (2014), Hong et al. (2015)
	Lettuce	*Lactuca sativa*		Leaf	Mg↓	
					P↓ & Fe↓	
SnO_2-NPs	Tomato	*Solanum lycopersicum*	Soil	Leaf & stem	K↓, Ca↓, Mg↓ & S↓	Antisari et al. (2015)

(*Continued*)

TABLE 2.2 (Continued)

Nanoparticles	Exposed plant-type	Scientific name	Route of application	Tissue part(s)	Influence on mineral uptake	References
Fe_2O_3-NPs TiO_2-NPs	Peanut	*Arachis hypogaea*	Soil	Grain	Cu↑	Rui et al. (2018)
FeS-NPs	Brown mustard	*Brassica juncea*	Foliar	Leaf	Fe↑	Rawat et al. (2017)
Fe_3O_4-NPs	Maize	*Zea mays*	Foliar	Leaf	Fe↑ & Ca↑	Jalali et al. (2016)
CeO_2-NPs	Barley	*Horduem vulgare*	Soil	Shoot	P↑, K↑, Ca↑, Mg↑, S↑, Fe↑, Zn↑ & Cu↑	Rico et al. (2015)
Mn-NPs	Wheat	*Triticum aestivum*	Soil	Shoot	P↓ & K↓	Dimpka et al. (2018)
			Foliar	Grain	Mn↑	
$NiFe_2O_4$-NPs	Barley	*Horduem vulgare*	Hydroponic	Leaf	Ni↑, K↑, Na & Fe↑	Tombuloglu et al. (2019)
Graphene oxide	Wheat	*Triticum aestivum*	Hydroponic	Shoot	Ca↓, Mg↓, Fe↓, Zn↓ & Cu↓	Zhang et al. (2016)
Graphene oxide quantum dots	Wheat	*Triticum aestivum*	Soil	Grain	Mg↓, Ca↓, Mn↓ Zn↓, Cu↓ & Mo↓	Li et al. (2018)
Ag-NPs	Wheat	*Triticum aestivum*	Soil	Grain	Fe↓, Cu↓ & Zn↓	Yang et al. (2018)

Ag-NPs	Radish	*Raphanus sativus*	Petri dish	Root	Ca↓, Mg↓, B↓, Cu↓, Mn↓ & Zn↓	Zuverza-Mena et al. (2016)
CuO-NPs	Cucumber	*Cucumis sativus*	Foliar	Fruit	Mo↓	Hong et al. (2016)
CeO_2-NPs					Zn↓	
CeO_2-NPs	Wheat	*Triticum aestivum*	Soil	Grain	Mn↓	Rico et al. (2014)
				Root	Fe↑, Na↑	
Si-NPs	Cucumber	*Cucumis sativus*	Soil	Root	K↑	Alsaeedi et al. (2019)
TiO_2-NPs	Pea	*Arachis hypogaea*	Soil	Root	Mn↓K↓ & Zn↓	Bellani et al. (2020)
				Shoot	Mn↓	
TiO_2-NPs	Tomato	*Solanum lycopersicum*	Hydroponic	Root	P↑	Tiwari et al. (2017)
				Leaf	Fe↓	
ZnO-NPs	Cilantro	Coriandrum sativum	Hydroponic	Root & shoot	Fe↑, Cu↑ & Zn↑	Sharifan et al. (2020)
MWCNTs	Maize	*Zea mays*	Hydroponic		Ca↑ & FeC	Tiwari et al. (2014)
TiO_2-NPs	Cucumber	*Cucumis sativus*	Soil	Fruit	P↑ & K↑	Servin et al. (2013)
CeO_2-NPs	Rice	*Oryza sativa*	Soil	Grain	Fe↓ & S↓	Rico et al. (2013)

(*Continued*)

TABLE 2.2 (Continued)

Nanoparticles	Exposed plant-type	Scientific name	Route of application	Tissue part(s)	Influence on mineral uptake	References
CuO-NPs	Bean	*Phaseolus vulgaris*	Soil	Shoot	Cu↑ & Na↑	Dimkpa et al. (2014a)
					Fe↓, Mn↓, Zn↓ & Ca↓	
TiO_2-NPs	Tomato	*Solanum lycopersicum*	Soil	Stem	Ca↑, Mg↑, P↑& S↑	Antisari et al. (2015)
				Leaf	P↑ & S↑.	
CeO_2-NPs	Tomato	*Solanum lycopersicum*	Soil	Root	Ca↑, K↑, Mg↑ & P↑	Antisari et al. (2015)
				Stem	Ca↑, Mg↑ & Na↓	
				Leaf	K↓, Na↓, P↓ & S↓	
CeO_2-NPs	Spinach	*Spinacia oleracea*	Foliar	Leaf	Zn↓ & Ca↓	Zhang et al. (2019)
ZnO-NPs	Bean	*Phaseolus vulgaris*	Soil	Shoot	Fe↓ & Mn↓	Dimkpa et al. (2014a)
CeO_2-NPs	Barley	*Hordeum vulgare*	Soil	Kernel	K↓ & S↓	Pošcic et al. (2016)
TiO_2-NPs					Zn↑, Mg↑ & Ca↑	
Fe_3O_4-NPs	Soybean	*Glycine max*	Soil	Root & leaf	P↑, K↑, Ca↑ & Mn↑	Torres et al. (2019)

Note: ↓- decrease; ↑- increase.

(Bakshi et al., 2019). Similarly, Tiwari et al. (2017) also showed that TiO_2-NPs promoted nutrient (P, S, and Mg) uptake and accumulation in tomato with about threefold increase of P in roots and leaves. Phosphorus is reported to be a limiting factor in plant nutrition and must be available in high concentration in plants for enhanced growth (Tiwari et al., 2017). TiO_2-NPs have also been reported to increase P uptake in several studies on different plant species such as lettuce (Pilon-Smits et al., 2009) and wheat (Rafique et al., 2018). In peas (*Pisum sativum*), Bellani et al. (2020) reported the alteration of mineral nutrition as Mn (50% in the shoots and 80% in the roots), P, Zn, K, (respectively, 35%, 40%, 80 in the roots) were reduced, and an increase of N in the shoots. This was also reported by the authors to portend possible consequences on the quality of the crop. Recently, Wang et al. (2020b) found that mineral elements such as Mn, Mg, and Zn were reduced in taproots of carrots exposed to TiO_2-NPs with different surface charges, although the authors did not make comparison on leaf elemental contents. However, their results were similar to those obtained by Jacob et al. (2013) on beans and wheat.

2.3.1.3 ZnO-NPs

The importance of ZnO-NPs use as nanofertilizers for improved crop performance and nutrient acquisition has been reported in literature. Peralta-Videa et al. (2014) studied the effects of ZnO-NPs on soybean cultivated to maturity in soil and found that the NP-effect on mineral acquisition in the plants was differential. The authors noted that micronutrients like Cu and Mo were increased in nodules and pods, while leaf Fe and root Cu accumulation were reduced. The significant increase in Mo which happens to be a major component of plant sulfite oxidases and nitrate reductases may assist soybean plants in nitrogen assimilation. The authors also found that ZnO-NPs interfered in accumulation of macronutrients such as K and Mg by reducing their concentrations in roots and nodules. So, from literatures now it is evident that ZnO-NPs can improve acquisition and grain fortification of N in cereal plants. When ZnO-NPs, either fresh or weathered were used for nutrient acquisition and fortification of wheat grains, Dimpka et al. (2018) found that there was preferential partitioning of P and N in the grain, and lesser accumulation in root and shoot. But Mn, contrarily was greatly accumulated in shoot of wheat plants with low translocation to grains. The authors also found that Fe was predominantly accumulated in the roots with insignificant amounts in shoots and grains (Dimpka et al., 2018). In the use of ZnO-NPs to alleviate stress from drought in sorghum, Dimpka et al. (2019) found that the ZnO-NPs improved N and K acquisition in shoots and grains. Also, the use of ZnO-NPs to mitigate drought stress in wheat plants, Dimkpa et al. (2020) found that ZnO-NPs increased grain Zn of wheat by 29% and reduced grain Fe significantly. The surface coating of ZnO-NPs has been reported to influence the NPs effects on crop plants. In red kidney beans (*Phaseolus vulgaris*) cultivated to fruiting under different surface-coated ZnO-NPs, Medina-Velo et al. (2017a,b) reported reduction of shoot Cu content of bean plants grown in soil amended with uncoated ZnO-NPs. Medina-Velo et al. (2017b) also reported that surface coating of ZnO-NPs negatively influenced seed accumulation of K and P concentrations whereas Mg, Ca concentrations were enhanced with differential modulation of micronutrients such as Fe, Cu, Mn, Mo, and Ni. The influence of both ZnO-NPs on the cucumber fruit mineral contents was studied by Zhao et al. (2014), and the authors found that ZnO-NPs was able to change the profile of fruit micronutrients (Mo, Mn, Zn, and Cu). Nano-ZnO was found to reduce the contents of Mo and Cu in the fruits whereas fruit Mn content increased. It is important to note, from the study (Zhao et al., 2014) that Mg was the only macronutrient that was increased in the cucumber fruits.

2.3.1.4 CeO_2-NPs

Cerium oxide NPs (CeO_2-NPs) is a member of the lanthanide family and has been reported in the literature to affect mineral contents in plant tissues. The interactions of CeO_2-NPs with mineral elements in plants could be linked to their stability which permits CeO_2-NPs to remain unchanged (nanoparticulate form) in a wide range of environmental media (Cornelis et al., 2011). In soybean, Peralta-Videa et al. (2014) found that CeO_2-NPs interfered with the uptake and tissue bioaccumulation of both macro and micronutrients. For macronutrients, the authors reported that CeO_2-NPs affected the uptake of Mg, K, and P in soybean roots and also Ca bioaccumulation in leaves and pods whereas P was increased in the pods. As for micronutrients in the study, Zn and Fe concentrations in all organs were reportedly unaltered while Na uptake and subsequent translocation to stems/leaves and bioaccumulation in pods were all reduced (Peralta-Videa et al., 2014). Similarly, CeO_2-NPs reduced Cu and Mo contents in the roots, stem and pods of soybean plants (Peralta-Videa et al., 2014). Also, common bean plants (*P. vulgaris*) exposed to foliar CeO_2-NPs presented pods with reduced Mo, Mn, and S concentrations while none of P, Ca, Fe, and Cu were altered (Xie et al., 2019). However, CeO_2-NPs did not affect the homeostasis of mineral elements in tomato plant (Barrios et al., 2016). In another study, fruit Mo concentrations of cucumber exposed to high CeO_2-NPs concentrations of 400 and 800 $mg\ kg^{-1}$ were reduced while Mg which transports via apoplastic pathway was the only macronutrient that increased in fruit (Zhao et al., 2014). Studies have also shown that CeO_2-NPs can alter mineral nutrients in plants at high concentration regardless of route of exposure. When CeO_2-NPs was foliar-applied to cucumber, Hong et al. (2016) found that mineral contents (Fe, Mn, and B) of cucumber leaves was significantly reduced at elevated concentrations, and at the same time reduced fruit Zn content by ~25%. In wheat grown to maturity (grain production), CeO_2-NPs did not alter Zn, Cu, and Ca concentrations, however increased the root concentrations of P, K, Na, S, Mn, and Fe (Rico et al., 2014). In grains, the authors found marked reduction only in S and Mn concentrations. The authors suggested that CeO_2-NPs may have positively modulated the elemental contents in roots without detrimental impact on the nutritional content of wheat grains. From literature, CeO_2-NPs seemed to have profound negative effect on grain S concentration of cereals while K concentration is usually promoted in grains (Rico et al., 2013, 2014, 2015). Similarly, in another study on a cereal crop barley, the beneficial effect of CeO_2-NPs was profound as remarkable increases in Mg, K, P, Ca, Cu, S, Fe, and Zn concentrations in grains were reported (Rico et al., 2015), and this presents good agronomic use for barley production. The enriched mineral nutrients in barley grains was ascribed to the ability of CeO_2-NPs to modulate elemental uptake by barley leaves which possibly became beneficial during grain formation (Rico et al., 2015).

2.3.2 Metal nanomaterials

2.3.2.1 Cu-NPs

In a study on a culinary species oregano (*Origanum vulgare*), Cu-NPs was found to modify Mg, Fe, Mn and Ca concentrations in roots and leaves by increasing the root uptake of Ca and Mg, and reducing Mn concentrations in both roots and leaves (Du et al., 2018). In similar pattern, shoot concentrations of Mg and Ca were increased while concentrations of Zn and Mn were reduced. The authors suggested that Cu-NPs exerted less effects on Ca and Mg in oregano. However, differential modulations are reported in literatures for leaf Zn contents of edible plants, though Cu and Zn shared same transporter. Some authors have also reported significant alteration of Fe concentrations in pea plants (Ogunkunle et al., 2016, 2017, 2019). Similarly, Apodaca et al. (2017) reported the reduction of Fe contents in seeds of Cu-NPs-exposed kidney bean by ~29% in relation to untreated plants.

2.3.2.2 Ag-NPs

One of the metal/metal oxide NPs that negatively affect acquisition of mineral nutrients and modify their distributions in plants is Ag-NPs. Zuverza-Mena et al. (2016) examined the impact of Ag-NPs on the nutritional quality of radish seedlings and found that Ag-NPs affected the accumulation of mineral nutrients by reducing the uptake of macroelements Ca and Mg and microelements B, Mn, Cu, and Zn. This indicated that Ag-NPs can impair the absorption of mineral elements in seedlings of radish. In wheat plant (*Triticum aestivum*), Yang et al. (2018) observed that mineral contents in grains after four-month exposure period to Ag-NPs (200 and 2000 mg kg^{-1}) significantly reduced the contents of micronutrients Fe, Cu, and Zn. Similarly, Martinez-Fernandez et al. (2016) investigated the hydraulic conductivity and nutrient uptake of *Helianthus annuus* as affected by root adsorption of γ-Fe_2O_3-NPs (50, 100 mg L^{-1}). It was also found that the concentrations of Ca, K, Mg, and S in shoot were reduced as compared to control plants. In addition, foliar application of Ag-NPs was reported to decrease Ca, P, B, Zn, Fe, Cu, K, Mn, Mg and Na in different organs of tomato seedlings with implications on nutrient deficiency (Shams et al., 2013).

2.3.2.3 Ca-NPs

Calcium NP (Ca-NP) is one of the metal NPs that act as nanofertilizer by supplying mineral nutrients to plants and also assist in unbounding minerals in soil for plant uptake (Azeez et al., 2020). Azeez et al. (2020) have found that supplementing soil with Ca-NPs enhanced the nutritional quality of *Moringa oleifera* tissues by significantly increasing the macro and micronutrient contents (Fe, K, Cu, Zn, Mg and Ca). This influence on mineral nutrients of *M. oleifera* was ascribed to the modulative role of Ca-NPs on mineral absorption and the ability to stabilize water absorption balance in *M. oleifera.*

2.3.2.4 Si-NPs

Silicon is a vital element involved in plant protection against some biotic and abiotic stresses like the salinity and heavy metals toxicity (Ma and Takahashi, 2002), so incorporating Si as base constituent of NPs has found use in crop production. Amorphous Si-NPs have been found to improve the productivity of drought-stressed cucumber by increasing the uptake and accumulation of K in stems and leaves (Alsaeedi et al., 2019). For alleviation of metal stress, Wang et al. (2015) investigated the role of foliar-applied Si-NPs prepared from Na_2SiO_3 on Cd-stressed rice, and found that Si-NPs could enhance the uptake and bioaccumulation of Fe, Mg, and Zn in rice tissues. Similarly, in a study to alleviate Cr stress in pea seedlings, Si-NPs was found to improve uptake by roots and shoot of Ca, Mg, K, B, Cu, Mn, and Fe (Tripathi et al., 2015).

2.3.3 Carbon-based nanomaterials

Carbon nanotubes such as multiwalled (MWCNTs) and singlewalled (SWCNTs) carbon nanotubes have been reported to portend both beneficial and detrimental effects on crop plants depending on size, concentration, and ultrasonication among other factors (Ma et al., 2010; Mukherjee et al., 2016). However, little studies are available in literature on the impact on mineral elements accumulation and modulation of food crops. MWCNTs have been reported to increase Fe and Ca concentrations by 2- and 1.6-fold, respectively, in maize grown for seven days in agar medium (Tiwari et al., 2014). Li et al. (2018) have shown that grapheme oxide (GO) significantly reduced the concentrations of Na, Ni, B, and Al in the grains of wheat plant.

Similarly, a previous study has also demonstrated that graphene at 500 mg L^{-1} in water cultivation caused significant nutrient reduction (K, N, Mg, Ca, Cu, Zn, and Fe) in the shoots of wheat (Zhang et al., 2016). And since mineral elements are precursors for some metabolites and important for the growth and quality of crop plants (Rasmussen et al., 2008), it is possible that graphene portends detrimental impacts on crop production.

2.4 Mechanisms of change in tissue minerals as influenced by nanoparticles

There are no straightforward documented mechanisms in the literature for the modulation of minerals in crop plants exposed to nanomaterials. However, some strategies have been identified from published literature and were described next (Fig. 2.1).

2.4.1 Transporter proteins

Transporter plays a major role in the modulation of mineral elements in crop plants. In CuO-NPs-exposed plants, level of mineral accumulation is attributed to regulatory role played by K uptake HKT1 transporters (Schachtman and Schroeder, 1994). And for P that usually experiences reduction in plant

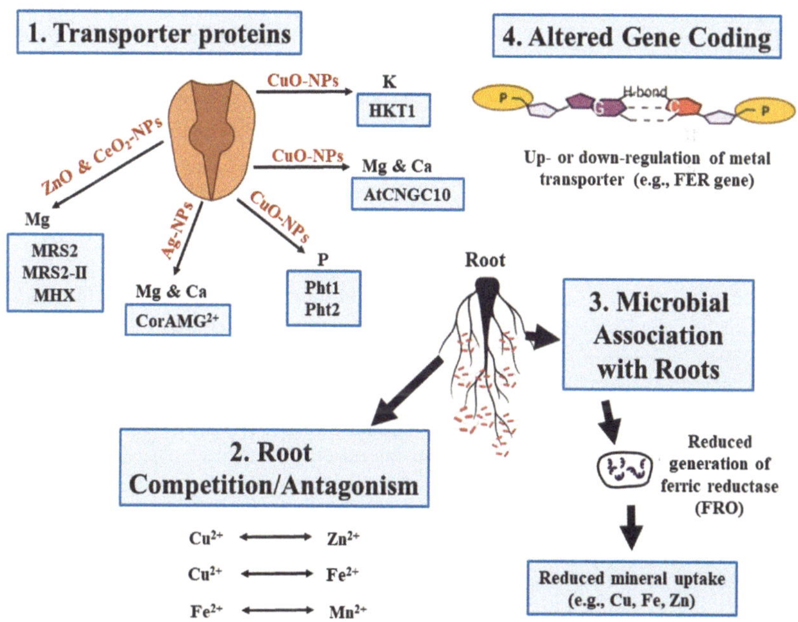

FIGURE 2.1 Schematic representation of mechanisms involved in modulation of mineral elements in nanoparticle-exposed plants.

exposed to CuO-NPs, Zuverza-Mena et al. (2015) ascribed these observations to the membrane transporters blockage or the formation of Cu-phosphates compounds. Further to that, Pht1 transporters in roots are specific for the uptake of P, and Pht2 transporters control the translocation of P to aboveground parts of plant (Buchner et al., 2004), so it is a possibility that CuO-NPs downregulate Pht1 molecules (Deng et al., 2020). In most studies, Ca and Mg are accumulated in tissues of plants exposed to CuO-NPs and these elements have similar transporter- AtCNGC10 (Guo et al., 2010), so CuO-NPs may induce AtCNGC10 expression, thereby facilitating Ca^{2+} and Mg^{2+} transport inside cells.

The mechanism responsible for Ag-NPs translocation in plants has not been well-documented, however, some studies indicated that Ag-NPs transport across cells by either active or passive transporters (such as aquaporins or potassium transporters) through the plasmodesmata and membrane transporters (Akter et al., 2018; Castro-González et al., 2019; Geisler-Lee et al., 2013; Noori et al., 2017). Generally, the transport of Ag across membranes creates positivity in the cell cytoplasm and subsequently modifies the electrochemical potential of the membrane. The modification of the electrochemical potential affects transport of other chemical materials such as nutrients (Noori et al., 2020). Therefore the positivity created by Ag in cells is expected to alter the expression of cation transporters as well as anion transporters, thereby influencing the uptake of minerals by plants. Furthermore, Ag is a membrane disruptor that affect cellular homeostasis, so disruption affects uptake of essential elements and possibly down regulate the genes encoding for metal transporters (Magesky and Pelletier, 2015; Zuverza-Mena et al., 2016). In Ag-NPs-exposed plants, Ca concentrations are usually decreased due to the decreased expression of Ca channel protein, thereby affecting its uptake. Since Mg can be absorbed through transporters $CorAMg^{2+}$ (Maathuis, 2009), there is the likelihood of physical blockage of the channels by Ag-NPs at high concentration, thereby reducing Mg uptake and absorption by plant (Zuverza-Mena et al., 2016).

There has not been a well-defined mechanism for mineral elements accumulation in plants in response to ZnO-NPs. But it is generally believed that ZnO-NPs can promote the expression of aquaporins or channels such as the Mrs2 that is responsible for Mg^{2+} transporter to the shoot (Karley and White, 2009). This mechanism has recently been hypothesized for increased uptake of Mg^{2+} in CeO_2-NP-exposed cucumber where the fruit significantly accumulated Mg ion (Zhao et al., 2014). Similarly, Ca accumulation in plants as a response to ZnO-NPs is ascribed to the size-mediated activation of Ca^{2+} channels that modulate Ca concentration (White and Broadley, 2009) while Nramp transporters are responsible for Mn uptake by plants (Pitman, 2005). Molybdenum, which is one of the micronutrients, requires specific transporters for its distribution in plants after root uptake. Nie et al. (2014) opined that intracellular Mo distribution in plants is coordinated by the specific transporter MOT1 and inter-organ Mo transfer is linked to MOT2. CeO_2-NPs have been reported to modulate mineral elements in plants directly or indirectly through certain ion transporters (Xie et al., 2019). For instance, Mg was found to be transported along the rca channel in roots of wheat plants (Pineros and Tester, 1997). Additional, CeO_2-NPs are also capable to initiate the expression of aquaporins or such channel as the Mrs2 (Mg^{2+} transporter to the shoot), thereby enhancing translocation of Mg^{2+} to the aboveground biomass in plants (Karley and White, 2009; Zhao et al., 2014). This was also confirmed by Li et al. (2017) that asserted that Mg delivery to edible portions of plants is promoted by the MHX, Mrs2; and the Mrs2−11 transporters. Furthermore, P modulation was ascribed to Ce as reported in CeO_2-NPs-exposed cucumber and soybean (Gardea-Torresdey et al., 2014). The uptake mechanisms of boron have been mentioned to include the passive transport via uncharged boric acid molecule and active transport into plant xylem via boron transporter 1 (BOR 1) (Miwa and Fujiwara, 2010; Wimmer and Eichert, 2013).

2.4.2 Metal competition/antagonism

The process of competition or antagonism in the uptake of metallic nutrients by roots is also a process that directs the accumulation of mineral elements in crop plants that are exposed to NPs. Yruela (2009) found that antagonism exists between Cu and Fe levels and these metals could also modulate competitive uptake, that is one could competitively inhibit the other from uptake by roots thereby affecting nutritional level in crops. The mechanism for this competitive inhibition between Fe and Cu was speculated to be caused by altered function of ferric reductase (FRO) in the roots of bean plant exposed to CuO-NPs which led to decreased Fe accumulation in bean shoots (Dimkpa et al., 2014b). The activity of reduced FRO usually result in less generated-Fe^{2+} that will eventually be transported into the plant by Itr-type transporter which is associated with FRO (Sinclair and Krämer, 2012). In another study, Dimkpa et al. (2014c) reported a strong competition between Mn^{2+} and Fe^{2+} in bean plant exposed to ZnO-NPs. In addition, there is competition in uptake channel between Cu and both Fe and Zn, hence the usual antagonistic/reduction in Fe and Zn concentrations of tissues exposed to Cu/CuO-NPs (Hong et al., 2015, 2016; Stegemeier et al., 2015). In fact, antagonism among several divalent metals in soil-plant systems has been reported in the literature such as competition between Zn^{2+} and Fe^{2+} (Alam et al., 2001; Dimpka et al., 2014a; Ghasemi-Fasaei and Ronaghi, 2008; Wang et al., 2011). Another explanation that could suffice for the competitive/antagonistic relationship is the nature of surface areas of metal NPs which are usually high with strong binding potential for mineral elements such as Fe^{2+}, K^{+}, and Ca^{2+} (Song et al., 2019). This allows for strong bonds between NPs and cations in rhizosphere and reduces availability for roots of crop plants.

2.4.3 Soil microbial intervention

Microbial intervention in the rhizosphere of plant exposed to NPs is another mechanism that may control or modulate mineral accumulation in crop plants. Root colonization by associated soil bacterium may be involved in the secretion of siderophores modulating uptake of minerals by NP-exposed roots. For instance, Dimpka et al. (2014a) observed that root colonization of bean roots by *Pseudomonas chlororaphis* O6(PcO6) was involved in the increased secretion of siderophore in the bean root surfaces and simultaneously decreased ferric reductase (FRO) activity after exposure to ZnO-NPs. In bean plant, FRO is an embedded component of the cytoplasmic membrane and known to be involved in the uptake of Fe by reducing Fe^{3+} to Fe^{2+} extracellularly prior to internalization by the root cells. Similarly, Dimpka et al. (2014c) observed that root FRO activity in ZnO-NPs-exposed bean plant reduced by 31% and that the presence of root-associated bacterium, *P. chlororaphis* O6 (PcO6) increased the FRO activity and the generation of siderophore by 17%. However, the microbial intervention decreased uptake of Zn and Fe, by 58% and 18%, respectively. This mechanism has also been reported to apply to other mineral elements such as Cu (Kim and Guerinot, 2007; Mukherjee et al., 2006). Additionally, microbial siderophores have also been reported to modulate or influence the uptake and accumulation of Fe into dicots (Dimkpa et al., 2009; Vansuyt et al., 2007).

2.4.4 Altered gene expressions

Plants are known to adjust their metabolism in order to regulate the uptake of metals by up- or downregulating the level of expression of metal transporters (Ma et al., 2016). The regulation of FER gene which is important for the synthesis of Fe-containing proteins has been reported in metal-based NP-exposed terrestrial plants (Ma et al., 2016). Dimkpa et al. (2014a) reported that ZnO-NPs altered the uptake of Mn and Fe in bean shoot and ascribed this to the altered transcription levels of genes that encodes for metal ion transporters or perhaps specific enzyme activities both of which could greatly impact the Fe and Mn uptake from the substrates. Copper has also been found to inhibit the expression of genes coding for the reductase proteins (Kim and Guerinot, 2007).

2.5 Implications of elemental changes to human health and generational seed vitality

The deficiency of mineral elements in food crops portends serious implications on human health, especially for people living in the rural areas (Elemike et al., 2019). It is reported that any imbalance in nutritional elements such as Na, P, and Ca could portend serious consequences to human/animal nutrition and health (White et al., 2012). In fact, Kemi et al. (2010) proposed that Ca/P ratio in diets must be 1.0 and lower dietary Ca/P ratio could possibly affect Ca homeostasis in humans. Therefore the application of engineered NPs for improved mineral nutrition as a sustainable approach to overcome the challenges of bioavailability and mineral uptake in food crop production is important. For instance, TiO_2-NPs has been found to promote contents of mineral elements in seeds and grains of plants. In rice exposed to 20–100 nm TiO_2-NPs for 130 days, Du et al. (2017) found that some mineral elements such as P, Mg, Ca, Mn, and Zn in grains increased when exposed to NPs. Similarly, Sathiyabama and Manikanan (2021) recently found that foliar application of chitosan-NPs was able to increase the grain mineral contents (Fe, Zn, Mn, P, Ca, and Mg) of finger millet plant. Therefore increase in mineral contents of seeds/grains/fruits when exposed to NPs will be beneficial to plants and humans from nutritional perspective. However, some NPs have potential to create nutrient imbalance in food crop, thereby affecting food quality and safety. For instance, CeO_2-NPs have the potential to negatively affect the balance of some essential elements such as Mo, B, Mn, Fe, Ca and S in crop plants with nutritional implications on the seed quality and human health (Barrios et al., 2019; Hernandez-Viezcas et al., 2013; Peralta-Videa et al., 2014; Priester et al., 2012). Soybean seed which is an important source of Mg for human diet (Moraghan et al., 2006) has been reported to be significantly reduced by CeO_2-NPs. In bean plants exposed to ZnO-NPs for 45 days, the nutritional quality of seeds was altered as levels of Zn, Fe, Mg, Ca, Fe, and Mn in produced seeds were reportedly reduced by NPs application in relation to organic-matter enriched soil (Medina-Velo et al., 2017b). Similarly, the mineral elements such as Fe, Cu, and Zn of wheat grains were reduced upon exposure to Ag-NPs treatments, suggesting low crop quality (Yang et al., 2018). This will consequentially pose health hazards as estimated daily intake of these minerals (Fe, Cu and Zn) in human were further reduced and therefore portends serious health implications (Table 2.3).

Seed vitality of food crops could be affected in terms of dietary nutritional status and germination efficiency for subsequent planting. In fact, seed vitality may become a major concern for farmers as NP-exposed crops could produce subsequent generational offspring with low seed vitality if alteration of normal mineral homeostasis of produced seeds occurs. Song et al. (2019) reported that NPs with high surface areas and strong binding sites could form complexes with cations such as K^+, Ca^{2+}, and Fe^{3+} and lower the phytoavailability/uptake of macro- or micronutrient. Nanoscale ceria is one of the most studied NPs on seed vitality across generational cycle, and in fact, crop plant exposure to CeO_2-NPs have been suggested to impact subsequent generations in terms of growth, development, and seed mineral acquisition (Rico et al., 2017, 2019; Wang et al., 2013). There was also a report that CeO_2-NPs can present intergenerational effects on wheat grain nutrients through reduced seed quality when exposed to CeO_2-NPs. Rico et al. (2017)

TABLE 2.3 **Estimated daily intake (EDI) of some mineral elements in wheat grains exposed to Ag-NPs and implication on human health.**

Treatments		Zn	Fe	Cu
Control	Estimated grain concentration[a] ($mg\ kg^{-1}$)	20	57	5
	Estimated daily intake (EDI) ($mg\ kg^{-1}\ bw\ day^{-1}$)[b]	0.049	0.137	0.012
20 $mg\ kg^{-1}$	Estimated grain concentration ($mg\ kg^{-1}$)	20	54	5
	EDI ($mg\ kg^{-1}\ bw\ day^{-1}$)	0.049	0.130	0.012
200 $mg\ kg^{-1}$	Estimated grain concentration ($mg\ kg^{-1}$)	15	47	4
	EDI ($mg\ kg^{-1}\ bw\ day^{-1}$)	0.036	0.113	0.010
2000 $mg\ kg^{-1}$	Estimated grain concentration ($mg\ kg^{-1}$)	8	4	3
	EDI ($mg\ kg^{-1}\ bw\ day^{-1}$)	0.019	0.010	0.007
Recommended daily intake ($mg\ kg^{-1}\ bw\ day^{-1}$)		1.0[c,d]	0.8[e]	0.5[c,d]

Note: Variables used in computing the estimated daily intake (EDI): global average body weight = 62 kg (Walpole et al., 2012); recommended daily intake of legume/grains = 0.176 $kg\ day^{-1}$ and conversion factor = 0.86 (Zhuang et al., 2009).
[a] *Value adapted from Yang et al. (2018).*
[b] *USEPA (2000).*
[c] *FAO/WHO (1993).*
[d] *FAO/WHO (2010).*
[e] *WHO (1983).*

reported significant reduction in Mn, Ca, K, Mg, and P concentrations in wheat grains at first and second-generational exposures to low CeO_2-NPs dosage. However, there was no concentration-dependent trend in the response to the NPs. Wang et al. (2013) reported that second generation of seedlings of tomato plants from previously CeO_2-NPs-exposed seeds displayed altered development, though mineral contents of fruits were not determined. Rico et al. (2017) also found that the first generation exposure of wheat plants to CeO_2-NPs negatively affected the nutrient profile of the second-generation grains of the plant. The authors reported that the wheat grains from both first- and second-generation plants contained lower minerals such as Mn, Ca, K, Mg, and P, though not concentration-dependent responses. This will naturally portend serious human health issue as such crops will be nutrient-deficient and expose humans to poor nutrition. Contrarily for metal oxide-NPs such as ZnO-NPs, Medina-Velo et al. (2018) have shown minimal effects on transgenerational seed quality in beans (*P. vulgaris*). This is in fact beneficial agronomically, as exposure did not alter mineral contents (K, P, S, Mg, Fe, Mn, B and Cu) in second-generation seeds of bean plant. In a recent study (Rico et al., 2020), wheat plant exposed to CeO_2-NPs for three generations showed a decrease of 13%–16% in Fe concentration in the grains. The authors suggested that the changes in concentration of mineral elements indicate potential alterations in grain quality of exposed progeny or/and negative impact on ecological functioning such as nutrient cycling and ecosystem stability.

2.6 Summary

There has been a great decline in environmental quality and crop productivity due to climate change, conventional farming practices, and misuse of soil resources. This has necessitated the introduction of nanotechnology in agriculture to boost the sustainable production of food for the growing world population. The unique characteristics of NPs such as large surface area, holding capacity, efficient reactivity, and variable size allow them to be considered and used as smart delivery systems in agriculture, which have brought about improvement in nutrient uptake, quality, and yield, and disease management of crop plants. However, NPs also reduce nutrient uptake and accumulation in plant roots, shoots, pods, leaves or seeds, thereby negatively affecting crop quality and yield, and human nutrition. Reports on the impact of NMs application on plant biochemical components and yield are inconsistent and still not well understood. There is, therefore, still significant research that needs to be done to examine the influence of various NPs on seed quality and plant nutrients uptake/availability.

Limited researches have been carried out on the generational impacts of nanomaterials on food crops, especially the consequences on nutritional quality and possible health implications via dietary consumption. There are scanty data on attempts to extrapolate from the accumulation of mineral elements due to nanomaterial exposures in crop plants to potential dietary exposure in humans (viz-a-viz tolerable upper intake levels). Therefore further research is required to better understand the relationship between specific NP characteristics, their uptake by plants, and their potential biological effects, such as the mineral nutrient levels on generational and transgenerational progeny, and possible human dietary risk that may arise. This knowledge could then be utilized to design NP-based systems that address the long-term risk assessment to promote sustainable agriculture (by enhancing crop nutrient quality and promote human health through quality dietary intake).

References

Adhikari, T., Biswas, A.K., Kundu, S., 2010. Nanofertilizer—a new dimension in agriculture. Indian. J. Fertil. 6 (8), 22–24.

Akter, M., Sikder, M.T., Rahman, M.M., Ullah, A.K.M.A., Hossain, K.F.B., Banik, S., et al., 2018. A systematic review on silver nanoparticles-induced cytotoxicity: physicochemical properties and perspectives. J. Adv. Res. 9, 1–16. Available from: https://doi.org/10.1016/j.jare.2017.10.008.

Alam, S., Kamei, S., Kawai, S., 2001. Effect of iron deficiency on the chemical composition of the xylem sap of barley. Soil. Sci. Plant. Nutr. 47, 643–649.

Alsaeedi, A., El-Ramady, H., Alshaal, T., El-Garawany, M., Elhawat, N., Al-Otaibi, A., 2019. Silica nanoparticles boost growth and productivity of cucumber under water deficit and salinity stresses by balancing nutrients uptake. Plant. Physiol. Biochem. 139, 1–10.

Antisari, L.V., Carbone, S., Gatti, A., Vianello, G., Nannipieri, P., 2015. Uptake and translocation of metals and nutrients in tomato grown in soil polluted with metal oxide (CeO_2, Fe_3O_4, SnO_2, TiO_2) or metallic (Ag, Co, Ni) engineered nanoparticles. Environ. Sci. Pollut. Res. 22 (3), 1841–1853. Available from: https://doi.org/10.1007/s11356-014-3509-0.

Apodaca, S., Tan, W., Dominguez, O.E., Hernandez-Viezcas, J.A., Peralta-Videa, J.R., Gardea-Torresdey, J.L., 2017. Physiological and biochemical effects of nanoparticulate copper, bulk copper, copper chloride, and kinetin in kidney bean (Phaseolus vulgaris) plants. Sci. Total. Environ. 599–600, 2085–2094. Available from: https://doi.org/10.1016/j.scitotenv.2017.05.095.

Ashfaq, M., Verma, N., Khan, S., 2017. Carbon nanofibers as a micronutrient carrier in plants: efficient translocation and controlled release of Cu nanoparticles. Env. Sci. Nano 4, 138–148.

Azeez, L., Adejumo, A.L., Ogunbode, M.S., Lateef, A., 2020. Influence of calcium nanoparticles (CaNPs) on nutritional qualities, radical scavenging attributes of Moringa oleifera and risk assessments on human health. J. Food Meas. Charact. . Available from: https://doi.org/10.1007/s11694-020-00465-6.

Bakshi, M., Line, C., Bedolla, D.E., Stein, R.J., Kaegi, R., Sarret, G., et al., 2019. Assessing the impacts of sewage sludge amendment containing nano-TiO_2 on tomato plants: a life cycle study. J. Hazard. Mater. . Available from: https://doi.org/10.1016/j.jhazmat.2019.02.036.

Barrios, A.C., Medina-Velo, I.A., Zuverza-Mena, N., Dominguez, O.E.D., Peralta-Videa, J.R., Gardea-Torresdey, J.L., 2019. Nutritional quality assessment of tomato fruits after exposure to uncoated and citric acid coated cerium oxide nanoparticles, bulk cerium oxide, cerium acetate and citric acid. Plant. Physiol. Biochem. 110, 100–107.

Barrios, A.C., Rico, C.M., Trujillo-Reyes, J., Medina-Velo, I.A., Peralta-Videa, J.R., Gardea-Torresdey, J.L., 2016. Effects of uncoated and citric acid coated cerium oxide nanoparticles, bulk cerium oxide, cerium acetate, and citric acid on tomato plants. Sci. Total. Environ. 563–564, 956–964. Available from: https://doi.org/10.1016/j.scitotenv.2015.11.143.

Bashaw, J., 2012. Hazardous substances, CERCLA, and nanoparticles-can the three be reconciled? Int. J. Dose Response 10 (3), 397–404.

Bellani, L., Siracusa, G., Giorgetti, L., Di Gregorio, S., Castiglione, M.R., Spano, C., et al., 2020. TiO_2 nanoparticles in a biosolid-amended soil and their implication in soil nutrients, microorganisms and Pisum sativum nutrition. Ecotoxicol. Environ. Saf. 190. Available from: https://doi.org/10.1016/j.ecoenv.2019.110095.

Borm, D.R., Paul, J.A., Stephan, H., Thomas, K., Heinz, F., Ken, D., et al., 2006. The potential risks of nanomaterials: a review carried out for ECETOX. Part. Fibre Toxicol. 3, 11. Available from: https://doi.org/10.1186/1743-8977-3-11.

Buchner, P., Takahashi, H., Hawkesford, M.J., 2004. Plant sulphate transporters: co-ordination of uptake, intracellular and long-distance transport. J. Exp. Bot. 55, 1765–1773. Available from: https://doi.org/10.1093/jxb/erh206.

Castro-González, C.G., Sánchez-Segura, L., Gómez-Merino, F.C., Bello-Bello, J.J., 2019. Exposure of stevia (Stevia rebaudiana B.) to silver nanoparticles in vitro: transport and accumulation. Sci. Rep. 9 (1), 1–10. Available from: https://doi.org/10.1038/s41598-019-46828-y.

Cornelis, G., Ryan, B., McLaughlin, M.J., Kirby, J.K., Beak, D., Chittleborough, D., 2011. Solubility and batch retention of CeO_2 nanoparticles in soils. Environ. Sci. Technol. 45, 2777–2782.

Corral-Diaz, B., Peralta-Videa, J.R., Alvarez-Parrilla, E., Rodrigo-García, J., Morales, M.I., Osuna-Avila, P., 2014. Cerium oxide nanoparticles alter the antioxidant capacity but do not impact tuber ionome in Raphanus sativus (L). Plant. Physiol. Biochem. 84, 277–285.

Deng, C., Wang, Y., Cota-Ruiz, K., Reyes, A., Sun, Y., Peralta-Videa, J., et al., 2020. Bok choy (Brassica rapa) grown in copper oxide nanoparticles-amended soils exhibits toxicity in a phenotype-dependent manner: translocation, biodistribution and nutritional disturbance. J. Hazard. Mater. 398. Available from: https://doi.org/10.1016/j.jhazmat.2020.122978.

Dimkpa, C.O., Andrews, J., Fugice, J., Singh, U., Bindraban, P.S., Elmer, WH., Gardea-Torresdey, J.L., White, J.C., 2020. Facile coating of urea with low-dose ZnO nanoparticles promotes wheat performance and enhances Zn uptake under drought stress. Front. Plant Sci. 11. Available from: https://doi.org/10.3389/fpls.2020.00168.

Dimkpa, C., Weinand, T., Asch, F., 2009. Plant-rhizobacteria interactions alleviate abiotic stress conditions. Plant. Cell Env. 32, 1682–1694.

Dimpka, C.O., Hansen, T., Stewart, J., Mclean, J.E., Britt, D.W., Anderson, A.J., 2014a. ZnO nanoparticles and root colonization by a beneficial pseudomonad influence essential metal responses in bean (Phaseolus vulgaris). Nanotoxicology 9. Available from: https://doi.org/10.3109/17435390.2014.900583.

Dimpka, C.O., McLean, J.E., Britt, D.W., Anderson, A.J., 2014b. Nano-CuO and interaction with nano-ZnO or soil bacterium provide evidence for the interference of nanoparticles in metal nutrition of plants. Ecotoxicol 24, 119–129.

Dimpka, C.O., Hansen, T., Stewart, J., McLean, J.E., Britt, D.W., Anderson, A.J., 2014c. ZnO nanoparticles and root colonization by a beneficial pseudomonad influence essential metal responses in bean (Phaseolus vulgaris). Nanotoxicology . Available from: https://doi.org/10.3109/17435390.2014.900583.

Dimpka, C.O., Singh, U., Bindraban, P.S., Elmer, W.E., Gardea-Torresdey, J.L., White, J.C., 2018. Exposure to weathered and fresh nanoparticle and ionic Zn in soil promotes grain yield and modulates nutrient acquisition in wheat (Triticum aestivum). J. Agric. Food Chem. . Available from: https://doi.org/10.1021/acs.jafc.8b03840.

Dimpka, C.O., Singh, U., Bindraban, P.S., Elmer, W.E., Gardea-Torresdey, J.L., White, J.C., 2019. Zinc oxide nanoparticles alleviate drought-induced alterations in sorghum performance, nutrient acquisition, and grain fortification. Sci. Total. Environ. 688, 926–934. Available from: https://doi.org/10.1016/j.scitotenv.2019.06.392.

Du, W., Tan, W., Peralta-Videa, J.R., Gardea-Torresdey, J.L., Ji, R., Yin, Y., et al., 2017. Interaction of metal oxide nanoparticles with higher terrestrial plants: physiological and biochemical aspects. Plant. Physiol. Biochem. 110, 210–225.

Du, W., Tan, W., Yin, Y., Ji, R., Peralta-Videa, J.R., Guo, H., et al., 2018. Differential effects of copper nanoparticles/microparticles in agronomic and physiological parameters of oregano (Origanum vulgare). Sci. Total. Environ. 618, 306–312.

Elemike, E.E., Uzoh, I.M., Onwudiwe, D.C., Babalola, O.O., 2019. The role of nanotechnology in the fortification of plant nutrients and improvement of crop production. Appl. Sci. 9, 499. Available from: https://doi.org/10.3390/app9030499.

FAO/WHO, 1993. Evaluation of certain food additives and contaminants. Forty-first report of the Joint FAO/WHO Expert Committee on Food Additives. World Health Organization, WHO Technical Report Series, No. 837. Geneva.

FAO/WHO, 2010. Summary report of the seventy-third meeting of JECFA. Joint FAO/WHO Expert Committee on Food Additives. Geneva.

Fogel, R., Limson, J., 2016. Developing biosensors in developing countries: South Africa as a case study. Biosensors 6 (1), 5. Available from: https://doi.org/10.3390/bios6010005.

Gardea-Torresdey, J.L., Rico, C.M., White, J.C., 2014. Trophic transfer, transformation, and impact of engineered nanomaterials in terrestrial environments. Environ. Sci. Technol. 48, 2526–2540.

Geisler-Lee, J., Wang, Q., Yao, Y., Zhang, W., Geisler, M., Li, K., et al., 2013. Phytotoxicity, accumulation and transport of silver nanoparticles by Arabidopsis thaliana. Nanotoxicology 7 (3), 323–337. Available from: https://doi.org/10.3109/17435390.2012.658094.

Ghasemi-Fasaei, R., Ronaghi, A., 2008. Interaction of iron with copper, zinc, and manganese in wheat as affected by iron and manganese in a calcareous soil. J. Plant. Nutr. 31, 839–848.

Göran, L, Hubert, R, Gert, R, Birgit, SK, Neil, G, Jean-Philippe, P, et al. (2010). Considerations on a definition of nanomaterial for regulatory purposes. In: Joint Research Centre of European Commission Reference Report.

Guo, K.M., Babourina, O., Christopher, D.A., Borsic, T., Rengel, Z., 2010. The cyclic nucleotide-gated channel AtCNGC10 transports Ca^{2+} and Mg^{2+} in Arabidopsis. Physiol. Plant. 139, 303–312. Available from: https://doi.org/10.1111/j.1399-3054.2010.01366.x.

Hernandez-Viezcas, J.A., Castillo-Michel, H., Andrews, J.C., Cotte, M., Rico, C.M., Peralta-Videa, J.R., et al., 2013. In situ synchrotron fluorescence mapping and coordination of CeO_2 and ZnO nanoparticles in soil cultivated soybean (Glycine max). ACS Nano 7, 1415–1423.

Hong, J., Rico, C.M., Zhao, L., Adeleye, A.S., Keller, A.A., Peralta-Videa, J.R., et al., 2015. Toxic effects of copper-based nanoparticles or compounds to lettuce (Lactuca sativa) and alfalfa (Medicago sativa). Environ. Sci. Process. Impacts 17, 177–185. Available from: https://doi.org/10.1039/C4EM00551A.

Hong, J., Wang, L., Sun, Y., Zhao, L., Niu, G., Tan, W., et al., 2016. Foliar applied nanoscale and microscale CeO_2 and CuO alter cucumber (Cucumis sativus) fruit quality. Sci. Total. Environ. 563–564, 904–911.

Ion, A.C., Ion, I., Culetu, A., 2010. Carbon-based nanomaterials: environmental applications. Univ. Politehn. Buchar. 38, 129–132.

Jacob, D.L., Borchardt, J.D., Navaratnam, L., Otte, M.L., Bezbaruah, A.N., 2013. Uptake and translocation of Ti from nanoparticles in crops and wetland plants. Int. J. Phytoremediat. 15, 142–153. Available from: https://doi.org/10.1080/15226514.2012.683209.

Jalali, M., Ghanati, F., Modarres-Sanavi, A.M., 2016. Effect of Fe_3O_4 nanoparticles and iron chelate on the antioxidant capacity and nutritional value of soil-cultivated maize (Zea mays) plants. Crop. Pasture Sci. 67 (6), 621–628.

Jat, S.K., Bhattacharya, Sharma, M.K., 2020. Nanomaterial based gene delivery: a promising method for plant genome engineering. J. Mater. Chem. B 19, 4165–4175. Available from: https://doi.org/10.1039/D0TB00217H.

Karley, A.J., White, P.J., 2009. Moving cationic minerals to edible tissues: potassium, magnesium, calcium. Plant. Biol. 12, 291–298.

Kemi, V.E., Karkkainen, M.U., Rita, H.J., Laaksonen, M.M., Outila, T.A., Lamberg-Allardt, C.J., 2010. Low calcium:phosphorus ratio in habitual diets affects serum parathyroid hormone concentration and calcium metabolism in healthy women with adequate calcium intake. Br. J. Nutr. 103, 561–568.

Kim, S.A., Guerinot, M.L., 2007. Mining iron, iron uptake and transport in plants. FEB Lett. 581, 2273–2280.

Li, H., Wang, N., Ding, J., Liu, C., Du, H., Huang, K., et al., 2017. The maize CorA/MRS2/MGT-type Mg transporter, ZmMGT10, responses to magnesium deficiency and confers low magnesium tolerance in transgenic Arabidopsis. Plant. Mol. Biol. 95, 269–278. Available from: https://doi.org/10.1007/s11103-017-0645-1.

Li, X., Mu, L., Li, D., Ouyang, S., He, C., Hu, X., 2018. Effects of the size and oxidation of grapheme oxide on crop quality and specific molecular pathway. Carbon 140, 352–361.

Liu, R., Lal, R., 2015. Potentials of engineered nanoparticles as fertilizers for increasing agronomic productions. Sci. Total. Environ. 514, 131–139.

Lopez-Moreno, M.L., Aviles, L.L., Perez, N.G., Irizarry, B.A., Perales, O., Cedeno-Mattei, Y., et al., 2016. Effect of Cobalt Ferrite ($CoFe_2O_4$) nanoparticles on the growth and development of lycopersicon lycopersicum (tomato plants). Sci. Total. Environ. 550, 45–52.

Lu, C.M., Wen, J., Wu, G., Tao, M., Lu, C.M., Zhang, C.Y., et al., 2002. Research on the effect of nanometer materials on germination and growth enhancement of Glycine max and its mechanism. Soybean. Sci. 21 (3), 168–171.

Lv, J., Zhang, S., Luo, L., Zhang, J., Yang, K., Christie, P., 2015. Accumulation, speciation and uptake pathway of ZnO nanoparticles in maize. Environ. Sci. Nano 2, 68–77. Available from: https://doi.org/10.1039/c4en00064a.

Ma, C., Liu, H., Guo, H., Musante, C., Coskun, S.H., Nelson, B., et al., 2016. Defense mechanisms and nutrient displacement in Arabidopsis thaliana upon exposure to CeO_2 and In_2O_3 nanoparticles. Environ. Sci.: Nano. . Available from: https://doi.org/10.1039/C6EN00189K.

Ma, X., Geiser-Lee, J., Deng, Y., Kolmakov, A., 2010. Interactions between engineered nanoparticles (ENPs) and plants: phytotoxicity, uptake and accumulation. Sci. Total. Env. 408 (16), 3053–3061.

Ma, J.F., Takahashi, E., 2002. Soil, fertilizer, and plant silicon research in Japan. Elsevier Science, Amsterdam.

Maathuis, F.J.M., 2009. Physiological functions of mineral macronutrients. Curr. Opin. Plant. Biol. 12, 250–258. Available from: https://doi.org/10.1016/j.pbi.2009.04.003.

Magesky, A., Pelletier, E., 2015. Toxicity mechanisms of ionic silver and polymer-coated silver nanoparticles with interactions of functionalized carbon nanotubes on early development stages of sea urchin. Aquat. Toxicol. 167, 106–123. Available from: https://doi.org/10.1016/j.aquatox.2015.07.011.

Marschner, H., Marschner, P., 2012. Marschner's Mineral Nutrition of Higher Plants, 3rd edn. Elsevier/Academic Press, London.

Martinez-Fernandez, D., Barroso, D., Komarek, M., 2016. Root water transport of Helianthus annuus L. under iron oxide nanoparticle exposure. Env. Sci. Pollut. Res. 23, 1732–1741.

Medina-Velo, I.A., Barrios, A.C., Zuverza-Mena, N., Hernandez-Viezcas, J.A., Chang, C.H., Ji, Z., et al., 2017a. Comparison of the effects of commercial coated and uncoated ZnO nanomaterials and Zn compounds in kidney bean (Phaseolus vulgaris) plants. J. Hazard. Mater. 332, 214–222.

Medina-Velo, I.A., Dominguez, O.E., Ochoa, L., Barrios, A.C., Hernández-Viezcas, J.A., White, J.C., et al., 2017b. Nutritional quality of bean seeds harvested from plants grown in different soils amended with coated and uncoated zinc oxide nanomaterials. Environ. Sci. Nano 4, 2336–2347.

Medina-Velo, I.A., Zuverza-Mena, N., Tamez, C., Ye, Y., Hernandez-Viezcas, J.A., White, J.C., et al., 2018. Minimal transgenerational effect of ZnO nanomaterials on the physiology and nutrient profile of Phaseolus vulgaris. ACS Sustain. Chem. Eng. .

Miller, A.J., 2014. Plant Mineral Nutrition. John Wiley & Sons, Chichester. Available from: https://doi.org/10.1002/9780470015902.a0023717.

Miwa, K., Fujiwara, T., 2010. Boron transport in plants: coordinated regulation of transporters. Ann. Bot. 105, 1103–1108. Available from: https://doi.org/10.1093/aob/mcq044.

Mohammadi, R., Maali-Amiri, R., Mantri, N.L., 2014. Effect of TiO_2 nanoparticles on oxidative damage and antioxidant defense systems in chickpea seedlings during cold stress. Rus. J. Plant. Physiol. 61, 768–775.

Moraghan, J.T., Etchevers, J.D., Padilla, J., 2006. Contrasting accumulations of calcium and magnesium in seed coats and embryos of common bean and soya bean. Food Chem. 95, 554–561. Available from: https://doi.org/10.1016/j.foodchem.2004.10.060.

Mukherjee, A., Majumdar, S., Servin, A.D., Pagano, L., Dhankher, O.P., White, J.C., 2016. Carbon nanomaterials in agriculture: a critical review. Front. Plant. Sci. 7, 172. Available from: https://doi.org/10.3389/fpls.2016.00172.

Mukherjee, I., Campbell, N.H., Ash, J.H., Connolly, E.L., 2006. Expression profiling of the Arabidopsis ferric chelate reductase (FRO) gene family reveals differential regulation by iron and copper. Planta 223, 1178–1190.

Nie, Z., Hu, C., Liu, H., Tan, Q., Sun, X., 2014. Differential expression of molybdenum transport and assimilation genes between two winter wheat cultivars (Triticum aestivum). Plant. Physiol. Biochem. 82, 27–33.

Noori, A., Ngo, A., Gutierrez, P., Theberge, S., White, J.C., 2020. Silver nanoparticle detection and accumulation in tomato (Lycopersicon esculentum). J. Nanopart. Res. 22, 131. Available from: https://doi.org/10.1007/s11051-020-04866-y.

Noori, A., White, J.C., Newman, L.A., 2017. Mycorrhizal fungi influence on silver uptake and membrane protein gene expression following silver nanoparticle exposure. J. Nanopart. Res. 19 (2), 66. Available from: https://doi.org/10.1007/s11051-016-3650-4.

Ogunkunle, CO, Bornmann, B, Wagner, R, Fatoba, PO, Frahm, R, Luetzenchirchen-Hecht, D., 2016. Biotransformation evidence of copper nanoparticles in cowpea (Vigna unguiculata) by XANES. In: Sternemann, C., Wagner, R., Lutzenkirchen-Hecht, D. (Eds.), 12th DELTA User Meeting and Annual Report 2016, Dortmund, Germany. https://doi.org/10.13140/RG.2.2.11579.62249.

Ogunkunle, C.O., Bornmann, B., Wagner, R., Fatoba, P.O., Frahm, R., Luezenkirchen-Hecht, D., 2019. Copper uptake, tissue partitioning and biotransformation evidence by XANES in cowpea (Vigna unguiculata L) grown in soil amended with nano-sized copper particles. Environ. Nanotechnol. Monit. Manag. 12. Available from: https://doi.org/10.1016/j.enmm.2019.100231.

Ogunkunle, C.O., Folarin, O., Olorunmaiye, K.S., Fatoba, P.O., 2017. Bioaccumulation of copper and tissue partitioning of selected mineral elements in German round pea (Kleine Rheinlanderin) under the influence of copper nanoparticles. Annals. Food Sci. Technol. 18 (4), 653–660.

Ogunkunle, C.O., Oyedeji, S., Okoro, H.K., Adimula, V., 2021. Interaction of nanoparticles with soil. In: Amrane, A., Mohan, D., Nguyen, T.A., Assad, A.A., Yasin, G. (Eds.), Nanomaterials for Soil Remediation. Elsevier, Netherland, pp. 101–132. , chapter 6.

Pandey, R., 2015. Mineral nutrition in plants. In: Bahadur, et al., (Eds.), Plant Biology and Biotechnology: Plant Diversity, Organization, Functions and Improvement. Springer, India. Available from: https://doi.org/10.1007/978-81-322-2286-6_20.

Patel, N., Jha, A., Gautam, H.K., 2014. Agronanotechnology for plant fungal disease management: a review. Int. J. Curr. Microbiol. Appl. Sci. 3 (10), 71–84.

Peralta-Videa, J.R., Hernandez-Viezcas, J.A., Zhao, L., Diaz, B.C., Ge, Y., Priester, J.H., et al., 2014. Cerium dioxide and zinc oxide nanoparticles alter the nutritional value of soil cultivated soybean plants. Plant. Physiol. Biochem. 80, 128–135.

Pérez-de-Luque, A., 2017. Interaction of nanomaterials with plants: what do we need for real applications in agriculture? Front. Environ. Sci. 5, 12. Available from: https://doi.org/10.3389/fenvs.2017.00012.

Pilon-Smits, E.A., Quinn, C.F., Tapken, W., Malagoli, M., Schiavon, M., 2009. Physiological functions of beneficial elements. Curr. Opin. Plant. Biol. 12, 267–274.

Pineros, M., Tester, M., 1997. Calcium channels in higher plant cells: selectivity, regulation and pharmacology. J. Exp. Bot. 48, 551–557.

Pitman, J.K., 2005. Managing the manganese: molecular mechanisms of manganese transport and homeostasis. N. Phytol. 167, 733–742.

Pošcic, F., Mattiello, A., Fellet, G., Miceli, F., Marchiol, L., 2016. Effects of cerium and titanium oxide nanoparticles in soil on the nutrient composition of barley (Hordeum vulgare L.) kernels. Int. J. Environ. Res. Public. Health 13. Available from: https://doi.org/10.3390/ijerph13060577.

Priester, J.H., Ge, Y., Mielke, R.E., Horst, A.M., Cole Moritz, S., Espinosa, K., et al., 2012. Soybean susceptibility to manufactured nanomaterials: evidence for food quality and soil fertility interruption. Proc. Natl. Acad. Sci. U.S.A. 109 (37), E2451–E2456.

Rafique, R., Zahra, Z., Virk, N., Shahid, M., Pinelli, E., Park, T.J., et al., 2018. Dose-dependent physiological responses of Triticum aestivum L. to soil applied TiO_2 nanoparticles: alterations in chlorophyll content, H_2O_2 production, and genotoxicity. Agric. Ecosyst. Environ. 255, 95–101.

Raliya, R., Biswas, P., Tarafdar, J.C., 2015. TiO_2 nanoparticle biosynthesis and its physiological effect on mung bean (Vigna radiata L.). Biotechnol. Rep. 5, 22–26. Available from: https://doi.org/10.1016/j.btre.2014.10.009.

Raliya, R., Dimpka, C., Biswas, P., 2018. Nano fertilizer for precision and sustainable agriculture: current state and future perspectives. J. Agric. Food Chem. 66 (26), 6487–6503.

Raliya, R., Franke, C., Chavalmane, S., Nair, R., Reed, N., Biswas, P., 2016. Quantitative understanding of nanoparticle uptake in watermelon plants. Front. Plant. Sci. 7, 1288. Available from: https://doi.org/10.3389/fpls.2016.01288.

Rasmussen, S., Parsons, A.J., Fraser, K., Xue, H., Newman, J.A., 2008. Metabolic profiles of Lolium perenne are differentially affected by nitrogen supply, carbohydrate content, and fungal endophyte infection. Plant. Physiol. 146, 1440–1453.

Rawat, M., Nayan, R., Negi, B., Zaidi, M.G.H., Arora, S., 2017. Physio-biochemical basis of iron-sulfide nanoparticle induced growth and seed yield enhancement in Brassica juncea. Plant. Physiol. Biochem. 118, 274–284.

Rico, C.M., Abolade, O.M., Wagner, D., Lottes, B., Rodriguez, J., Biagioni, R., et al., 2019. Wheat exposure to cerium oxide nanoparticles over three generations reveals transmissible changes in nutrition, biochemical pools, and response to soil N. J. Hazard. Mater. . Available from: https://doi.org/10.1016/j.jhazmat.2019.121364.

Rico, C.M., Barrios, A.C., Tan, W., Rubenecia, R., Lee, S.C., Varela-Ramirez, A., et al., 2015. Physiological and biochemical response of soil-grown barley (Hordeum vulgare L.) to cerium oxide nanoparticles. Env. Sci. Pollut. Res. 22 (14), 10551–10558.

Rico, C.M., Lee, S.C., Rubenecia, R., Mukherjee, A., Hong, L., Peralta-Videa, J.R., et al., 2014. Cerium oxide nanoparticles impact yield and modify nutritional parameters in wheat (Triticum aestivum L.). J. Agric. Food Chem. 62, 9669–9675.

Rico, C.M., Morales, M.I., Barrios, A.C., McCreary, R., Hong, J., Lee, W.Y., et al., 2013. Effect of cerium oxide nanoparticles on the quality of rice (Oryza sativa L.) grains. J. Agric. Food Chem. 61, 11278–11285.

Rico, C.M., Wagner, D., Abolade, O., Lottes, B., Coates, K., 2020. Metabolomics of wheat grains generationally-exposed to cerium oxide nanoparticles. Sci. Total. Environ. 712, 136487. Available from: https://doi.org/10.1016/j.scitotenv.2019.136487.

Rico, C.M., Johnson, M.G., Marcus, M.A., Andersen, C.P., 2017. Intergenerational responses of wheat (Triticum aestivum L.) to cerium oxide nanoparticles exposure. Env. Sci. Nano 4 (3), 700–711. Available from: https://doi.org/10.1039/c7en00057j.

Rui, M., Ma, C., White, J.C., Hao, Y., Wang, Y., Tang, X., et al., 2018. Metal oxide nanoparticles alter peanut (Arachis hypogaea L.) physiological response and reduce nutritional quality: a life cycle study. Environ. Sci.: Nano . Available from: https://doi.org/10.1039/c8en00436f.

Sathiyabama, M., Manikanan, A., 2021. Foliar applivcation of chitosan nanoparticle improves yield, mineral content and boost innate immunity in finger millet plants. Carbohydr. Polym. 258. Available from: https://doi.org/10.1016/j.carbpol.2021.117691.

Schachtman, D.P., Schroeder, J.I., 1994. Structure and transport mechanism of a high affinity potassium uptake transporter from higher plants. Nature 370, 655–658. Available from: https://doi.org/10.1038/370655a0.

Schmidt, W., Bartels, M., Tittel, J., Fühner, C., 1997. Physiological effects of copper on iron acquisition processes in Plantago. New Phytol. 135 (4), 659–666. Available from: https://doi.org/10.1046/j.1469-8137.1997.00691.x.

Sekhon, B.S., 2014. Nanotechnology in agri-food production: an overview. Nanotechnol. Sci. Appl. 7, 31–53.

Servin, A.D., Morales, M.I., Castillo-Michel, H., Hernandez-Viezcas, J.A., Munoz, B., Zhao, L., et al., 2013. Synchrotron verification of TiO_2 accumulation in cucumber fruit: a possible pathway of TiO_2 nanoparticle transfer from soil into the food chain. Environ. Sci. Technol. 47, 11592–11598.

Shams, G., Ranjbar, M., Amiri, A.A., Khodarahmpour, Z., 2013. The effect of 35 nm silver nanoparticles on antagonistic and synergistic mineral elements in leaves and fruit of tomato (Lycopersicon esculentum Mill.). Int. J. Agric. Crop. Sci. 5, 439–500.

Sharifan, H., Moore, J., Ma, X., 2020. Zinc oxide (ZnO) nanoparticles elevated iron and copper contents and mitigated the bioavailability of lead and cadmium in different leafy greens. Ecotoxicol. Environ. Saf. 191, 110177. Available from: https://doi.org/10.1016/j.ecoenv.2020.110177.

Sinclair, S.A., Krämer, U., 2012. The zinc homeostasis network of land plants. Biochim. Biophys. Acta 1823, 1553–1567.

Solanki, P., Bhargava, A., Chhipa, H., Jain, N., Panwar, J., 2015. Nano-fertilizers and their smart delivery system. In: Rai, M., Ribeiro, C., Mattoso, L., Duran, N. (Eds.), Nanotechnologies in Food and Agriculture. Springer, Cham, pp. 81–101. Available from: https://doi.org/10.1007/978-3-319-14024-7_4.

Song, C., Ye, F., Zhang, H., Hong, J., Hua, C., Wang, B., et al., 2019. Metal(loid) oxides and metal sulfides nanomaterials reduced heavy metals uptake in soil cultivated cucumber plants. Environ. Pollut. . Available from: https://doi.org/10.1016/j.envpol.2019.113354.

Stegemeier, J.P., Schwab, F., Colman, B.P., Webb, S.M., Newville, M., Lanzirotti, A., et al., 2015. Speciation matters: bioavailability of silver and silver sulfide nanoparticles to alfalfa (Medicago sativa). Environ. Sci. Technol. 49, 8451.

Sun, D., Hussain, H.I., Yi, Z., Siegele, R., Cresswell, T., Kong, L., et al., 2014. Uptake and cellular distribution, in four plant species, of fluorescently labeled mesoporous silica nanoparticles. Plant. Cell Rep. 33, 1389–1402. Available from: https://doi.org/10.1007/s00299-014-1624-5.

Tiwari, D., Dasgupta-Schubert, N., Cendejas, L.V., Villegas, J., Montoya, L.C., García, S.B., 2014. Interfacing carbon nanotubes (CNT) with plants: enhancement of growth, water and ionic nutrient uptake in maize (Zea mays) and implications for nanoagriculture. Appl. Nanosci. 4, 577–591.

Tiwari, M., Sharma, N.C., Fleischmann, P., Burbage, J., Venkatachalam, P., Sahi, S.V., 2017. Nanotitania exposure causes alterations in physiological, nutritional and stress responses in tomato (Solanum lycopersicum). Front. Plant. Sci. 8, 1–12.

Tombuloglu, H., Slimani, Y., Tombuloglu, G., Munirah Almessiere, M., Baykal, A., Ercan, I., et al., 2019. Tracking of $NiFe_2O_4$ nanoparticles in barley (Hordeum vulgare L.) and their impact on plant growth, biomass, pigmentation, catalase activity, and mineral uptake. Environmental. Nanotechnol. Monit. Manag. 11–100223. Available from: https://doi.org/10.1016/j.enmm.2019.100223.

Torres, A.D., Govea-Alcaide, E., Masunaga, S.H., Rosabal, L.F., Effenberger, F., Rossi, L.M., et al., 2019. Impact of Fe_3O_4 nanoparticle on nutrient accumulation in common bean plants grown in soil. SN Appl. Sci. . Available from: https://doi.org/10.1007/s42452-019-0321-y.

Tripathi, D.K., Singh, V.P., Prasad, S.M., Chauhan, D.K., Dubey, N.K., 2015. Silicon nanoparticles (SiNp) alleviate chromium (VI) phytotoxicity in Pisum sativum (L.) seedlings. Plant. Physiol. Biochem. . Available from: https://doi.org/10.1016/j.plaphy.2015.07.026.

Trujillo-Reyes, J., Majumdar, S., Botez, C.E., Peralta-Videa, J.R., Gardea-Torresdey, J.L., 2014. Exposure studies of core-shell Fe/Fe_3O_4 and Cu/CuO NPs to lettuce (Lactuca sativa) plants: are they a potential physiological and nutritional hazard. J. Hazard. Mater. 267, 255–263.

USEPA, 2000. Risk based concentration table. United States Environmental Protection Agency, Washington, DC.

Vanderroost, M., Ragaert, P., Devlieghere, F., DeMeulenaer, B., 2014. Intelligent food packaging: the next generation. Trends Food Sci. Technol. 39, 47–62.

Vansuyt, G., Robin, A., Briat, J.F., Curie, C., Lemanceau, P., 2007. Iron acquisition from Fe-pyoverdine by Arabidopsis thaliana. Mol. Plant. Microbe Interact. 20, 441–4417.

Velasco, M.N., 2009. Optical biosensors for probing at the cellular level: a review of recent progress and future prospects. Semin. Cell Develop. Biol. 20 (1), 27–33.

Wang, H., Kou, X., Pei, Z., Xial, J.Q., Shan, X., Xing, B., 2011. Physiological effects of magnetite (Fe3O4) nanoparticles on perennial ryegrass (Lolium perenne L.) and pumpkin (Cucurbita mixta) plants. Nanotoxicology 5, 30–42.

Walpole, S.C., Prieto-Merino, D., Edwards, P., Cleland, J., Stevens, G., Roberts, I., 2012. The weight of nations: an estimation of adult human biomass. BMC Public Health 12, 439. Available from: https://doi.org/10.1186/1471-2458-12-439.

Wang, L., Deng, C., Cota-Ruiz, K., Peralta-Videa, J.R., Sun, Y., Rawat, S., et al., 2020a. Improvement of nutrient elements and allicin content in green onion (Allium fistulosum) plants exposed to CuO nanoparticles. Sci. Total. Environ. 725. Available from: https://doi.org/10.1016/j.scitotenv.2020.138387.

Wang, Q., Ebb, S.D., Chen, Y., Ma, X., 2013. Trans-generational impact of cerium oxide nanoparticles on tomato plants. Metallomics 5, 753. Available from: https://doi.org/10.1039/c3mt00033h.

Wang, S., Wang, F., Gao, S., 2015. Foliar application with nano-silicon alleviates Cd toxicity in rice seedlings. Env. Sci. Pollut. Res. Int. 22, 2837–2845. Available from: https://doi.org/10.1007/s11356-014-3525-0.

Wang, Y., Deng, C., Cota-Ruiz, K., Tan, W., Reyes, A., Peralta-Videa, J.R., et al., 2020b. Effects of different surface-coated $nTiO_2$ on full-grown carrot plants: impacts on root splitting, essential elements, and Ti uptake. J. Hazard. Mater. 402. Available from: https://doi.org/10.1016/j.jhazmat.2020.123768.

White, P.J., Broadley, M.R., Gregory, P.J., 2012. Managing the nutrition of plants and people. Appl. Environ. Soil. Sci. . Available from: https://doi.org/10.1155/2012/104826.

White, P.J., Broadley, M.R., 2009. Biofortification of crop with seven mineral elements often lacking in human diets- iron, zinc, copper, calcium, magnesium, selenium and iodine. N. Phytologist 182, 49–84.

WHO, 1983. Evaluation of the joint FAO/WHO Expert Committee on Food additives (JECFA). World Health Organisation. Geneva. . Available from: https://apps.who.int/food-additives-contaminants-jecfa-database/chemical.aspx?chemID = 2859Accessed 02/03/2021.

Wimmer, M.A., Eichert, T., 2013. Review: mechanisms for boron deficiency-mediated changes in plant water relations. Plant. Sci 20, 25–32. Available from: https://doi.org/10.1016/j.plantsci.2012.12.012.

Xie, C., Ma, Y., Yang, J., Zhang, B., Luo, W., Feng, S., et al., 2019. Effects of foliar applications of ceria nanoparticles and $CeCl_3$ on common bean (Phaseolus vulgaris). Environ. Pollut. 250, 530–536.

Yang, J., Jiang, F., Ma, C., Rui, Y., Rui, M., Muhammad, A., et al., 2018. Alteration of crop yield and quality of wheat upon exposure to silver nanoparticles in a life cycle study. J. Agric. Food Chem. . Available from: https://doi.org/10.1021/acs.jafc.7b04904.

You, C., Bhagawati, M., Brecht, A., Piehler, J., 2009. Affinity capturing for targeting proteins into micro and nanostructures. Anal. Bioanal. Chem. 393 (6-7), 1563–1570. Available from: https://doi.org/10.1007/s00216-008-2595-6.

Yruela, I., 2009. Copper in plants, acquisition, transport and interactions. Funct. Plant. Biol. 36, 409–430.

Yue, L., Ma, C., Zhan, X., White, J.C., Xing, B., 2017. Molecular mechanisms of maize seedling response to La_2O_3 NP exposure: water uptake, aquaporin gene expression and signal transduction. Environ. Sci. Nano. 4.

Zarei, F., Negahdari, B., Eatemadi, A., 2018. Diabetic ulcer regeneration: stem cells, biomaterials, growth factors. Artif. Cell Nanomed. Biotechnol. 46 (1), 26–32.

Zhang, H., Lu, L., Zhao, X., Zhao, S., Gu, X., Du, W., et al., 2019. Metabolomics reveals the "invisible" responses of spinach plants exposed to CeO_2 nanoparticles. Environ. Sci. Technol. 53, 6007–6017.

Zhang, P., Zhang, R., Fang, X., Song, T., Cai, X., Liu, H., et al., 2016. Toxic effects of grapheme on the growth and nutritional level of wheat: short and long term exposure studies. J. Hazard. Mater. 317, 543–551.

Zhao, L., Peralta-Videa, J.R., Rico, C.M., Hernandez-Viezcas, J.A., Sun, Y., Niu, G., et al., 2014. CeO_2 and ZnO nanoparticles change the nutritional qualities of cucumber (Cucumis sativus). J. Agric. Food Chem. 62, 2752–2759.

Zhuang, P., McBride, M.B., Xia, H., Lia, N., Lia, Z., 2009. Health risk from heavy metals via consumption of food crops in the vicinity of Dabaoshan mine, South China. Sci. Total. Env. 407, 1551–1561.

Zuverza-Mena, N., Armendariz, R., Peralta-Videa, J.R., Gardea-Torresdey, J.L., 2016. Effects of silver nanoparticles on radish sprouts: root growth reduction and modifications in the nutritional value. Front. Plant. Sci. 7, 90. Available from: https://doi.org/10.3389/fpls.2016.00090.

Zuverza-Mena, N., Medina-Velo, I.A., Barrios, A.C., Tan, W., Peralta-Videa, J.R., Gardea-Torresdey, J.L., 2015. Copper nanoparticles/compounds impact agronomic and physiological parameters in cilantro (Coriandrum sativum). Environ. Sci. Process. Impacts 17, 1783–1793. Available from: https://doi.org/10.1039/C5EM00329F.

Chapter 3

Spontaneous plant species responses to engineered nanoparticles

Luca Marchiol, Laura Pilotto, Daniel Lizzi and Guido Fellet
Department of Agrifood, Environmental and Animal Science (DI4A), University of Udine, Udine, Italy

Chapter Outline

3.1 Engineered nanoparticles and plants

Nanoscience and nanotechnology are rapidly evolving in different applications because of the potential to revolutionize human life. Engineered nanoparticles (ENPs) have one dimension between about 1 and 100 nm and novel properties due to their size (Boyes et al., 2017). Considerable headways have been made for the applications of ENPs and nano-enabled products in medicine, energy, electronics, innovative materials, and many others (Roco, 2020; Sun, 2019).

The flip side of nanotechnology is the release of tons of ENPs into the environment (Mortimer and Holden, 2019). Discharged ENPs contaminate soil, move to surface and groundwater, and enter the biota posing risks to the food chains. Simultaneously, nanomaterials directly discharged or eventually

Plant Exposure to Engineered Nanoparticles. DOI: https://doi.org/10.1016/B978-0-323-85032-2.00006-3

released in wastewater effluents can be transported to aquatic systems. According to the models describing the fluxes of ENPs in the environment, we know that soil and water are the end point of nanomaterials (Holden et al., 2016).

The environmental fate of ENPs is influenced by several factors, depending on their native properties (size and shape, hydrodynamic behavior, crystalline phase, surface charge, and others) and the conditions of the end-point medium. In the native or transformed form, due to subsequent interactions with the mineral and organic components present in the solid matrices and the water, the ENPs come into contact with the environmental biotic component. However, we still have patchy knowledge regarding the impacts of these materials on biota (Reddy Pullagurala et al., 2019). In fact, the toxicity of ENMs and long-term effects on ecosystems are still new topics.

When we consider that plants represent about 82% of living biomass on Earth (Bar-On et al., 2018) and that primary producers support the food chains, then it is clear that from an ecological point of view, it is of the utmost importance to understand the relationships between ENPs and plants. In particular, studying the behavior and fate of ENPs within plants is of great significance for exploring (1) ENPs uptake, translocation, and storage in plant tissues, (2) mechanisms of plant toxicity, and (3) life cycle risk assessment of ENPs and risks of transfer to the trophic chain.

The research has been paid almost exclusively to food crops, while the spontaneous plant species have been almost neglected. Although this was primarily justified by the potential risks for ENPs human exposure, the potential negative impact of ENPs on primary producers could have severe consequences on food webs and ecosystem services (Hawthorne et al., 2014). Therefore, it should not be deemed less significant.

Experiments carried out on crops demonstrated that the chemical and physical properties of ENPs (e.g., size, shape, structure, composition, concentration, and others), the environmental conditions, the plant species, and age contribute to determining the effects on plants (Spielman-Sun et al., 2019). It is not advisable to generalize the crops to other plants living on natural ecosystems, neither fertilized nor irrigated, and potentially more exposed to ENPs fluxes having a longer life-cycle than crops.

Studies on the relationships between plants and nanomaterials have started relatively recently. There is still no thorough and systematic knowledge on the ENPs uptake and toxicity mechanisms on plants. This fact poses a serious concern due to the global increase in the discharge of these nanomaterials into the environment and plants' ecological roles within the food chains.

Studies have been conducted to investigate the flow of nanomaterials into aquatic and terrestrial environments. This chapter reports a literature overview regarding the relationships between ENPs and wild aquatic and terrestrial species.

3.2 Engineered nanoparticles in wetlands ecosystems

Wetlands are among the most productive ecosystems in the world, comparable to rain forests and coral reefs. They provide essential ecological services, including protecting and improving water quality, wildlife habitats, storing floodwater, and maintaining surface water flow during dry periods. Wetland primary productivity depends upon the type of wetland and the vegetation found there and on hydrology, climate, and environmental variables such as soil type and nutrient availability. An average value for the aboveground primary productivity of wetlands in temperate zones is about 2500 gC m^{-2} $year^{-1}$ (Fennessy and Cronk, 2018). Since wetlands are one of the central carbon sinks globally, their ecology's negative impacts may result in detrimental C sequestration capacity. Recently some concerns rise regarding the impacts of ENPs on wetlands since the reduction of biomass production can increase global C emissions (Erwin, 2009).

Aquatic plants are more exposed to ENP flows than terrestrial plants. They do not have the buffer effect provided by soil which impedes or slow down ENPs movements. So, they are in constant direct contact with dissolved substances, will have substantial potential for exposure and accumulation. The fate of ENPs in aqueous environments is tremendously influenced by their chemical composition, stability and solubility, particle number and concentration, mass, size distribution, specific surface area, surface charge/ zeta potential, and the presence of capping agents; stability; and solubility (Anawar and Strezov, 2019). In brief, after having reached the aquatic environment, the ENPs may aggregate with suspended organic particles, sediment particles, dissolved organic matter, or other colloids. However, owing to hydrodynamic pulses, resuspension movements of ENPs upward to surface water may occur (Dale et al., 2015). For a deeper insight into such complex processes, refer to Peng et al. (2017).

Macrophytes are the larger primary producers living in aquatic environments. They are represented in seven plant divisions: *Cyanobacteria*, *Chlorophyta*, *Rhodophyta*, *Xanthophyta*, *Bryophyta*, *Pteridophyta*, and *Spermatophyta*. This chapter focuses on the last two divisions, indicated as vascular macrophytes (Chambers et al., 2008). They have different growth types: floating species, submerged, and emerged species since the different ecology significantly influences the nutrition strategies of such plants.

The literature overview regarding the relationships between aquatic plants and the ENPs is organized according to this criterion.

3.2.1 Engineered nanoparticles and floating plants

Floating plants have a peculiar morphology and ecology. The roots of floating plants are not anchored in sediments but float into the water column while absorbing nutrients. Plants grow on the water surface and move with winds and water currents.

Before deepening the literature analysis on this topic, it must be emphasized that the plants belonging to the genus *Lemna* have been studied with some attention compared to other floating plant species. The reason lies in the fact that they have been indicated as target organisms for ecotoxicology by the Organization for Economic Cooperation and Development (OECD, 2006).

In Tables 3.1–3.3 the literature references for the genus *Lemna* are reported. According to the OECD methods on *Lemna minor* L. (duckweed), several investigations were carried out to observe the plants' response to different nanomaterials. In particular, there were reported studies regarding the effects of silver nanoparticles (Ag-NPs), alumina nanoparticles (Al-NPs), copper oxide nanoparticles (CuO-NPs), selenium nanoparticles (Se-NPs), titanium oxide nanoparticles (TiO_2-NPs), zinc oxide nanoparticles (ZnO-NPs), and single-walled carbon nanotubes (SWCNTs).

The first couple of studies investigated the phytotoxic impact of Ag-NPs on *L. minor*. The first paper was provided by Gubbins et al. (2011). According to the OECD protocols, specimens of *L. minor* exposed to Ag-NPs with different sizes (16 and 84 nm) were grown. The nanoparticles were supplied to plants for 14 days at concentrations 0, 5, 10, 20, 40, 80, and 160 mg L^{-1}. Treatment effects were evident after the exposure to both smaller and larger Ag-NPs, even at low concentration (5 mg L^{-1}); this impact became progressively more acute with a longer exposure time. The principal evidence of Ag-NPs phytotoxicity was a growth reduction of *L. minor* in lower frond number and dry matter accumulation (Table 3.1). A similar study was carried out by Üçüncü et al. (2014). In this case, *L. minor* plants were treated with Ag-NPs having a size range in the range 5–20 nm (0, 8, 16, 32, 96, and 128 μg L^{-1}) for a shorter time (7 days). Shortage in dry matter accumulation and reduction of frond number in plant specimens were recorded even at lower Ag-NPs concentrations (8–32 mg L^{-1}). The authors suggested that *L. minor* may be used for the bioremediation of Ag-NPs in wastewater.

Juhel et al. (2011) studied the impact of alumina nanoparticles (Al-NPs) on the growth, morphology, and photosynthetic activity of *L. minor*. Cultures of duckweed were grown in magenta boxes containing half-strength Hutner's medium (Biswas et al., 2010) and 0–1000 mg L^{-1} Al-NPs. After 7 days, there were observed end-point parameters according to OECD (2000). Al-NPs showed a stimulatory effect on plant growth and biomass accumulation. An increase in root length and number of fronds per colony was recorded, as well. Furthermore, an enhancement of photosynthetic efficiency was observed. The authors hypothesized that the enhancement of biomass accumulation in *L. minor* is associated with increased efficiencies in the light reactions of photosynthesis (Table 3.1). However, similar evidence was reported some years later by Giraldo et al. (2014) while studying the photosynthesis light reaction in chloroplasts of *Arabidopsis thaliana* treated with SWCNTs.

TABLE 3.1 Observed effects of engineered nanoparticles on spontaneous aquatic species: floating plants.

Species	Engineered nanoparticles	Materials and experimental conditions	Results	References
Lemna minor	Ag-NPs	• Citrate capped Ag-NPs, 16- (NP1) and 84 nm (NP2); Treatments: NP1, NP2, $AgNO_3$ at 0, 5, 10, 20, 40, 80 and 160 mg L^{-1}; hydroponic conditions, 14 d plant growth.	• Plant growth inhibition after exposure to small and larger Ag-NPs even at low concentrations (5 mg L^{-1}); more acute effect in longer exposure time.	Gubbins et al. (2011)
	Ag-NPs	• Ag-NPs 5- and 20 nm at 8, 16, 32, 96 and 128 μg L^{-1}; plants were grown in 100 mL filtered greenhouse water (pH 8.8–9.2), 24 h/0 h light/dark.	• Shortage of dry matter accumulation and reduction of frond numbers.	Üçüncü et al. (2014)
	Al-NPs	• Al-NPs 20 nm at 0, 10, 100 mg L^{-1}; plants were grown into Magenta boxes containing half-strength Hutner's medium; 16 h/8 h light/dark for 7 d.	• Enhanced biomass accumulation; Increased root length and number of fronds per colony; Enhancement of photosynthetic efficiency.	Juhel et al. (2011)
	CuO-NPs	• CuO-NPs 25 nm at 0, 0.2, 0.4, 0.6, 0.8, 1.2, 1.6, 2 and 2.5 mg L^{-1}; 16 h/8 h light/dark.	• Total frond area most sensitive endpoint to evaluate acute phytotoxicity of CuO-NPs.	Song et al. (2015)
	CuO-NPs	• CuO-NPs 40 nm, bulk CuO, $CuCl_2$ at 0, 10, 50, 100, 150, 200 mg L^{-1}; 500 mL beakers containing 200 mL 1/10 modified Steinberg medium, pH 6.5; 16 h/8 h light/dark.	• CuO-NPs more toxic than bulk CuO in *L. minor*.	Song et al. (2016)
	Se-NPs	• 20–30 nm (i) L-cysteine capped Se-NPs and (ii) tannic acid capped Se-NPs at 0 and 80 mg L^{-1}.	• *L. minor* was poisoned by both capped Se-NPs and $Se4^{+}$.	Tarrahi et al. (2017)

(*Continued*)

TABLE 3.1 (Continued)

Species	Engineered nanoparticles	Materials and experimental conditions	Results	References
	SWCNT	• SWCNT 2.5 mg L^{-1} in the water column; aquatic/subaquatic/terrestrial environment mesocosm (Lowry et al., 2012). 10-months incubation.	• Limited mobility and low bioavailability of SWCNTs. No bioaccumulation of SWNT in aquatic plants.	Schierz et al. (2014)
	TiO_2-NPs	• TiO_2-NPs 5–10 nm and bulk TiO_2 at 0, 10, 50, 100, 200, 1000, 2000 mg L^{-1}; glass troughs containing 1/10 modified Steinberg medium; 16 h/8 h light/dark. for 7 d.	• Enhancement of plant growth at 200 mg L^{-1}; severe damage and decrease in phytochemicals such as chlorophyll and enzymes at 500 mg L^{-1}.	Song et al. (2012)
	ZnO-NPs	• ZnO-NPs and coated ZnO-NPs (1 wt% Silane coupling agent); particle average size 25.9 ± 7.8 6 nm at 0, 1, 10 mg L^{-1}; Plants grown in glass containers, in 250 mL half-strength Hutner's medium for 7 d; 16 h/8 h light/dark.	• Dissolved Zn play a central role in causing ZnO-NPs toxicity.	Chen et al. (2016)
	ZnO-NPs	• ZnO-NPs 30–40 nm at 0, 0.3, 3.0, 30 mg L^{-1}; greenhouse conditions, fluorescent lamps 12 h/2 h light/dark for 7 d.	• *L. minor* able to assimilate up to 98.6% of ZnO-NPs could be seen as fairly tool to remove nano-pollutants from water.	Ergen and Tunca (2018)
	ZeSe-NPs	• ZeSe-NPs 30 nm at 0, 1, 5, 10, 20, 40, 80 mg L^{-1}; plants were grown hydroponically in 30% Steimberg solution for 14 d; 16 h/8 h light/dark.	• Higher toxicity by Zn^{2+} in comparison with ZnSe-NPs.	Tarrahi et al. (2018)

Song et al. (2015) compared the effect of copper oxide nanoparticles (CuO-NPs), bulk copper oxide, and Cu ions on *L. minor*. Among the three Cu treatments supplied at 0, 10, 50, 100, 150, and 200 mg L^{-1}, CuO-NPs exerted the highest growth impairment on duckweed, followed by copper ions and copper oxide. Dose-response-related inhibiting effects caused by the copper ions were found at all end-points studied. In contrast, the particles only showed dose–response related inhibiting effects on the total frond area (Table 3.1).

In a later study by Song et al. (2016), *L. minor* was exposed to different concentrations of CuO-NPs, bulk CuO, and ionic Cu released from $CuCl_2$ in culture media. In addition to plant growth, chlorophyll content, peroxidase, catalase, superoxide dismutase enzyme activity, and malondialdehyde content were studied. Copper nanoparticles inhibited plant growth at a lower concentration than bulk CuO. As expected CuO-NPs was more toxic than bulk CuO, and responding to such toxicity, plants accumulated more reactive oxygen species (ROS) than in the same concentration of bulk CuO. At the concentration of 50 mg L^{-1}, plant cells were seriously damaged (Table 3.1).

Few papers regarding the effects of nanometric forms of selenium on higher plants exist. Among these, a study by Tarrahi et al. (2017) investigated the response of *L. minor* in the presence of selenium nanoparticles (Se-NPs) coated with L-cysteine and tannic acid. Fluorescence microscopy and TEM observations proved that Se-NPs entered plant roots. The assessments of nonenzymatic antioxidant activities (phenolic and flavonoids contents) and enzymatic antioxidant activities (SOD, POD, and CAT) were carried out to evaluate the impact of Se-NPs on plant metabolism. It was concluded that Se-NPs inhibited duckweed growth and photosynthetic activity. A stimulating anti-oxidative response as a defense strategy to remove ROS was hypothesized (Table 3.1).

Carbon-based nanomaterials are allotropic forms of C, having a huge potential for high-performance applications in several fields (Kumar and Kumbhat, 2016). SWCNTs are included among the various allotropic forms of C most commonly used in nanotechnologies (Zaytseva and Neumann, 2016). Of course, SWCNTs were also considered in nano-ecotoxicology studies. Schierz et al. (2014) carried out earlier study addressing the fate and transport of SWCNTs in a mesocosm mimicking emergent freshwater wetland. Several macrophytes, and among them *L. minor* were studied to evaluate the uptake of SWCNTs. Observations regarding these materials' fate in the water column and the accumulation in the sediments were carried out. In the experimental conditions, the SWCNTs showed limited mobility and low bioavailability. No bioaccumulation of SWNT in aquatic plants or vertebrates was observed over the 10-month incubation (Table 3.1).

Titanium dioxide nanoparticles (TiO_2-NPs) are among the most often used nanomaterials worldwide (Baranowska-Wójcik et al., 2020). For that reason, the presence of TiO_2-NPs in freshwater, sediments, and soils was reported (Meesters et al., 2016). Song et al. (2012) examined the effects of TiO_2-NPs on *L. minor*. Plants were exposed to TiO_2-NPs and bulk TiO_2 at 0, 10, 50, 100, 200, 1000, and 2000 mg L^{-1} for 7 days. TiO_2-NPs stimulated the growth of *L. minor* up to a concentration of 200 mg L^{-1}. In contrast, at higher concentrations, alterations in antioxidant defense-enzyme (POD, SOD, and CAT) activity occurred, resulting in serious cell damages (Table 3.1).

As a general rule, ENPs encounter various chemical and physical transformations in the environment (Read et al., 2016), which may affect their toxicological characteristics. Zinc oxide nanoparticles (ZnO-NPs) have been reported to be toxic to many aquatic organisms. However, it is disputed whether this is per se caused by ZnO-NPs or due to ionic Zn dissolution. To clarify that, Chen et al. (2016) investigated the effects on *L. minor* exposed to uncoated and coated ZnO-NPs in a growth medium with acidic (pH 4.5) and alkaline conditions (pH 8). It was demonstrated that ZnO-NPs toxicity is primarily caused by dissolved Zn, whereas phytotoxicity was absent in alkaline conditions. Besides, particle coating reduces the dissolution of ionic Zn; this reinforces the hypothesis that ZnO-NPs toxicity is mainly due to dissolved Zn than the presence of ZnO-NPs (Table 3.1).

Another study dedicated to the relationships between duckweed and ZnO-NPs was carried out by Ergen and Tunca (2018). This paper aimed to evaluate the toxic effect of ZnO-NPs and analyze the plant removal of ZnO-NPs in aqueous medium. Specimens of *L. minor* were treated with ZnO-NPs at 0, 0.3, 3.0, 30 mg L^{-1} for 7 days. Growth rate and growth inhibition rates for plant fronds were calculated according to OECD (2002), whereas the ZnO-NPs percentage removal efficiency was calculated according to Tanhan et al. (2007). Time-dependent and concentration-dependent removal efficiencies were statistically significant. *L. minor* could assimilate up to 98.6% of ZnO-NPs and could be a fair tool to remove nano-pollutants from water (Table 3.1).

Hybrid nanomaterials are defined as unique chemical conjugates of organic and/or inorganic materials. That is, these are mixtures of two (or more) inorganic components, two (or more) organic components, or at least one of both types of components (Gregorczyk and Knez, 2016). An investigation regarding the effects of the nano-hybrid zinc selenide nanoparticles (ZnSe-NPs) on growth parameters and physiological indices of *L. minor* was carried out by Tarrahi et al. (2018). The relative frond number, frond size, fresh weight, and dry weight were observed to evaluate the impacts of ZnSe-NPs on plants. As for plant physiology, the activities of some antioxidant enzymes [superoxide dismutase (SOD) and catalase (CAT)], as well as the contents of nonenzymatic antioxidants (phenols and flavonoids) photosynthetic pigments content were determined. The results showed a remarkable decline in growth parameters and photosynthetic activity due to ZnSe-NPs phytoxicity (Table 3.1).

Besides the remarkable amount of studies done on *L. minor*, two other species belonging to the genus *Lemna* have been studied (Table 3.2). ENPs phytotoxicity in aquatic environments is due to ENPs concentration and particle solubilization, leading to the release of toxic ions. Perreault et al. (2014) evaluated the effects of CuO-NPs on *Lemna gibba* L. Plant specimens were exposed for 48 h to CuO-NPs or soluble copper. It was evident that an inhibition of photosynthetic activity occurred due to a decrease in electron transport in PSII. Although the bioaccumulation of CuO-NPs in plant tissues was observed, the phytotoxic effect in *L. gibba* was due to the release of soluble Cu ions from CuO-NPs. The last paper we report for *Lemna* genus, carried out by Kim et al. (2011), is the oldest one. According to the OECD test guideline 221 (OECD, 2002), it investigated the tolerance of *Lemna paucicostata* Hegelm. 6746 to different concentrations of 50 nm Ag-NPs and 2–3 nm TiO_2-NPs. The plant growth rate indicated that Ag-NPs caused severe growth inhibition at low concentration (>1 mg L^{-1}) whereas TiO_2-NPs treatment showed signals of toxicity at concentrations greater than 250 mg L^{-1}.

With similar approaches, studies were also conducted on other species belonging to the group of floating plants. In aquatic environments, these are typically present plants belonging to the division *Pteridophyta* (ferns), free-sporing vascular plants. The effects of ENPs on such a group of macrophytes were studied on two *Salviniaceae*: *Azolla caroliniana* Willd. (Glenn and Klaine, 2013; Glenn et al., 2012) and *Salvinia natans* L. (Asztemborska et al., 2018), exposed to Ag-NPs and TiO_2-NPs, respectively. *A. caroliniana* floats a top of the water column, and the downward roots take up nutrients and suspended ENPs. In a first investigation, Glenn et al. (2012) supplied *A. caroliniana* with 250 μg L^{-1} of citrate capped Au-NPs (4 and 18 nm) and verified the root absorption of Au-NPs within 24 h after treatment. Results showed that Au-NPs of both sizes entered the root tissues, but the smaller ones (4 nm) were present in more significant quantities, demonstrating a size-dependent Au-NPs uptake (Table 3.3).

Typically, macrophytes are found in shallow water ecosystems that are rich in natural organic matter. Thus, the previous study developed by Glenn and Klaine (2013) investigated the factors that can influence the Au-NPs bioavailability. In particular, specimens of *A. caroliniana* were exposed to 4, 18, and 30 nm Au-NPs in the presence or absence of dissolved organic carbon (DOC). Data showed that Au-NPs bioavailability was a result of different parameters. Size and species-dependent absorption were observed. DOC's presence promotes the formation of a Au-NPs/organic matter complex that resulted in a decrease of Au-NPs absorption by *A. caroliniana* (Table 3.3).

The other *Pteridophyta*—*S. natans*—was studied more recently by Asztemborska et al. (2018). The approach of this study was more complex than those presented so far. In addition to investigating the bioaccumulation

TABLE 3.2 **Observed effects of engineered nanoparticles on spontaneous aquatic species: floating plants.**

Species	Engineered nanoparticles	Materials and experimental conditions	Results	References
Lemna gibba	CuO-NPs	• CuO-NPs 30–40 nm and $CuSO_4$ at 0, 0.4, 0.7, 1.0, 2.1, 4.5 g L^{-1}; plants grown in controlled conditions in freshwater inorganic culture medium, 16 h/8 h light/dark for 48 h.	• Toxicity of CuO-NPs modified by agglomeration and deposition in natural organic colloids.	Perreault et al., (2014)
Lemna paucicostata	Ag-NPs, TiO_2-NPs	• Ag-NPs 50 nm at 0, 0.1, 1, 10, 20, 40, 50, 100, 200 mg L^{-1}, TiO_2-NPs 2–3 nm at 0, 31, 50, 62, 100, 250, 500 mg L^{-1}. Plants grown in 250 mL flask containing 100 mL AAP (algal assay procedure) medium, pH 7.5. 16 h/8 h light/dark for 7 d.	• Growth inhibition by low concentration Ag-NPs; phytotoxicity of TiO_2-NPs >250 mg L^{-1}.	Kim et al. (2011)

TABLE 3.3 Observed effects of engineered nanoparticles on spontaneous aquatic species: floating plants.

Species	Engineered nanoparticles	Materials and experimental conditions	Results	References
Azolla caroliniana	Au-NPs	• Citrate-Au-NPs 4- and 18 nm at 250 μg L^{-1}; greenhouse conditions; plants sampled after 1, 3, 6, 12, 18, 24 h of static Au-NPs exposition.	• Both 4- and 18-nm Au-NPs absorbed into the roots. Smallest Au-NPs were present in higher numbers.	Glenn et al. (2012)
	Au-NPs	• Citrate-Au-NPs 4-, 18- and 30 nm at 250 μg L^{-1}; presence/absence of dissolved organic carbon (DOC).	• Confirmed size-dependent Au-NPs absorption; Au-NPs/DOC complex reduce Au-NPs absorption.	Glenn and Klaine (2013)
Salvinia natans	TiO_2-NPs, ZnO-NPs	• TiO_2-NPs 25 nm and ZnO-NPs 50 nm at 0.1 and 1 g L^{-1}; hydroponic conditions and presence Zn and Sr 0.01 mg L^{-1} for 8 d.	• ENMs can modify the bioaccumulation of ionic Zn and Sr. Possibly negative effects on nutrient uptake.	Asztemborska et al. (2018)
Eichhornia crassipes	CuO-NPs	• CuO-NPs 25 nm at 0, 50, 100, 200 mg L^{-1} and bulk CuO at 200 mg L^{-1}; plants grown in tap water, 12 h/12 h light/dark for 8 d.	• Growth inhibition and phytotoxicity of CuO-NPs at concentration > 50 mg L^{-1}.	Liu et al. (2014)
	Fe_3O_4-NPs	• Fe_3O_4-NPs 40–80 nm at 0, 50, 100, 200 mg L^{-1}, Fe_3O_4-NPs 500 nm at 200 mg L^{-1}, Fe^{3+} 0.8 mg L^{-1}; hydroponics for 21 d.	• Root growth inhibition, root tip cell disorder. Plant physiology was also affected.	Ding et al. (2019)

(*Continued*)

TABLE 3.3 (Continued)

Species	Engineered nanoparticles	Materials and experimental conditions	Results	References
Landoltia punctata	CuO-NPs	• CuO-NPs 10–15 nm, bulk CuO, $CuCl_2$ at 0, 0.2, 0.6, 1.0 mg L^{-1}; culture dishes in growth chamber for 14 d.	• Inhibitory effect of CuO-NPs; Cu content in fronds exposed to CuO-NPs four times higher than in fronds exposed to soluble Cu.	Shi et al. (2011)
	CuO-NPs	• CuO-NPs 22 nm, humic acid (HA) and fulvic acid (FA); growth chamber, 16 h/8 h light/dark for 7 d.	• HA, and to a lesser extent, FA, decrease the toxicity of CuO-NPs.	Rippner et al. (2018)
Spirodela polyrhiza	Ag-NPs	• GA-Ag-NPs 6 nm 0, 0.5, 1, 1.5, 10 mg L^{-1}, PVP-Ag-NPs 20 nm, 0, 10, 20 mg L^{-1} and 1 μ Ag particles 10 mg L^{-1}; hydroponics, 16 h/8 h light/dark for 72 h.	• Oxidative stress and damages to chloroplast structure caused by Ag-NPs but not by particle released Ag ions.	Jiang et al. (2014)
	TiO_2-NPs	• TiO_2-NPs 8 nm, 0, 0.05, 0.1, 1, 5, 10 mg L^{-1}; hydroponics, 16 h/8 h light/dark for 20 d.	• Root uptake of TiO_2-NPs; growth parameters negatively affected by different TiO_2-NPs concentrations.	Movafeghi et al. (2018)
Spirodela polyrhiza, Wolffia arrhiza	CuO-NPs	• CuO-NPs 25 nm at 0, 0.2, 0.4, 0.6, 0.8, 1.2, 1.6, 2 and 2.5 mg L^{-1}; 16 h/8 h light/dark.	• Total frond area most sensitive end-point to evaluate acute phytotoxicity of CuO-NPs.	Song et al. (2015)

of TiO_2-NPs and ZnO-NPs in *S. natans*, the authors carried out observations on Zn^{2+} and Sr^{2+} absorption potential by nanoparticles, hypothesizing their use for the removal of metal ions present in the water. It was demonstrated that bioaccumulation of Zn^{2+} and Sr^{2+} by *S. natans* was modified in the presence of TiO_2-NPs and ZnO-NPs. In some conditions, a relevant percentage of Zn^{2+} can be removed from a solution through sorption using ZnO-NPs (Table 3.3).

The free floating *Eichhornia crassipes* (Mart.) Solms (water hyacinth) is an aquatic, very invasive species able to alter the composition of native vegetation by lowering light penetration and dissolved oxygen levels. On the other hand, being very common, it is a good target for studies regarding the impact of ENPs in aquatic ecosystems. Liu et al. (2014) carried out a study to determine whether the growth of *E. crassipes* could be affected by CuO-NPs. Water-grown plant specimens received a series of CuO-NPs concentrations (0, 50, 100, and 200 $mg\,L^{-1}$) and bulk CuO (200 $mg\,L^{-1}$) for 8 d. Plant growth was significantly affected by CuO-NPs concentrations higher than 50 $mg\,L^{-1}$. Severe alteration on the root cap cells resulted in root growth inhibition (Table 3.3). Further results regarding the impact of nanomaterials on *E. crassipes* were provided by Ding et al. (2019). In this case, there were compared the effects of different concentrations of Fe_3O_4-NPs (0, 50, 100, and 200 $mg\,L^{-1}$), bulk Fe_3O_4 (200 $mg\,L^{-1}$) and Fe^{3+} (0.8 $mg\,L^{-1}$). After a 21-day exposure at higher concentration, nanoparticles caused significant root growth inhibition and cell disorders. Plant physiology was affected as well. A decrease of chlorophyll content and catalase activity was recorded, whereas an increase of malondialdehyde content was observed (Table 3.3).

As already described, to investigate the effects of ENPs on plants, it is advisable to compare the plants' response to the nanometric and ionic forms of a given element. Another example of such an approach is provided by Shi et al. (2011) on *Landoltia punctata* (G. Mey.) Les & D.J. Crawford. Plant specimens were exposed to CuO-NPs, bulk CuO, and ionic Cu, released by $CuCl_2$. It was evidenced of an inhibitory effect of CuO-NPs, which were responsible for a four times higher Cu concentration in fronds than in plants exposed to soluble Cu. An extra Cu solubilization from CuO-NPs was due to root exudates and subsequent water acidification (Table 3.3). Such findings were more recently updated by Rippner et al. (2018). The experimental approach was implemented considering DOM's presence, provided as humic (HA) and fulvic (FA) acids. Growth inhibition of *L. punctata* was most significant from aqueous $CuCl_2$ and, for CuO-NPs, increased with decreasing CuO-NPs particle size due to Cu dissolution. However, the current results suggest that HA, and to a lesser extent, FA, decreases the toxicity of CuO-NPs and ionic Cu to *L. punctata*. The Cu-DOM complex formation likely results in a reduction of Cu availability (Table 3.3).

Jiang et al. (2014) used *Spirodela polyrrhiza* (L.) Schleid. as model species to study the oxidative stress and cell damages determined by Ag-NPs and whether that is due to Ag-NPs itself or by ionic Ag released from the nanoparticles. In treated plants, results indicated a dose-dependent increase in levels of ROS and other oxidase-enzymes. Also, TEM images showed alterations in the ultrastructure of chloroplasts not present in plants treated with Ag microparticles (Table 3.3). Again on *S. polyrrhiza*, Movafeghi et al. (2018) evaluated the effects of different concentrations of TiO_2-NPs. The root uptake of TiO_2-NPs was confirmed using epifluorescence microscopy. The increasing concentration of TiO_2-NPs led to a corresponding decrease in growth parameters, chlorophyll, and carotenoid contents, whereas antioxidant enzyme activities were enhanced. Finally, Song et al. (2016) studied the impact of CuO-NPs on *S. polyrrhiza*, and *Wolffia arrhiza* (L.) Horkel ex Wimm. With respect to *L. minor*, these species were respectively more sensitive and less sensitive to Cu-NPs. This evidence demonstrated the importance of developing a better ability to assess the effects caused by the entry of ENPs in aquatic ecosystems populated by different species (Table 3.3).

A group of floating species has their roots anchored in the substrate. A flexible stem connects the floating leaves with the root apparatus. The most representative plant species belonging to this group are the water lilies (*Nymphaceae*) and water lotus (*Nembulonaceae*) (Chambers et al., 2008). While several studies carried out on free-floating species exist, no reports regarding the effects of ENPs on such plants were found in literature.

3.2.2 Engineered nanoparticles and submerged plants

Submerged plants are rooted plants with most of their vegetative mass below the water surface (Chambers et al., 2008). These plants create valuable habitats for fish and small invertebrates and provide food for other animals. Studies on the impact of ENPs in aquatic environments also concern the most representative species belonging to this group (Table 3.4).

Lu et al. (2010) carried out a study on the distribution and accumulation of CeO_2-NPs in simulated aquatic ecosystems, which included water, sediments collected in a natural reservoir, the submerged macrophyte *Ceratophyllum demersum* L., a shellfish (*Cipangopaludina chinensis*) and a fish (*Pseudorasbora parva*). Radioactive ceria nanoparticles $^{141}CeO_2$-NPs having an average diameter of 7 nm were added to simulated ecosystems. The distribution and accumulation characteristics of ceria nanoparticles in various aquatic organisms were different (Table 3.4). The macrophyte *C. demersum* showed a high ability of accumulation of $^{141}CeO_2$-NPs from water. Again, on this species, in aquatic microcosms, Ostroumov et al. (2014) studied the effects of exposure to Au-NPs. Specimens of *C. demersum* were treated with different concentrations of 20 nm Au-NPs in aquatic microcosms. Under the experimental conditions, symptoms of phytotoxicity appeared after 17 d of Au-NPs exposure. After 24 d about 50% of *C. demersum* shoots died (Table 3.4).

TABLE 3.4 Observed effects of engineered nanoparticles on spontaneous aquatic species: submerged plants.

Species	Engineered nanoparticles	Materials and experimental conditions	Results	References
Ceratophyllum demersum	CeO_2-NPs	• $^{141}CeO_2$-NPs 7 nm; final *n* CeO_2-NPs concentration in water not clarified. Simulated aquatic ecosystem; 12 h/12 h light/dark for 15 d. Plants sampled 1, 2, 4, 8, 12 h, 1, 3, 6 d after treatment.	• High ability of accumulation of nanoceria. Most CeO_2-NPs accumulated at the water-sediment interface.	Lu et al. (2010)
	Au-NPs	• Au-NPs 20 nm at 0, 5 additions of 2 mL 6×10^{-6} M, 5 additions of 6 mL 1.8×10^{-5} M, respectively at 0, 3, 8, 17, 25 d of incubation in aquatic microcosms (experiment duration 28 d).	• Demonstrated Au-NPs phytoxicity. At highest concentration 50% plants died.	Ostroumov et al. (2014)
Egeria densa, Myriophyllum simulans	Au-NPs	• Au-NPs, 4 nm and 18 nm. Plants exposed to Au-NPs at 250 $\mu g\ L^{-1}$ for 24 h and harvested after 1, 3, 6, 12, 18, and 24 h of exposition.	• *E. densa* did not absorb Au-NPs of either size; only Au-NPs 4 nm were absorbed by *M. simulans*. Particle uptake confirmed by EDX-spectroscopy.	Glenn et al. (2012)
	Au-NPs	• Citrate-Au-NPs 4 nm, 18 nm and 30 nm at 250 $\mu g\ L^{-1}$; presence/absence of dissolved organic carbon (DOC).	• Confirmed size-dependent Au-NPs absorption; *n* Au/DOC complex reduce Au-NPs absorption.	Glenn and Klaine (2013)
Egeria densa	CeO_2-NPs	• CeO_2-NPs 3.8 and 185 nm dosed weekly to wetland mesocosms over 9 months	• CeO_2-NPs poorly bioavailable	Geitner et al. (2018)

(Continued)

TABLE 3.4 (Continued)

Species	Engineered nanoparticles	Materials and experimental conditions	Results	References
Elodea canadensis	TiO_2-NPs	• TiO_2-NPs 0, 12 mmol L^{-1} + P treatments (0, 32, 64, 128 μmol as NH_2PO_4 H_2O); greenhouse hydroponics for 15 d.	• A negative correlation between P and Ti content in plant tissues suggests that in eutrophic environments Ti uptake could be reduced	Jacob et al. (2013)
	SWCNT	• SWCNT 2.5 mg L^{-1} in the water column; aquatic/subaquatic/terrestrial environment mesocosm . 10-months incubation.	• Limited mobility and low bioavailability of SWCNTs. No bioaccumulation of SWNT in aquatic plants.	Schierz et al. (2014)
	TiO_2-NPs, ZnO-NPs	• TiO_2-NPs 25–100 nm and ZnO-NPs 50 and 100 nm at 0.1 and 1 g L^{-1}; hydroponic conditions and presence Zn and Sr 0.01 mg L^{-1} for 8 d.	• ZnO-NPs (1 g L^{-1}) reduced bioaccumulation of Zn^{2+} (−68%) and Sr^{2+} (49%).	Asztemborska et al. (2018)
Hydrilla verticillata	ZnO-NPs	• ZnO-NPs < 50 nm at 0, 0.01, 0.1, 1, 10, 100, 1000 mg L^{-1}; mesocosm containing sand and gravel and Hoagland solution (without zinc sulfate). 5 weeks growth.	• Dissolution rate of ZnO-NPs inversely related to the concentration; plants showed adaptation to highest ZnO-NPs concentration over time-	Song and Lee (2016)

The invasive macrophytes *Egeria densa* Planch. and *Myriophyllum simulans* Orch., being able to adapt in different ecological conditions, are distributed worldwide (Cabrera Walsh et al., 2013; Chambers et al., 2008). The already cited studies by Glenn et al. (2012) and Glenn and Klaine (2013) also observed the effects of Au-NPs on *E. densa* and *M. simulans*. Glenn et al. (2012) treated the macrophyte species with 250 mg L^{-1} Au-NPs (4 and 18 nm). Under the same experimental conditions, *E. densa* and *M. simulans* behaved differently than the pteridophyte *A. caroliniana*. Only small Au-NPs were absorbed by *E. densa*, whereas in the tissues of *M. simulans* Au-NPs were not detected. This demonstrates that nanomaterials' properties and species-specific variables strongly influence the relationships between plants and ENPs (Table 3.4). Similar findings come from the study by Glenn and Klaine (2013). They studied the same species in an experimental system in which, in addition to the variables (1) Au-NPs size and (2) plant species, (3) the presence of DOC was added. The Au-NPs plant uptake was significantly influenced by the experimental factors' interactions confirming that nanoparticles bioavailability is a very complex process, the understanding of which requires careful and in-depth study (Table 3.4).

The effects of exposure to CeO_2-NPs of *E. densa* were studied by Geitner et al. (2018). In an artificial freshwater wetland mesocosm was investigated the long-term fate and transport of CeO_2-NPs (3.8 and 185 nm), weekly supplied over nine months. Actually, the experiment's duration is because the study was mainly aimed to clarify the fate of CeO_2-NPs and not at the evaluation of effects on macrophytes, which could have been studied with a much shorter specific study. However, the results demonstrated that the more significant fraction of CeO_2-NPs remained in the upper layers of sediments, and therefore, the bioavailability of the nanoparticles was very low (Table 3.4).

Another invasive macrophyte considered in studies regarding the impact of ENPs in wetlands, due to its worldwide distribution, is *Elodea canadensis* Michx. Jacob et al. (2013) investigated the toxicity of TiO_2-NPs in different plants representative of terrestrial and aquatic environments. Hydroponically grown specimens of *E. canadensis* were exposed to 20 mg L^{-1} TiO_2-NPs for two weeks. Results showed that TiO_2-NPs altered the uptake patterns of Fe, Mn, and Mg, and P. Treated plants showed a strong negative correlation between P and Ti leading to hypothesis that *E. canadensis* in eutrophic environments could reduce Ti uptake (Table 3.4).

A study previously cited about *L. minor*, included also *E. canadensis* (Jacob et al., 2013). As reported, experimental results showed that over ten months of incubation, SWCNTs were accumulated in surficial sediments. No bioaccumulation of SWNT in macrophytes or vertebrates was observed (Table 3.4). The study by Asztemborska et al. (2018) put together the floating macrophyte *S. natans* and the rooted submerged *E. canadensis*. The paper aims were (i) to determine the sorption capacity of TiO_2-NPs (25 and

100 nm) and ZnO-NPs (50 and 100 nm) and (ii) to quantify the bioaccumulation of Zn^{2+} and Sr^{2+}. The results have shown that ENPs can adsorb metal ions present in the water. In *E. canadensis*, Zn and Sr ions' bioaccumulation was reduced by ZnO-NPs at 1 g L^{-1} concentration of about 68% and 49%, respectively (Table 3.4).

With the study of Song and Lee (2016), the section on submerged species ends. Further, the aquatic ecosystems are represented by two different species: the submerged macrophyte *Hydrilla verticillata* (L.f.) Royle and *Phragmites australis* (Cav.) Trin. ex Steud., belonging to emerged macrophytes. *H. verticillata* is an aquatic weed known to be a strong heavy metal accumulator in contaminated wetlands (Srivastava et al., 2006). A mesocosm experiment was carried out to test the phytotoxic effects of ZnO-NPs on growth and physiology of *H. verticillata*. Since the early development stages, the highest ZnO-NPs concentration (1000 mg L^{-1}) determined a growth reduction. However, some weeks later, plants showed better growth indicating that *H. verticillata* exposed to high ZnO-NPs concentrations can recover after a period of adaptation (Table 3.4).

3.2.3 Engineered nanoparticles and emerged plants

Among the plant species populating wetland ecosystems, relevant examples of adaptation to aquatic environments exist. However, there is also a group of plants very similar to terrestrial species. Emerged plants are macrophytes rooted in the soil or sediments beneath the water surface, while the aerial part consisting of leaves, stems, and flowers emerging above the water surface. Typically, the most common emerged plants are herbaceous species belonging to *Poaceae*, *Cyperaceae*, and *Juncaceae*.

Among the early studies regarding the impact of ENMs on biota in natural environments, the investigation carried out by Yin et al. (2012) must certainly be mentioned. In particular, it studied the impact of Ag-NPs on *Carex crinita* Lam., *Carex lurida* Wahlenb., *Carex scoparia* Schkuhr ex Willd., *Carex vulpinoidea* Michx., *Eupatorium fistulosum* Barratt, *Juncus effusus* L., *Lobelia cardinalis* L., *Lolium multiflorum* L., *Panicum virgatum* L., *Phytolacca americana* L. and *Scirpus cyperinus* L., typically present in wetland environments. Plants' responses to Ag-NPs exposure were observed (1) after direct exposure of germinating seeds to nanoparticles and (2) in plants grown in Ag-NPs amended soil in a greenhouse experiment. Seed germination was carried out in Petri dishes supplied with PVP-Ag-NPs, GA-Ag-NPs, and $AgNO_3$ at 0, 1, 10, 40 mg L^{-1}. In the greenhouse experiment, each pot was sown with a seed mixture of 7 species: *C. crinita*, *C. lurida*, *C. scoparia*, *C. vulpinoidea*, *E. fistulosum*, *L. multiflorum* and *P. americana*. After seeding, soil was sprayed with 250 mL of a 40 mg L^{-1} of PVP-Ag-NPs, GA-*n* Ag, and $AgNO_3$. After 50 d, plant biomass was harvested. In seed direct exposure experiment, PVP-Ag-NPs did not affect germination. In contrast,

40 mg Ag L^{-1} GA-Ag-NPs exposure significantly reduced the germination rate of *S. cyperinus*, *J. effusus* and *P. americana*, whereas it was enhanced in *E. fistulosum*. Silver nitrate enhanced the germination rate of five species. Ag exposure affected the root growth much more than the leaf growth. In the soil exposure experiment, germination effects were less pronounced (Table 3.5). In the pot experiment, the plant growth response differed by taxa with *L. multiflorum* growing more rapidly under both GA-Ag-NPs and $AgNO_3$ exposures. In contrast, all other species plant growth was affected by GA-Ag-NPs exposure. $AgNO_3$ did not reduce the growth of any species while PVP-Ag-NPs negatively affected only the growth of *P. americana* (Table 3.5).

Apart from the floating species *L. minor* and the submerged one *E. canadensis*, Schierz et al. (2014) also provided data regarding the impact of SWCNTs on *C. lurida*, *Juncus effuses*, *Lobelia elongata* and *P. virgatum*. It is a work carried out in an artificial aquatic ecosystem complete with sediments and a fish species (*Gambusia holbrooki*, Eastern mosquito fish) representing a consumer. The absence of SWCNTs bioaccumulation reported for *L. minor* and *E. canadensis* was confirmed for emerged macrophytes as well (Table 3.5). The same macrophytes and *E. densa* were cited in the already mentioned study by Geitner et al. (2018). A longer-term (9 months) experiment was carried out in an artificial wetland mesocosm to study the fate of CeO_2-NPs. Very poor CeO_2-NPs accumulation was recorded in plants due to the low bioavailability of nanoceria (Table 3.5).

We conclude this review by citing a couple of works conducted to evaluate the effects of *P. australis* exposure to ZnO-NPs. The first one was already presented since it was carried out to compare the effects of ZnO-NPs on *H. verticillata* and *P. australis* (Song and Lee, 2016). Contrary to *H. verticillata*, *P. australis* began to show significantly reduced growth due to Zn-NPs treatments after a few weeks without evidence of adaptation and recovery. The plant metabolism was affected as well with a decrease in chlorophyll content and reduction of antioxidant enzyme activity (Table 3.5). Lastly, Caldelas et al. (2020) carried out pot experiments to investigate ZnO-NPs uptake and phytotoxicity in *P. australis*. Plants were exposed to different materials: bulk ZnO, ZnO-NPs <100 nm, ZnO-NPs <50 nm, and Zn nanowires of 50 nm diameter at concentrations of 0, 0.1, 1, 10, 100, and 1000 mg L^{-1}. Results showed that ZnO-NPs in *P. australis* reduces shoot and root growth, transpiration, and photosynthetic rate while inducing nutrient uptake and distribution (Table 3.5).

3.3 Engineered nanoparticles in terrestrial plants

So far, literature data are available regarding the adverse effects on plant growth and metabolism due to exposure to ENPs (Miralles et al., 2012; Zhang et al., 2015; Zuverza-Mena et al., 2017). On the other hand,

TABLE 3.5 Observed effects of engineered nanoparticles on spontaneous aquatic species: emerged plants.

Species	Engineered nanoparticles	Materials and experimental conditions	Results	References
Carex crinita *Carex lurida* *Carex scoparia* *Carex vulpinoidea* *Eupatorium fistulosum* *Juncus effusus* *Lobelia cardinalis* *Lolium multiflorum* *Panicum virgatum* *Phytolacca americana* *Scirpus cyperinus*	Ag-NPs	• Polyvinylpyrrolidine (PVP)-coated Ag-NPs 20 nm, Gum Arabic (GA)-coated-Ag-NPs 6 nm, $AgNO_3$; • Seed germination in Petri dishes with PVP-Ag-NPs, GA-Ag-NPs, $AgNO_3$ at 0, 1, 10, 40 mg L^{-1}; • (ii) Pot experiment; after seeding soil sprayed with PVP-Ag-NPs, GA-Ag-NPs and $AgNO_3$, 250 mL at 40 mg L^{-1}, plan biomass harvested 50 d after seeding.	• PVP- Ag-NPs had no effects on germination by PVP- Ag-NPs; GA- Ag-NPs 40 mg Ag L^{-1} reduced germination of three species. $AgNO_3$ 40 mg Ag L^{-1} enhanced germination of five species; • Growth of *L. multiflorum* stimulated by GA-Ag-NPs and $AgNO_3$; other species negatively affected by GA- Ag-NPs. PVP-Ag-NPs reduced the growth of *P. americana*.	Yin et al. (2012)
C. lurida *Juncus effuses* *Lobelia elongate* *P. virgatum*	SWCNT	• SWCNT 2.5 mg L^{-1} in the water column; aquatic/subaquatic/terrestrial environment mesocosm. 10-months incubation.	• Limited mobility and low bioavailability of SWCNTs. No bioaccumulation of SWNT in aquatic plants.	Schierz et al. (2014)
Carex lurida, *Juncus effuses*, *Lobelia elongate*, *P. virgatum*	CeO_2-NPs	• CeO_2-NPs 3.8 and 185 nm dosed weekly to wetland mesocosms over 9 months	• CeO_2-NPs poorly bioavailable.	Geitner et al. (2018)
Phragmites australis	ZnO-NPs	• ZnO-NPs < 50 nm at 0, 0.01, 0.1, 1, 10, 100, 1000 mg L^{-1}; mesocosm containing sand and gravel and Hoagland solution (without zinc sulfate). 5 weeks growth.	• High ZnO-NPs concentrations induce significant phytotoxicity; low concentrations caused unpredictable phytotoxicity.	Song and Lee (2016)
P. australis	ZnO-NPs	• (i) bulk ZnO, (ii) ZnO-NPs <100 nm, (iii) ZnO-NPs <50 nm, (iv) Zn nanowires Ø 50 nm x 300 nm; greenhouse pot experiment.	• Effects of ZnO-NPs exposure are mostly due to dissolved Zn^{2+} uptake.	Caldelas et al. (2020)

convincing demonstrations of the positive effects of ENPs were published when provided as fertilizers or employed as carriers of plant nutrients (Lowry et al., 2019; Marchiol et al., 2020). Such contradictory indications are the result of research works that used heterogeneous methodological contexts. For example, studies aiming to test the effects of ENPs in plants carried out in different conditions (e.g., hydroponics, synthetic soil, potting soil or agricultural soil) provided very different results. However, as time went more solid protocols are being consolidated. It should also be mentioned that ENPs with new features and properties are continuously released due to a high-speed evolution of the production methods (Kah et al., 2019). Studies regarding ENPs uptake and biotransformation inside plant tissues, the response of plant metabolism, as well as risks of ENPs transfer to the trophic chain, are currently in progress and are expected to fill the current gaps within the relationships between ENPs and plants (Reddy Pullagurala et al., 2019) (Table 3.6).

The existing literature on ENPs-plants relationships is nearly entirely focused on food crops, this being primarily justified considering the human exposure to ENPs. However, the impacts of ENPs on the whole primary producers could indeed have dire consequences on food webs and ecosystem services, and therefore, they should not be considered less critical (Hawthorne et al., 2014).

As stated earlier, a minor part of the published papers concerns non-food plant species. Within this field, a substantial line of research regarding the relationships between ENPs and plants is focused on *A. thaliana* L. (Milewska-Hendel et al., 2019; Montes et al., 2017; Zhang et al., 2019). Although it is a very useful model species for genomic and metabolomics studies, *A. thaliana* is not representative from an ecological point of view. The second field of study based on the relationships between ENPs and higher plants concerns the biosynthesis of nanoparticles. In addition to top-down and bottom-up methods based on chemical and physical processes, biosynthesis processes of nanoparticles mediated by fungi, bacteria, and plants have also been observed (Iravani, 2011). The combination of biomolecules found in plant extracts (e.g., enzymes, proteins, amino acids, vitamins, polysaccharides, and others) with metal salt solutions promotes metals' bioreduction. This process, being conducted at room pressure and temperature, is very cost-effective and can be easily scaled up for metal nanoparticles' large-scale production (Mittala et al., 2013).

As for terrestrial ecosystems, to the best of our knowledge, *Pinus sylvestris* L. and *Quercus robur* L. are the only non-food plant species investigated for exposure to ENPs. Therefore, referring exclusively to studies carried out on aquatic macrophytes, it is obvious that this field's work has yet to begin. The response of such tree species to Ag-NPs and CuO-NPs was studied by Aleksandrowicz-Trzcińska et al. (2019) in a pot experiment carried out in greenhouse conditions in a completely randomized experimental scheme.

TABLE 3.6 Observed effects of engineered nanoparticles on spontaneous terrestrial plant species.

Species	Engineered nanoparticles	Materials and experimental conditions	Results	References
Pinus sylvestris Quercus robur	Ag-NPs, CuO-NPs	• Ag-NPs, CuO-NPs 50 nm at 0, 5, 25 and 50 mg L^{-1} provided in four spray treatments of 100 mL m^{-2}; greenhouse pot trial; one-year-old seedlings grown on a peat–perlite mix for six months.	• Limited toxicity due to nanoparticles and only present with higher concentrations.	Aleksandrowicz-Trzcińska et al. (2019)
Holcus lanatus Diplotaxis tenuifolia Silene flos-cuculi	CeO_2-NPs	• CeO_2-NPs 25 nm and 50 nm at 0, 20, 200 mg L^{-1}; germination trial;	• Seed germination not affected by CeO_2-NPs. Moderate stimulating effect on root length.	Lizzi et al. (2020)
H. lanatus D. tenuifolia	CeO_2-NPs	• CeO_2-NPs 25 nm and 50 nm at 0 and 200 mg kg^{-1}; pot trial in greenhouse conditions for 60 d.	• CeO_2-NPs taken up by plant roots and then translocated. Stimulating effects on plant growth. Nutritional imbalances.	Lizzi et al. (2021a)
Silene flos-cuculi	CeO_2-NPs	• CeO_2-NPs 25 nm and 50 nm at 0 and 200 mg kg^{-1}; pot trial in greenhouse conditions for 60 d. Treatments: (i) D1, before seeding, (ii) D2, 20 d after seeding, (iii) D3 40 d after seeding	• Root uptake and translocation of CeO_2-NPs; Ce accumulation in roots and leaves. Shortage in biomass of roots and stems.	Lizzi et al. (2021b)

P. sylvestris and *Q. robur* were chosen since they are widespread, being easily adaptable to different ecological conditions. The aerial part of plant seedlings was sprayed with 100 mL m^{-2} in different concentrations (0, 5, 25, and 50 mg L^{-1}). The plant growth was slightly affected by Ag-NPs and CuO-NPs. Some signals of stress were evidenced by TEM observation on the ultrastructure of chloroplasts. *Q. robur* appeared to be more sensitive than *P. sylvestris*. However, comprehensive experimental data indicated that Ag-NPs and CuO-NPs toxicity on *P. sylvestris* and *Q. robur* was limited and present only at concentrations-effect.

3.4 Effects of CeO_2-NPs on three wild plant species

Given the scarcity of investigations on the effects of ENPs on spontaneous terrestrial plants, a Ph.D. project was developed at the University of Udine (Italy). Some experiments were planned to evaluate the influence of CeO_2-NPs on the growth of the spontaneous monocot *Holcus lanatus* (L.), and the dicots *Diplotaxis tenuifolia* (L.) DC and *Silene flos-cuculi* (L.).

Nanoceria was considered because this nanomaterial is among the top 10 most produced ENPs in the world (Piccinno et al., 2012). Its presence in different environmental matrices and ecosystems has already been demonstrated (Meesters et al., 2016). The plant species have been chosen since they are common and widespread in natural systems, highly competitive, and easily adaptable to different ecological conditions.

H. lanatus L. is a hairy, tufted, fibrous-rooted and meadow soft perennial grass, growing between 50 and 100 cm tall, belonging to Poaceae family (Stace, 1997). It has a wide climatic range and occurs over a wide range of soil types and fertility conditions (Thompson and Turkington, 1988). This species has long green leaves; the ligule is around 4 mm long, and the stems present pink veins at the bases; it produces a large number of seeds that could rapidly colonize disturbed grounds. The absence of rhizomes characterizes *H. lanatus*, and it spreads by developing new shoots and roots at its nodes and forming a blanket of runners on the soil surface.

D. tenuifolia L. DC. is a perennial flowering species growing in temperate climates, found in different habitats and ruderal plant associations in abandoned and disturbed fields (Hall et al., 2012). It is an erect plant with branching stems that could reach half a meter in height; this species has long lobed aromatic leaves, and at the top of the branches, it develops yellow flowers with rounded petals. They produce straight siliques containing small brown seeds.

S. flos-cuculi L. is an herbaceous perennial plant belonging to *Caryophyllaceae* family, and it is native and distributed throughout Europe. However, it also grows in North America, Central and Northern Asia, and Eastern Canada (Jalas and Suominen, 1986). It is found in open habitats, along roads, and in wet meadows and pastures. *S. flos-cuculi* forms vegetative rosettes with numerous flower stems that could be from 20 to 90 cm tall. The stems have barbed hairs that make the plant rough to the touch; stems grow over the foliage and end with pink flowers. The fruits consist of small capsules containing many dark seeds dispersed mechanically (Biere, 1991).

The study started with germination experiments. The seeds of *H. lanatus*, *D. tenuifolia*, and *S. flos-cuculi* were exposed to 25 and 50 nm CeO_2-NPs, provided in concentrations included in the range 0.2–2000 mg L^{-1}. Two pot experiments were subsequently carried out in greenhouse conditions to observe the possible effects of CeO_2-NPs over the plant's life cycle. The experiments were designed with different goals. In the first one, seeds of *H. lanatus* and *D. tenuifolia* were sown in a substrate previously amended with CeO_2-NPs, and observations were done on growing plants to verify possible signals of phytotoxicity. The second study was designed to reproduce the conditions of plants exposed to repeated pulses of CeO_2-NPs from an anthropogenic source. Therefore, plants of *S. flos-cuculi* were supplied with CeO_2-NPs suspensions also during their growth cycle.

3.4.1 Procedure

3.4.1.1 *Cerium oxide nanoparticles characterization*

The CeO_2-NPs having an average particle size of 25 nm, density of 7.13 g mL^{-1} at 25°C, and purity of 99.95% (81.25% of Ce) were purchased from Sigma-Aldrich. The characterization of 99.95% indicated the following parameters: Z—average size 126.7 ± 1.0 nm, ζ potential 39.2 ± 1.1 mV, and relative polydispersity index 0.17 ± 0.01. The size and average shape were measured with transmission electron microscopy (FEI Tecnai F20).

3.4.1.2 *Germination trial*

Seeds of *D. tenuifolia* were provided by Sementi Bruni (Corbetta, Milan, Italy) while seeds of *H. lanatus* and *S. flos-cuculi* were purchased from SemeNostrum (Udine, Italy). Seeds were placed in Petri dishes containing filter paper soaked with 10 mL of deionized water (control) and 20 and 200 mg mL^{-1} of CeO_2-NPs suspensions previously sonicated for 10 minutes to prevent particle aggregation. The duration of the experiment was two weeks. Each treatment was replicated three times. Germination was calculated as the ratio of germinated seeds out of the total number of seeds in each Petri dish.

3.4.1.3 *Pot experiments*

The pot trials were carried out in a semi-controlled greenhouse facility at the University of Udine's experimental farm (Italy). Compo Sana organic potting mix (pH = 6.8) was used as a growing substrate. In the first pot experiment

(Pot-1), the substrate was amended with two water suspensions containing 25 and 50 nm CeO_2-NPs to reach a final substrate concentration of 200 mg L^{-1} CeO_2-NPs. Before soil amendment, the suspensions were stirred and sonicated for 30 min to avoid aggregation. Tap water was used as control. The amended substrate was carefully mixed and stored in the dark at 15°C for three days to equilibrate. Then 2 L plastic pots were filled with control and amended substrate. Seeds of *H. lanatus* and *D. tenuifolia* were put about 2.5 cm deep in the substrate. The pots were placed in full sunlight, at 18°C/27°C (night/day), with a relative humidity of 60%. During the trial, irrigation was provided as needed based on the water holding capacity of the substrate. Plants were harvested 60 days after sowing (Fig. 3.1). The plant biomass was separated into roots, stems, and leaves which were oven-dried at 60°C for 48 h and weighed.

As already mentioned, the design approach for the second experiment (Pot-2) soil was different. The potting soil was amended with 25 nm CeO_2-NPs water suspensions to reach a final substrate concentration of 20 and 200 mg kg^{-1} CeO_2-NPs, respectively. After soil equilibration, the pots were filled with 500 g soil. Repeated applications of the same CeO_2-NPs suspensions were performed after 20 and 40 days from seedling emergence (DSE). Therefore, the experimental scheme had (1) two CeO_2-NPs concentrations and (2) three CeO_2-NPs dose levels, and precisely D1 (before seeding), D2 (20 DSE), and D3 (40 DSE).

FIGURE 3.1 Pot experiments. (A-B) Plantlets of *H. lanatus*, *D. tenuifolia* and *S. flos-cuculi* at the beginning of Pot-1. (C) Comparison between plants of *H. lanatus* grown in the substrate untreated and treated with CeO_2-NPs (20–200 mg kg^{-1}). (D) Plants of *S. flos-cuculi* at the end of Pot-2.

3.4.1.4 *Internalization of CeO_2-NPs and Ce concentration in plant tissues*

The uptake of CeO_2-NPs by plant seedlings was verified on plant seedlings at the end of the germination experiments by enzymatic digestion (Jiménez-Lamana et al., 2016). The digesting enzyme used was Macerozyme R-10 enzyme—pectinase from Rhizopus sp. (Sigma Aldrich). The supernatants were analyzed using an sp-ICP-MS (NexION 350, PerkinElmer). According to the USEPA 3052 method (USEPA, 1995), the plant samples were acid digested. Cerium total content in both plantlets and adult plant specimens was determined using an ICP-MS (NexION 350, Perkin Elmer, USA).

3.4.2 Results and discussion

3.4.2.1 *Germination trial*

Treatments improved the germination percentage in all three species compared to controls. Indeed, germination increased more in *H. lanatus*, than in *D. tenuifolia* and *S. flos-cuculi* (Lizzi et al., 2020). Regarding the observed stimulating effect on germination by CeO_2-NPs, literature reports conflicting data. Low toxicity and reduction of seed germination were observed on *Lycopersicum esculentum* and *Zea mays* (López-Moreno et al., 2010a) and *Glycine max* (López-Moreno et al., 2010b). In contrast, germination rate of *Hordeum vulgare* was not different up to 2000 mg L^{-1} CeO_2-NPs (Mattiello et al., 2015).

Nanoceria treatments stimulate the root growth in all three species, with a clear increase of length, particularly in *L. flos-cuculi* and *D. tenuifolia*, if compared with control seedlings. Some differences in response to treatments are species-specific (Lizzi et al., 2020). Contradictory literature evidence also exists regarding the influence of CeO_2-NPs on root development in different plant growth stages. Positive effects on root growth were observed in *Zea mays*, *Cucumis sativus*, and *Lactuca sativa* (López-Moreno et al., 2010a, b; Ma et al., 2010). As occurred in our species, very high tolerance to CeO_2-NPs was reported for *C. sativus*, *Brassica oleracea*, *Brassica napus*, and *Raphanus sativus*, whose root growth was not affected up to 2000 mg L^{-1} CeO_2-NPs. In opposite, a delayed root development in treated *Medicago sativa* and *L. esculentum* was reported (López-Moreno et al., 2010a,b). In our study, CeO_2-NPs does not cause acute toxicity in the early stages of plant development. On the other hand, as for other rare earth elements, literature reports that Ce may positively promote the effects of phytohormones on germinating seeds (Ramos et al., 2016; Shtangeeva, 2014). Also, it seems that eventually, monocots species are more tolerant to Ce than dicot ones (Pošćić et al., 2017).

Samples of control and treated seedlings of *H. lanatus*, *D. tenuifolia*, and *S. flos-cuculi* were subjected to the enzymatic extraction and further

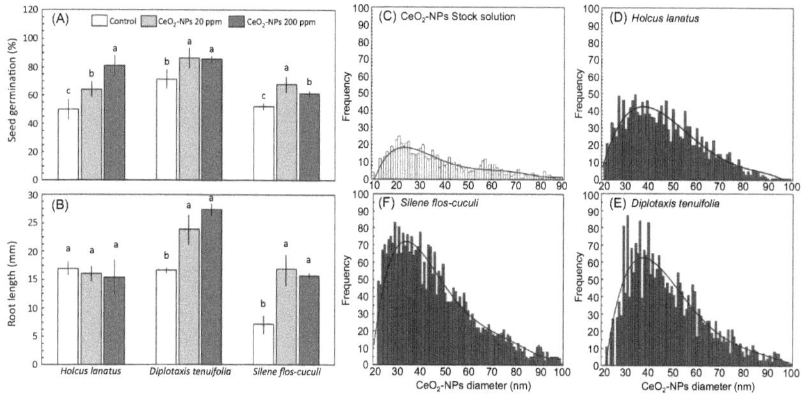

FIGURE 3.2 (A) Percentage of seed germination and (B) root length in seedlings of *H. lanatus*, *S. flos-cuculi* and *D. tenuifolia*, in control and treated with solutions of 25 nm CeO_2-NPs at 20 and 200 mg L^{-1}, respectively. The values are mean $\pm$ standard deviation of 3 replicates. Different letter indicates statistically significant difference between treatments at $p \leq 0.05$. Size distribution of CeO_2-NPs in (C) stock solutions and extracted from seedlings of (D) *H. lanatus*, (E) *D. tenuifolia*, and (F) *S. flos-cuculi* treated with 20 mg L^{-1} of CeO_2-NPs after enzymatic extraction.

analyzed by sp−ICP−MS to determine the size distribution of CeO_2-NPs. As expected, in control seedlings, CeO_2-NPs was not detected. In contrast, in all treated species, (1) the presence of internalized CeO_2-NPs was verified, and (2) the CeO_2-NPs have a different size distribution than stock solution suggesting aggregation phenomena between nanoparticles. Data from sp−ICP−MS analysis confirmed that CeO_2-NPs underwent agglomeration (Fig. 3.2). The increase of the median diameter of CeO_2-NPs was evident for seedlings treated with CeO_2-NPs 25 nm, being 41.7 nm, the average size of the nanoparticles extracted from seedlings (41 nm in *H. lanatus* and *S. flos-cuculi*, and 43 nm in *D. tenuifolia*). Besides, the most frequent CeO_2-NPs diameter is smaller in dicots *D. tenuifolia* and *S. flos-cuculi* (31 and 35 nm, respectively) than monocot *H. lanatus* (40 nm) (Lizzi et al., 2020).

3.4.2.2 *Pot trials*

The main objectives of Pot-1 are the following: (1) to observe the response of *H. lanatus* and *D. tenuifolia* to CeO_2-NPs and (2) verify whether that response was different depending on the size of the CeO_2-NPs provided (25 and 50 nm). The following biometric variables were detected on the fractions of *H. lanatus* and *D. tenuifolia*: root dry matter, number of stems per plant, stems dry matter, leaves dry matter, leaf area per plant, and total plant dry matter (Lizzi et al., 2021a).

An overall view of the results clearly shows that the response of the species to CeO_2-NPs was different. There was no significant change in the root

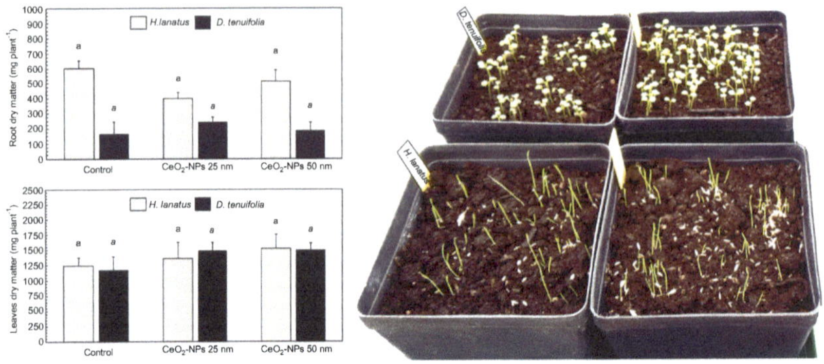

FIGURE 3.3 Pot-1 experiment. Root dry matter and leaves dry matter ± standard deviation of *H. lanatus* and *D. tenuifolia*. Comparison between control and plants grown in presence of 200 mg kg^{-1} 25 and 50 nm CeO_2-NPs. For each species the statistically significant difference ($p \leq 0.05$) between treatments is indicated by the letters using one-way ANOVA follow by Tukey's test. On the right, plant seedlings in the early stages of the experiment.

dry mass of *H. lanatus* ($p = 0.1154$) and *D. tenuifolia* ($p = 0.3096$) in response to the treatments. That is likely due to the potting soil's buffer effect, which at the not very high concentration of 200 mg kg^{-1}, CeO_2-NPs did not affect the root growth (Fig. 3.3). The aerial biomass production of plants responded to the presence of CeO_2-NPs. The effect of the "treatment" factor was statistically significant in the case of the number of stems per plant ($p = 0.0094$**), for the leaf area ($p = 0.0005$**) and the dry weight of the leaves ($p = 0.0482$*) (Lizzi et al., 2021a). The leaves' dry mass responded to the treatment. The ANOVA indicated a statistically significant effect of the factor "treatment" ($p = 0.0482$*) and the post-hoc test showed that 25 and 50 nm CeO_2-NPs stimulated a higher biomass production in leaves than control plants (1210 mg kg^{-1}), with an increase of 17% (1420 mg kg^{-1}) and 24% (1511 mg kg^{-1}), respectively. However, analyzing separately the results of the two species, it can be verified that data variability hides the effect of the size of CeO_2-NPs, which is not statistically significant for both *H. lanatus* and *D. tenuifolia* compared to control (Fig. 3.3).

Lizzi et al. (2021a) demonstrated that *D. tenuifolia* was not influenced by CeO_2-NPs, while *H. lanatus* responded to the treatment. The most evident effect of the treatments occurred on developing the aerial biomass of *H. lanatus*. Unexpectedly, that was observed in response to treatment with the 50 nm CeO_2-NPs. The larger nanoparticles stimulated a more significant number of stems per plant. In turn, greater tillering has increased aerial biomass accumulation and, therefore, a greater surface area of photosynthetic tissues was produced. The stimulating effect agrees with other observations made on autumn-spring cereals *T. aestivum* and *H. vulgare* (Rico et al., 2015; Zhao et al., 2015). However, going back to *H. lanatus*, it must be specified that the observed increase in leaf area did not correspond to the enhancement of the dry weight of the leaves as it would have been reasonable to expect. So, the newly formed leaves were thinner than those previously formed. According to similar observations on *H. vulgare*, it is likely that treated plants had a more extended vegetative period than the controls (Rico et al., 2015).

The fate and effects of ENPs in the soil-plant system are studied by treating plants with nanomaterials, eventually structured with several capping molecules, at different concentrations, sizes, and shapes. Plants were always treated with a single ENP dose at a given time (Servin and White, 2016). Regardless of the ENPs source, plants are likely exposed to ENPs over a much longer time, at relatively lower concentrations but repeated pulses of ENPs; this last aspect is what we developed in the second experiment (Pot-2). This study's main goal was to evaluate and compare the effects of a single dose and two and three repeated applications of 25 nm CeO_2-NPs at different concentrations on the growth of *S. flos-cuculi* (L.) (Lizzi et al., 2021b).

An overall view of the experimental data indicates that the CeO_2-NPs dose (D1, D2, and D3), CeO_2-NPs concentration (20 and 200 mg kg^{-1}), and their interaction influenced the biometric variables observed on *S. flos-cuculi*. Carefully observing the effects of treatments on the vegetative development of plants, some aspects of particular interest can be highlighted. As already mentioned, the root biomass dry weight, being the plant fraction directly exposed to the treatments, showed to be particularly sensitive to the experimental conditions. The root apparatus development responded positively to D1 (single dose of CeO_2-NPs provided to pot soil before seed germination). An increase of 29% (at 20 mg kg^{-1}) and 9% (at 200 mg kg^{-1}) in root biomass compared to the control was observed.

At the lowest concentration, the higher doses of CeO_2-NPs (D2 and D3) did not promote the same effect detected in D1. The weight of the root biomass returned to a level very close to that of the control plants. That also occurred for D2 at the highest concentration (200 mg kg^{-1}). Treatment D3 reduced by about 27% the root development compared to the control (Fig. 3.3). Additionally, a statistical analysis was performed by isolating the concentration factor, that is testing the effect of single and repeated administrations of CeO_2-NPs to plants within the same concentration level. In this case, considering D1 as the reference within each concentration, we evaluated the effect of the additional doses of CeO_2-NPs on the plant root biomass.

Whether the single CeO_2-NPs dose stimulated the production of root biomass, the second additional dose (D3), even though supplied to the plants at a late vegetative stage, resulted in a reduction in the root biomass. Compared to D1, we recorded a reduction in root dry matter of *S. flos-cuculi* by approximately 21% and 33%, for CeO_2-NPs 20 and 200 mg kg^{-1}, respectively.

As reported by Lizzi et al. (2021b), some other biometric variables were observed in plants. In particular, on the aboveground plant biomass, the number of stems and leaves for each plant was counted. The total leaf area per plant and the leaves dry matter were recorded as well. For these variables, the statistical analysis did not reveal significant effects of the treatments. In contrast, there was a very evident negative effect of CeO_2-NPs on dry matter accumulation in the stems of *S. flos-cuculi*. Regardless of the concentration and dose of CeO_2-NPs, the treatment's negative effect determined an average reduction of 75.5% in dry matter accumulation in the stems compared to the control. Finally, regarding the plants' total dry matter, albeit to a lesser extent than that observed for the weight of the stems, the effect of CeO_2-NPs on plant development is also visible on total biomass production. The negative effect of the treatments on the growth of *S. flos-cuculi* is statistically significant ($p = 0.00000$***), regardless of the CeO_2-NPs dose and even at the lower concentration of nanoparticles.

In conclusion, the growth of *S. flos-cuculi* was negatively affected by CeO_2-NPs. The root development in plants that received the lowest nanoparticle concentration has not undergone apparent alterations due to subsequent CeO_2-NPs doses. At the higher concentration, the dose-effect is much clearer. A slowdown in plant growth was observed as well. The number of plant stems did not change, but they were shorter than the controls'. Although no statistically significant evidence was found regarding the effects of treatments on leaf biomass, perhaps the relevant data variability detected in the treated plants compared to control ones was an early signal of plant stress.

3.5 Summary

In this chapter, a review of the literature relating to the impact of ENPs on spontaneous plants was conducted. Drawing upon the primary scientific literature search databases was found a body of papers; subsequently, they were sorted to distinguishing studies on aquatic plants from terrestrial ones.

Environmental risk assessment of ENPs requires accurate data on their environmental concentrations and ecotoxicity. At the same time, a systematic investigation regarding the effects on biota is needed. As far as higher plants are concerned, this review demonstrates that there is a significant knowledge gap. If we do not consider the species of agricultural interest, there is an impressive disproportion between the number of studies dedicated to aquatic species and spontaneous terrestrial species.

The impact of ENPs on ecosystems is predicted to be significant. However, several uncertainties regarding the weathering of ENPs and the effects of soil properties on nanomaterials still exist. On the other hand, current knowledge regarding ENPs-plants relationships indicates that the effects of ENMs properties and species-specific responses can combine.

No model plant species to extrapolate predictions on plants' behavior in response to a given nanostructure and its modifications is available. Plant-nanotoxicology is still somewhat backward compared to other areas, and systematic work will have to be carried out to achieve adequate knowledge on these aspects.

References

Aleksandrowicz-Trzcińska, M., Bederska-Błaszczyk, M., Szaniawski, A., Olchowik, J., Studnicki, M., 2019. The effects of copper and silver nanoparticles on container-grown Scots Pine (*Pinus sylvestris* L.) and Pedunculate Oak (*Quercus robur* L.) seedlings. Forests 10, 269.

Anawar, H.Md, Strezov, V., 2019. Transport, fate, and toxicity of the emerging and nanomaterial contaminants in aquatic ecosystems: removal by natural processes. In: Kumar Mishra, A., Anawar, H.Md, Drouiche, N. (Eds.), Emerging and Nanomaterial Contaminants in Wastewater. Elsevier, pp. 43–62.

Asztemborska, M., Bembenek, M., Jakubiak, M., Stęborowski, R., Bystrzejewska-Piotrowska, G., 2018. The Effect of nanoparticles with sorption capacity on the bioaccumulation of divalent ions by aquatic plants. Int. J. Environ. Res. 12, 245–253.

Baranowska-Wójcik, E., Szwajgier, D., Oleszczuk, P., Winiarska-Mieczan, A., 2020. Effects of titanium dioxide nanoparticles exposure on human health—a review. Biol. Trace Elem. Res. 193, 118–129.

Bar-On, Y.M., Phillips, R., Milo, R., 2018. The biomass distribution on Earth. Proc. Natl. Acad. Sci. USA 115, 16506–16511.

Biere, A., 1991. Parental effects in *Lychnis floscuculi* L.: seed size, germination and seedling performance in a controlled environment. J. Evol. Biol. 4, 447–465.

Biswas, D.K., Scannell, G., Akhmetov, N., Fitzpatrick, D., Jansen, M.A.K., 2010. 2,4,6-Trichlorophenol mediated increases in extracellular peroxidase activity in three species of Lemnaceae. Aquat. Toxicol. 100, 289–294.

Boyes, W.K., Thornton, B.L.M., Al-Abed, S.R., Andersen, C.P., Bouchard, D.C., Burgess, R.M., et al., 2017. A comprehensive framework for evaluating the environmental health and safety implications of engineered nanomaterials. Crit. Rev. Toxicol. 47 (9), 771–814.

Cabrera Walsh, G., Magalí Dalto, Y., Mattioli, F.M., Carruthers, R.I., Anderson, L.W., 2013. Biology and ecology of *Brazilian elodea* (*Egeria densa*) and its specific herbivore, Hydrellia sp., in Argentina. BioControl 58, 133–147.

Caldelas, C., Poitrasson, F., Viers, J., Araus, J.L., 2020. Stable Zn isotopes reveal the uptake and toxicity of zinc oxide engineered nanomaterials in *Phragmites australis*. Environ. Sci.: Nano 7, 1927–1941.

Chambers, P.A., Lacoul, P., Murphy, K.J., Thomaz, S.M., 2008. Global diversity of aquatic macrophytes in freshwater. Hydrobiologia 595, 9–26.

Chen, X., O'Halloran, J., Jansen, M.A.K., 2016. The toxicity of zinc oxide nanoparticles to *Lemna minor* (L.) is predominantly caused by dissolved Zn. Aquat. Toxicol. 174, 46–53.

Dale, A.L., Casman, E.A., Lowry, G.V., Lead, J.R., Viparelli, E., Baalousha, M., 2015. Modeling nanomaterial environmental fate in aquatic systems. Environ. Sci. Technol. 49, 2587–2593.

Ding, Y., Bai, X., Ye, Z., Ma, L., Liang, L., 2019. Toxicological responses of Fe_3O_4 nanoparticles on Eichhornia crassipes and associated plant transportation. Sci. Tot. Env. 671, 558–567.

Ergen, Ş.F., Tunca, E.Ü., 2018. Nanotoxicity modelling and removal efficiencies of ZnONP with consortium. Int. J. Phytoremed. 20, 16–26.

Erwin, K.L., 2009. Wetlands and global climate change: the role of wetland restoration in a changing world. Wetl. Ecol. Manage. 17, 71.

Fennessy, M.S., Cronk, J.K., 2018. Primary production and respiration: ecological processes in wetlands. In: Finlayson, C., et al., (Eds.), The Wetland Book. Springer, Dordrecht, pp. 315–322.

Geitner, N.K., Cooper, J.L., Avellan, A., Castellon, B.T., Perrotta, B.G., Bossa, N., et al., 2018. Size-based differential transport, uptake, and mass distribution of ceria (CeO_2) nanoparticles in wetland mesocosms. Env. Sci. Technol. 52, 9768–9776.

Giraldo, J.P., Landry, M.P., Faltermeier, S.M., McNicholas, T.P., Iverson, N.M., Boghossian, A.A., et al., 2014. Plant nanobionics approach to augment photosynthesis and biochemical sensing. Nat. Mater. 13 (4), 400–408.

Glenn, J.B., Klaine, S.J., 2013. Abiotic and biotic factors that influence the bioavailability of gold nanoparticles to aquatic macrophytes. Environ. Sci. Technol. 47, 10223–10230.

Glenn, J.B., White, S.A., Klaine, S.J., 2012. Interactions of gold nanoparticles with freshwater aquatic macrophytes are size and species dependent. Environ. Toxicol. Chem. 31 (1), 194–201.

Gregorczyk, K., Knez, M., 2016. Hybrid nanomaterials through molecular and atomic layer deposition: top down, bottom up, and in-between approaches to new materials. Prog. Mater. Sci. 75, 1–37.

Gubbins, E.J., Batty, L.C., Lead, J.R., 2011. Phytotoxicity of silver nanoparticles to *Lemna minor* L. Environ. Poll. 159 (6), 1551–1559.

Hall, M.K.D., Jobling, J.J., Rogers, G.S., 2012. Some perspectives on rocket as a vegetable crop: a review. Veg. Crop. Res. Bull. 76, 21–41.

Hawthorne, J., De la Torre Roche, R., Xing, B., Newman, L.A., Ma, X., Majumdar, S., et al., 2014. Particle-size dependent accumulation and trophic transfer of cerium oxide through a terrestrial food chain. Environ. Sci. Technol. 48, 13102–13109.

Holden, P.A., Gardea-Torresdey, J., Klaessig, F., Turco, R.F., Mortimer, M., Hund-Rinke, K., et al., 2016. Considerations of environmentally relevant test conditions for improved evaluation of ecological hazards of engineered nanomaterials. Env. Sci. Technol. 50 (12), 6124–6145.

Iravani, R., 2011. Green synthesis of metal nanoparticles using plants. Green. Chem. 13, 2638–2650.

Jacob, D.L., Borchardt, J.D., Navaratnam, L., Otte, M.L., Bezbaruah, A.N., 2013. Uptake and translocation of Ti nanoparticles in crops and wetland plants. Int. J. Phytorem. 15, 142–153.

Jalas, J., Suominen, J., 1986. Atlas Florae Europaeae (AFE)—distribution of vascular plants in Europe 7 (Caryophyllaceae (Silenioideae)). The Committee for Mapping the Flora of Europe and Societas Biologica Fennica Vanamo, Helsinki.

Jiang, H.S., Qiu, X.N., Li, G.B., Li, W., Yin, L.Y., 2014. Silver nanoparticles induced accumulation of reactive oxygen species and alteration of antioxidant systems in the aquatic plant Spirodela polyrhiza. Environ. Toxicol. Chem. 33, 1398–1405.

Jiménez-Lamana, J., Wojcieszek, J., Jakubiak, M., Asztemborska, M., Szpunar, J., 2016. Single particle ICP-MS characterization of platinum nanoparticles uptake and bioaccumulation by Lepidium sativum and Sinapis alba plants. J. Anal. At. Spectrom. 31, 2321–2329.

Juhel, G., Batisse, E., Hugues, Q., Daly, D., van Pelt, F.N.A.M., O'Halloran, J., et al., 2011. Alumina nanoparticles enhance growth of *Lemna minor*. Aquat. Toxicol. 105, 328–336.

Kah, M., Tufenkji, N., White, J.C., 2019. Nano-enabled strategies to enhance crop nutrition and protection. Nat. Nanotechnol. 14, 532–540.

Kim, E., Kim, S.-H., Kim, H.-C., Lee, S.G., Lee, S.J., Jeong, S.W., 2011. Growth inhibition of aquatic plant caused by silver and titanium oxide nanoparticles. Toxicol. Environ. Health Sci. 3 (1), 1–6.

Kumar, N., Kumbhat, S., 2016. Essentials in Nanoscience and Nanotechnology, first ed. John Wiley & Sons.

Liu, L.J., Xu, Y.A., Wang, Z.Y., 2014. Toxicity of CuO engineered nanoparticles to Eichhornia crassipes (2014). Adv. Mater. Res. 881–883, 976–979.

Lizzi, D., Mattiello, A., Piani, B., Fellet, G., Adamiano, A., Marchiol, L., 2020. Germination and early development of three spontaneous plant species to nanoceria (nCeO2) with different concentrations and particle sizes. Nanomaterials 10 (12), 2534.

Lizzi, D., Mattiello, A., Adamiano, A., Marchiol, L., 2021a. Influence of cerium oxide nanoparticles on two terrestrial wild plant species. Plants 10, 335.

Lizzi, D., Mattiello, A., Piani, B., Gava, E., Fellet, G., Marchiol, L., 2021b. Single and repeated applications of CeO_2 nanoparticles differently affect *Silene floscuculi* L. (Caryophyllaceae) growth: evidences from a microcosm study. Nanomaterials 11, 229.

López-Moreno, M.L., De La Rosa, G., Hernàndez-Viezcas, J.A., Peralta-Videa, J.R., Gardea-Torresdey, J.L., 2010a. X-ray absorption spectroscopy (XAS) corroboration of the uptake and storage of CeO_2 nanoparticles and assessment of their differential toxicity in four edible plant species. J. Agric. Food Chem. 58, 3689–3693.

López-Moreno, M.L., de la Rosa, G., Hernàndez-Viezcas, J.A., Castillo Michel, H., Botez, C.E., Peralta-Videa, J.R., et al., 2010b. Evidence of the differential biotransformation and genotoxicity of ZnO and CeO_2 nanoparticles on soybean (*Glycine max*) plants. Environ. Sci. Technol. 44, 7315–7320.

Lowry, G.V., Avellan, A., Gilbertson, L.M., 2019. Opportunities and challenges for nanotechnology in the agri-tech revolution. Nat. Nanotechnol. 14, 517–522.

Lowry, G.V., Espinasse, B.P., Badireddy, A.R., Richardson, C.J., Reinsch, B.C., Bryant, L.D., et al., 2012. Long-term transformation and fate of manufactured Ag nanoparticles in a simulated large scale freshwater emergent wetland. Environ. Sci. Technol. 46, 7027–7036.

Lu, K., Zhang, Z., He, X., Ma, Y., Zhou, K., Zhang, H., et al., 2010. Bioavailability and distribution and of ceria nanoparticles in simulated aquatic ecosystems, quantification with a radiotracer technique. J. Nanosci. Nanotechnol. 10 (12), 8658–8662.

Ma, X., Geiser-Lee, M.J., Deng, Y., Kolmakov, A., 2010. Interactions between engineered nanoparticles (ENPs) and plants: phytotoxicity, uptake and accumulation. Sci. Total. Environ. 408, 3053–3061.

Marchiol, L., Iafisco, M., Fellet, G., Adamiano, A., 2020. Nanotechnology support the next agricultural revolution: perspectives to enhancement of nutrient use efficiency. Adv. Agron. 161, 27–116.

Mattiello, A., Pošćić, F., Musetti, R., Giordano, C., Vischi, M., Filippi, A., et al., 2015. Evidences of genotoxicity and phytotoxicity in *Hordeum vulgare* exposed to CeO_2 and TiO_2 nanoparticles. Front. Plant. Sci. 6, 1043.

Meesters, J.A.J., Quik, J.T.K., Koelmans, A.A., Hendriks, A.J., van de Meent, D., 2016. Multimedia environmental fate and speciation of engineered nanoparticles: a probabilistic modeling approach. Environ. Sci. Nano 3, 715–727.

Milewska-Hendel, A., Zubko, M., Stróz, D., Kurczynska, E.U., 2019. Effect of nanoparticles surface charge on the *Arabidopsis thaliana* (L.) roots development and their movement into the root cells and protoplasts. Int. J. Mol. Sci. 20, 1650.

Miralles, P., Church, T.L., Harris, A.T., 2012. Toxicity, uptake, and translocation of engineered nanomaterials in vascular plants. Env. Sci. Technol. 46, 9224–9239.

Mittala, K., Chisti, Y., Banerjee, U.C., 2013. Synthesis of metallic nanoparticles using plant extracts. Biotechnol. Adv. 31, 346–356.

Montes, A., Bisson, M.A., Gardella, J.A., Aga, D.S., 2017. Uptake and transformations of engineered nanomaterials: critical responses observed in terrestrial plants and the model plant *Arabidopsis thaliana*. Sci. Tot. Env. 607–608, 1497–1516.

Mortimer, M., Holden, P.A., 2019. Fate of engineered nanomaterials in natural environments and impacts on ecosystems. In: Marmiroli, N., White, J., Song, J. (Eds.), Exposure to Engineered Nanomaterials in the Environment, first ed. Elsevier, Amsterdam, pp. 61–103.

Movafeghi, A., Khataee, A., Abedi, M., Tarrahi, R., Dadpour, M., Vafaei, F., 2018. Effects of TiO_2 nanoparticles on the aquatic plant Spirodela polyrrhiza: evaluation of growth parameters, pigment contents and antioxidant enzyme activities. J. Environ. Sci. 64, 130–138.

OECD, 2000. Revised Draft Guidance Document on Aquatic Toxicity Testing of Difficult Substances and Mixtures. Environment Directorate, Organisation for Economic Cooperation and Development.

OECD, 2002. Guidelines for the testing of chemicals Lemna sp. Growth inhibition test. Draft guideline 221. OECD Publishing, Paris.

OECD, 2006. Test No. 221: Lemna sp. Growth Inhibition Test, OECD Guidelines for the Testing of Chemicals, Section 2, OECD Publishing, Paris. https://doi.org/10.1787/9789264016194-en.

Ostroumov, S.A., Poklonov, V.A., Kotelevtsev, S.V., Orlov, S.N., 2014. Toxicity of gold nanoparticles for plants in experimental aquatic system. Mosc. Univ. Biol. Sci. Bull. 69 (3), 108–112.

Peng, C., Zhang, W., Gao, H., Li, Y., Tong, X., Li, K., et al., 2017. Behavior and potential impacts of metal-based engineered nanoparticles in aquatic environments. Nanomaterials 2017 (7), 21.

Perreault, F., Samadani, M., Dewez, D., 2014. Effect of soluble copper released from copper oxide nanoparticles solubilisation on growth and photosynthetic processes of *Lemna gibba* L. Nanotox. 8 (4).

Piccinno, F., Gottschalk, F., Seeger, S., Nowack, B., 2012. Industrial production quantities and uses of ten engineered nanomaterials in Europe and the world. J. Nanopart. Res. 14.

Poščić, F., Schat, H., Marchiol, L., 2017. Cerium negatively impacts the nutritional status in rapeseed. Sci. Total. Environ. 593–594, 735–744.

Ramos, S.J., Dinali, G.S., Oliveira, C., Martins, G.C., Moreira, C.G., Siqueira, J.O., et al., 2016. Rare earth elements in the soil environment. Curr. Pollut. Rep. 2, 28–50.

Read, D.S., Matzke, M., Gweon, H.S., Newbold, L.K., Heggelund, L., Ortiz, M.D., et al., 2016. Soil pH effects on the interactions between dissolved zinc, non-nano- and nano-ZnO with soil bacterial communities. Environ. Sci. Pollut. Res. 23, 4120–4128.

Reddy Pullagurala, V.L., Adisa, I.O., Rawat, S., White, J.C., Zuverza-Mena, N., Hernandez-Viezcas, J.A., et al., 2019. Fate of engineered nanomaterials in agroenvironments and impacts on agroecosystems. In: Marmiroli, N., White, J., Song, J. (Eds.), Exposure to Engineered Nanomaterials in the Environment, 1st ed Elsevier, Amsterdam, pp. 105–142.

Rico, C.M., Barrios, A.C., Tan, W., Rubenecia, R., Lee, S.C., Varela-Ramirez, A., et al., 2015. Physiological and biochemical response of soil-grown barley (*Hordeum vulgare* L.) to cerium oxide nanoparticles. Environ. Sci. Pollut. Res. 22, 10551–10558.

Rippner, D.A., Green, P.G., Young, T.M., Parikh, S.J., 2018. Dissolved organic matter reduces CuO nanoparticle toxicity to duckweed in simulated natural systems. Environ. Pollut. 234, 692–698.

Roco, M.C., 2020. Principles of convergence in nature and society and their application: from nanoscale, digits, and logic steps to global progress. J. Nanopart. Res. 22 (11), 321.

Schierz, A., Espinasse, B., Wiesner, M.R., Bisesi, J.H., Sabo-Attwood, T., Lee Ferguson, P., 2014. Fate of single walled carbon nanotubes in wetland ecosystems. Environ. Sci. Nano 1 (6), 574–583.

Servin, A.D., White, J.C., 2016. Nanotechnology in agriculture: next steps for understanding engineered nanoparticle exposure and risk. NanoImpact 1, 9–12.

Shi, J., Abid, A.D., Kennedy, I.M., Hristova, K.R., Silk, W.K., 2011. To duckweeds (*Landoltia punctata*), nanoparticulate copper oxide is more inhibitory than the soluble copper in the bulk solution. Environ. Poll. 159, 1277–1282.

Shtangeeva, I., 2014. Europium and cerium accumulation in wheat and rye seedlings. Water Air Soil. Pollut. 225, 1964–1973.

Song, U., Lee, S., 2016. Phytotoxicity and accumulation of zinc oxide nanoparticles on the aquatic plants *Hydrilla verticillata* and *Phragmites australis*: leaf-type-dependent responses. Environ. Sci. Pollut. Res. 23, 8539–8545.

Song, G., Gao, Y., Wu, H., Hou, W., Zhang, C., Ma, H., 2012. Physiological effect of anatase TiO_2 nanoparticles on *Lemna minor*. Environ. Toxicol. Chem. 31, 1–6.

Song, G., Hou, W., Gao, Y., Wang, Y., Lin, L., Zhang, Z., et al., 2016. Effects of CuO nanoparticles on *Lemna minor*. Bot. Stud. 57, 3.

Song, L., Vijver, M.G., Peijnenburg, W.J.G.M., 2015. Comparative toxicity of copper nanoparticles across three Lemnaceae species. Sci. Total. Environ. 518–519, 217–224.

Spielman-Sun, E., Avellan, A., Bland, G.D., Tappero, R.T., Acerbo, A.S., Unrine, J.M., et al., 2019. Nanoparticle surface charge influences translocation and leaf distribution in vascular plants with contrasting anatomy. Env. Sci. Nano 6, 2508–2519.

Srivastava, S., Mishra, S., Tripathi, R.D., Dwivedi, S., Gupta, D.K., 2006. Copper-induced oxidative stress and responses of antioxidants and phytochelatins in *Hydrilla verticillata* (L.f.) Royle. Aquat. Toxicol. 80, 405–415.

Stace, C., 1997. New flora of the British Isles. Cambridge University Press, Cambridge, UK.

Sun, H., 2019. Grand challenges in environmental nanotechnology. Front. Nanotechnol. 1, 2.

Tanhan, P., Kruatrachue, M., Pokethitiyook, P., Chaiyarat, R., 2007. Uptake and accumulation of cadmium, lead and zinc by Siam weed [*Chromolaena odorata* (L.) King & Robinson]. Chemosphere 68 (2), 323–329.

Tarrahi, R., Khataee, A., Movafeghi, A., Rezanejad, F., 2018. Toxicity of ZnSe nanoparticles to *Lemna minor*: evaluation of biological responses. J. Environ. Manag. 226, 298–307.

Tarrahi, R., Khataee, A., Movafeghi, A., Rezanejad, F., Gohari, G., 2017. Toxicological implications of selenium nanoparticles with different coatings along with Se4 + on *Lemna minor*. Chemosphere 181, 655–665.

Thompson, J.D., Turkington, R., 1988. The biology of Canadian weeds - *Holcus lanatus* L. Canadian. J. Plant. Sci. 68, 131–147.

Üçüncü, E., Özkan, A.D., Kursungöz, C., Ülger, Z.E., Ölmez, T.T., Tekinay, T., et al., 2014. Effects of laser ablated silver nanoparticles on *Lemna minor*. Chemosphere 108, 251–257.

USEPA, 1995. EPA Method 3052: Microwave assisted acid digestion of siliceous and organically based matrices in US Environmental Protection Agency. Test Methods for Evaluating Solid Waste, Washington, USA.

Yin, L., Colman, B.P., McGill, B.M., Wright, J.P., Bernhardt, E.S., 2012. Effects of silver nanoparticle exposure on germination and early growth of eleven wetland plants. PLoS One 7, e47674.

Zaytseva, O., Neumann, G., 2016. Carbon nanomaterials: production, impact on plant development, agricultural and environmental applications. Chem. Biol. Technol. Agric. 3, 17.

Zhang, P., Ma, Y., Zhang, Z., 2015. Interactions between engineered nanomaterials and plants: phytotoxicity, uptake, translocation, and biotransformation. In: Siddiqui, M., Al-Whaibi, M., Mohammad, F. (Eds.), Nanotechnology and Plant Sciences. Springer, Cham, pp. 77–99.

Zhang, W., Yu, Z., Rao, P., Lo, I.M.C., 2019. Uptake and toxicity studies of magnetic TiO_2-based nanophotocatalyst in *Arabidopsis thaliana*. Chemosphere 224, 658–667.

Zhao, L., Sun, Y., Hernandez-Viezcas, J.A., Hong, J., Majumdar, S., Niu, G., et al., 2015. Monitoring the environmental effects of CeO_2 and ZnO nanoparticles through the life cycle of corn (*Zea mays*) plants and in situ μ-XRF mapping of nutrients in kernels. Environ. Sci. Technol. 49, 2921–2928.

Zuverza-Mena, N., Martínez-Fernández, D., Du, W., Hernandez-Viezcas, J.A., Bonilla-Bird, N., López-Moreno, M.L., et al., 2017. Exposure of engineered nanomaterials to plants: insights into the physiological and biochemical responses-A review. Plant. Physiol. Biochem. 110, 236–264.

Chapter 4

Changes in metabolite profile of plants exposed to engineered nanomaterials

Cyren M. Rico[1], Polycarp C. Ofoegbu[1], Naum J. Kirwa[1], Dane C. Wagner[1], Oluwasegun Michael Abolade[1], Siying Jia[2] and Jie Hong[2]

[1]Department of Chemistry and Biochemistry, Missouri State University, Springfield, MO, United States, [2]College of Environment, Zhejiang University of Technology, Hangzhou, Zhejiang, China

Chapter Outline

4.1 Introduction

This chapter summarizes the advantages and major knowledge gaps in metabolome changes in plants after exposure to engineered nanoparticles (NPs). Both single endpoint and advance metabolomics techniques are discussed, but greater emphasis is given for in-depth discussion on advanced metabolomics analysis. The concerns that engineered NP are detrimental to the growth and productivity, nutritional quality (especially of food crops), and ecological

Plant Exposure to Engineered Nanoparticles. DOI: https://doi.org/10.1016/B978-0-323-85032-2.00003-8

functions and ecosystem processes (i.e., nutrient cycling, ecosystem stability) of plants have generated vigorous research on nanophytotoxicity. Several reviews have noted that the effects of NPs on plants are dependent on factors such as plant species, vegetative growth stage, mode and length of exposure, and NPs characteristics including size, shape, morphology, and coating (Rico et al., 2011; Gardea-Torresdey et al., 2014; Ma et al., 2018; Majumdar and Keller, 2020; Miralles et al., 2012). Short-term exposure studies revealed mixed responses in plant or seedling growth, photosynthetic efficiency, or oxidative stress (Rico et al., 2011), while long-term and generational exposure studies demonstrated benign or inconsequential effects on plant growth and productivity especially when tested at realistic environmental concentrations of NPs (Rico et al., 2017; Rico et al., 2020a,b). However, there are still questions on whether NPs will cause detrimental effects on plant integrity and ecological balance. In addition, NPs are now being used in agricultural inputs and implemented especially as fertilizers and pesticides (White and Gardea-Torresdey, 2018; White, 2016; Hong et al., 2021; Paramo et al., 2020), and whether such applications could cause harm to human and environmental health is a question that lingers in literature. The need to address these concerns resulted in adopting more sensitive tools particularly the "-omics technologies" (genomics, transcriptomics, proteomics, and metabolomics) in systems biology (Majumdar and Keller, 2020; Matich et al., 2019). This chapter mainly describes metabolomics in plants exposed to engineered NPs.

The repeatedly reported major finding in the literature is that plants undergo metabolic reprogramming to activate and boost the defense system in response to NPs stress. Metabolomics can identify a peculiar set of metabolites more sensitive to NPs exposure and potentially use these metabolites as biomarkers. Studies have reported metabolic perturbations in NP-treated plants, but these changes were too broad and unspecific to identify unique metabolites that can characterize NPs impacts on plants. Also, most studies were performed at early plant growth stage that it becomes a challenge to predict the long-term detrimental effects of NPs to terrestrial plants and the environment. Nevertheless, metabolomics is useful in broadening knowledge on plants responses to NPs, and extensive information on biochemical differences in plants following exposure to NPs has been obtained in the last few years.

4.2 Single endpoint analysis of plant metabolites

Earlier reviews on plant-NPs interactions have reported changes in plant metabolites such as carbohydrates, fatty acids, amino acids, and phytochemicals using single endpoint colorimetric or chromatographic techniques (Gardea-Torresdey et al., 2014; Hatami et al., 2016; Rico et al., 2011). Hatami et al. (2016) reviewed several published papers that measured primary and secondary metabolites using mostly colorimetric assays. There are also reports on single endpoint chromatographic analysis of metabolites in plants exposed to engineered NP. For example, Rico et al. (2013b) measured the relationship between antioxidant activity, lipid peroxidation, and fatty acid and lignin contents in rice seedlings exposed to CeO_2 NPs using single endpoint assays (Fig. 4.1). Kole et al. (2013) employed HPLC to assess modifications in anticancer and antidiabetic components in bitter melon (*Momordica charantia*) exposed to fullerol NPs; these researchers found that fullerol improved the phytochemical contents (i.e. cucurbitacin-B, lycopene, charantin, insulin) in bitter melon. There are also several studies that measured fatty acid and amino acid contents in wheat (*Triticum aestivum* L.) and barley (*Hordeum vulagare* L.) using HPLC or GC-MS to assess the impacts of CeO_2 or TiO_2 NPs in grain nutritional quality (Rico et al., 2020a; Rico et al., 2015; Rico et al., 2014; Pošćić et al., 2016). Recently, there has been an increase in studies that employed LC or GC instrumentations to evaluate metabolite changes in plants. In general, literature reports indicate that analyses of specific metabolites were performed to measure impacts in membrane damage and nutritional quality. For example, phytotoxic effects of CuO NPs on *Arabidopsis thaliana* were shown by enhanced production or reactive oxygen species which resulted in increased degrees of fatty acid saturation (Yuan et al., 2016). The researchers found that superoxide anion ($O_2^{\bullet -}$) induced conversion of saturated fatty acid C18:1, C18:2, and C16:3 to C16:0 (Yuan et al., 2016). Likewise, responses of eggplant (*Solanum melongena*) to *in vitro* exposure to metal oxides NPs (CuO, ZnO, NiO NPs) were measured by single endpoint colorimetric analyses of phenolic and flavonoid compounds (Baskar et al., 2018). Chapter 6 also partly discussed metabolite changes in plants exposed to engineered NP using single endpoint metabolite analysis.

Single endpoint assays are still useful and being reported in recent publications as an additional evidence to understand plant integrity; however, they suffer from several shortcomings including the need for large amount of samples, time intensive extraction procedures, light sensitive color changes,

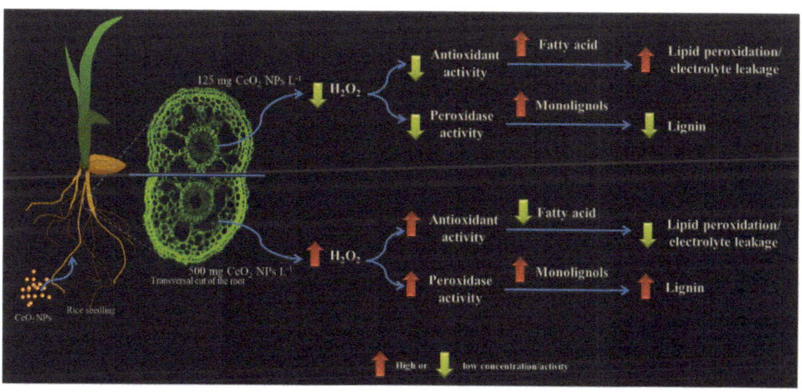

FIGURE 4.1 Single endpoint assays for selected metabolites in rice seedlings exposed to CeO_2 NPs. Data taken from Rico et al. (2013b).

poor detection limit, and requires multiple extractions as each protocol requires different extraction solutions and procedures for each endpoint. These traditional assays probe single process of nanophytotoxicity, and the results of which is limited for understanding biological processes and responses occurring in plants after exposure to NPs. Because of these limitations, metabolomics technique is rapidly becoming popular in the scientific community and increasingly being employed in nanophytotoxicity studies.

4.3 Metabolomics in plant-contaminant interaction studies

Metabolomics is an analytical technique that measures low-molecular weight metabolites in biological samples such as cells, fluids, and tissues using more advanced instrumentations (Hall, 2006; Dunn and Ellis, 2005; Jorge et al., 2016). Table 4.1 provides descriptions of relevant terminologies. These metabolites are small molecules ($\leq$1500 Da) that are involved as substrates, intermediates, or products of metabolic functions in organisms, and the complete collection of these metabolites in an organism is known as metabolome. Approaches to metabolomics could be either targeted or nontargeted (Hall, 2006; Dunn and Ellis, 2005; Jorge et al., 2016). Nontargeted (or global) analysis involves nondiscriminatory, broad-scale assessment of hundreds of metabolites in semiquantitative method that does not quantify the metabolites concentration in the samples. Nontargeted metabolomics is highly appropriate for generating testable hypothesis for future research. In contrast, targeted (or specific) approach to metabolomics can serve to test hypothesis and involves qualitative and quantitative analyses of one or few selected group of metabolites that have been identified from previous untargeted metabolomics analysis. Targeted metabolite analysis is a more rigorous approach than nontargeted metabolomics because it uses calibration curves for each metabolite and also needs to determine the recovery of each metabolite (Hall, 2006). Targeted metabolomics has been performed in cucumber leaves exposed to Cu NPs; in one study, 23 amino acids were successfully analyzed at high sensitivity (limit of detection $\geq$ 0.005 $ngmL^{-1}$) and good recovery rates (80%–120%) (Huang et al., 2018; Huang et al., 2019).

Metabolomics has been employed for many applications in plant studies such as metabolotyping, population screening, understanding of physiological processes, identifying biomarkers and bioactivity, plant product quality and breeding, and defining substantial equivalence between transgenic and wildtype plants (Hall, 2006; Jorge et al., 2016). The field of environmental metabolomics has also emerged by employing metabolomics technique toward understanding responses of terrestrial and aquatic organisms to environmental contaminants and stressors (Lankadurai et al., 2013; Matich et al., 2019; Gauthier et al., 2019; Long et al., 2020; Tang, 2011; Piasecka et al., 2019). More recently, metabolomics has become a powerful tool in assessing the responses of plants exposed to emerging contaminants and pollutants.

TABLE 4.1 Metabolomics-related terminologies.

Term	As defined by Hall (2006)	As defined by Dunn and Ellis (2005)
Metabolomics	The technology geared toward providing an essentially unbiased, comprehensive qualitative and quantitative overview of the metabolites present in an organism.	Nonbiased identification and quantification of all metabolites in a biological system. Sample preparation must not exclude metabolites, and selectivity and sensitivity of the analytical technique must be high.
Metabolome	The complete complement of small molecules present in an organism.	
Metabolite-targeted analysis	Following broad-scale metabolomics analysis, or based on previous knowledge, biochemical profiling can be performed in greater detail on selected groups of metabolites by using optimized extraction and dedicated separation/detection techniques.	Qualitative and quantitative analysis of one or a few metabolites related to a specific metabolic reaction. Extensive sample preparation and separation from other metabolites are required, and this approach is especially employed when low limits of detection are required. Generally, chromatographic separation is used followed by sensitive MS or UV detection.
Metabolic profiling	Identification and quantification of the metabolites present in an organism. For practical reasons, this is generally only feasible for a limited number of components, which are generally chosen on the basis of discriminant analysis or on molecular relationships based upon molecular pathways/networks.	Identification and quantification of a selected number of predefined metabolites, generally related to a specific metabolic pathway(s). Sample preparation and instrumentation are employed to isolate those compounds of interest from possible matrix effects prior to detection, normally with chromatographic separation prior to detection with MS.
Metabolic fingerprinting	High-throughput qualitative screening of the metabolic composition of an organism or tissue with the primary aim of sample comparison and discrimination analysis. Generally, no attempt is initially made to identify the metabolites present. All steps from sample preparation, separation, and detection should be rapid and as simple as is feasible. Often used as a forerunner to metabolic profiling.	High-throughput, rapid, global analysis of samples to provide sample classification. Quantification and metabolic identification are generally not employed. A screening tool to discriminate between samples of different biological status or origin. Sample preparation is simple and, as chromatographic separation is absent, rapid analysis times are small (normally 1 min or less).
Metabonomics	A nonplant term generally used to define the technology used to measure quantitatively the metabolic composition of body fluids following a response to pathophysiological stimuli or genetic modification.	Evaluation of tissues and biological fluids for changes in endogenous metabolite levels that result from disease or therapeutic treatments.

Specifically, metabolomics is quickly becoming a tool to monitor health and toxicity of plants following exposure to engineered NPs (Ruotolo et al., 2018; Majumdar and Keller, 2020; Li et al., 2019).

4.3.1 Metabolomics in nanophytotoxicity studies

Metabolomics application in nanophytotoxicity studies is a growing field. There are only about 30 reports considered in this review; however, the number of publications is expected to grow significantly in the coming years. Experts recognize major challenges needed to overcome to maximize the potential of this technique in elucidating mechanisms of action or biochemical pathways of toxicity in plants following NPs treatment (Hall, 2006; Majumdar and Keller, 2020; Matich et al., 2019). For example, the lack of comprehensive databases limits the number of compounds that can be identified and consequently leaves out numerous metabolites in the data mining analysis. The analytical instrumentations, although they have improved immensely in recent years, are also still inadequate to detect and identify vast majority of metabolites in plants. In addition, scientists have emphasized the importance of reporting precise details on exposure conditions such as plant age and growth stage, ambient growth conditions, exposure media, NPs characteristics, and others (Majumdar and Keller, 2020).

Metabolomics employs sensitive instrumentations capable of detecting subtle changes in metabolites composition of plants that are otherwise not measurable in more traditional analyses (especially colorimetric assays). In contrast to single endpoint assays, metabolomics protocol requires a facile extraction procedure that uses small amount of samples and yields ample amount of extracts that can be used several times over to simultaneously measure hundreds of metabolites even at very low concentration. Although plant metabolomics has its own challenges (including sampling, extraction, storage, instrumentations, data processing and interpretation, chemical diversity of metabolites, and high variability between them), it is nonetheless more effective in extraction and detection of several hundreds of metabolites in a simple procedure (Creydt et al., 2018; Hall, 2006; Majumdar and Keller, 2020; Lankadurai et al., 2013). For example, targeted metabolomics readily measured 13 antioxidants (including 2-hyroxycinnamic acid, 4-cinnamic acid, α-tocopherol, benzoic acid, caffeic acid, curcumin, ferulic acid, gallic acid, and reduced glutathione) and 23 amino acids (including ornithine, citrulline, homoserine and typical amino acids such as glycine, serine, alanine, and others) in the leaves of cucumber (*Cucumis sativus*) exposed to Cu NPs in two different but facile extraction methods (80:20 methanol and water with 2% formic acid for antioxidants and 0.5 M aqueous HCl for amino acids extractions) without a need for postextraction treatments (Huang et al., 2019; Huang et al., 2018). These compounds could not be analyzed in more traditional single endpoint methods employed in earlier nanophytotoxicity experiments covered in previous reviews (Rico et al., 2011; Gardea-Torresdey et al., 2014).

4.3.2 Metabolomics data procurement and processing

Analysis of the metabolome is quite challenging in both the type of analytical techniques to use and the identification of metabolites detected because of the sheer number of metabolites (there are $\geq$200,000 metabolites, a large number of which has not been identified yet) and huge differences in their concentrations. It employs instrumentations with superb sensitivity capable of detecting minute concentrations of metabolites (parts per trillion levels). The most widely used detection method is high-resolution MS (mass spectrometry) or tandem MS/MS, which is routinely hyphenated with GC (gas chromatography) or LC (liquid chromatography) depending on the type of extracts, while NMR is employed to a lesser extent probably because of sensitivity issues (Li et al., 2019; Majumdar and Keller, 2020). Metabolites are identified by comparison with databases such as NIST library and KEGG database.

Metabolomics generate huge amount of data, which requires bioinformatics software to mine, analyze, and present the complex data into useful reliable information. Several publications relied on free software MetaboAnalyst (http://www.metaboanalyst.ca/) for the analysis of metabolomics data. This online resource is a great tool for easy and direct analysis of large amount of data to give visual representations on perturbations in metabolite composition. This is often provided by multivariate analysis such as principal component analysis (PCA) or partial least-squares discriminant analysis (PLS-DA), variable importance projection (VIP), and pathway enrichment analysis. For example, Fig. 4.2A depicts the analysis and visualization of metabolomics data from generational exposure of wheat to CeO_2 NPs. PLS−DA analysis of wheat grains revealed a much stronger separation or difference between grains from control plants (C1C2C3 = control treatment in the 1st, 2nd, and 3rd generations) as opposed to grains from treated plants in three generations (T1T2T3 = exposed to CeO_2 NPs (500 mg kg^{-1}) in the 1st, 2nd, and 3rd generations). The discriminating metabolites responsible for the separation were the ones that had VIP scores $\geq$ 1 (Fig. 4.2B). In addition, pathway enrichment analysis identifies the biological pathways significantly affected by the altered metabolite compositions (Fig. 4.2C). Readers are directed to several reviews that discussed the extraction, detection, and data analysis involved in plant metabolomics (Hall, 2006; Matich et al., 2019; Piasecka et al., 2019; Li et al., 2019).

4.4 Applications of metabolomics in plant-NP interactions studies

The popularity of metabolomics lies not only on its versatility, but more importantly on the amount of information it can provide and the usefulness of those pieces of information to uncover metabolic responses specific to the

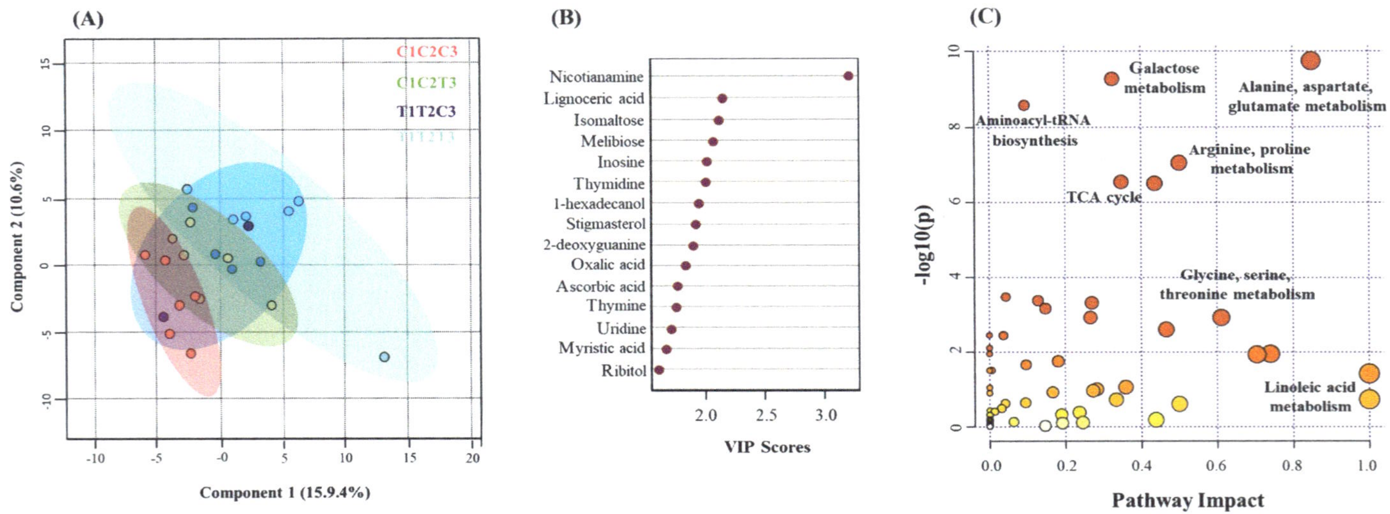

FIGURE 4.2 Metabolomics analysis of wheat grains generationally exposed to CeO_2 NPs: (A) partial least-square discriminant analysis (PLS-DA) of grains metabolomes, (B) metabolites with VIP scores ≥ 1 are responsible for discrimination between treatments, (C) biological pathways significantly impacted by the altered metabolite compositions as revealed by pathway enrichment analysis. C1C2C3 = wheat grains harvested from control treatment (0 mg kg^{-1}) for three generations; T1T2T3 = wheat grains harvested from CeO_2 NPs treatment (500 mg kg^{-1}) for three generations. Figures and information taken from Rico et al. (2020b).

time of plant growth and development. This is because the information obtained is the biological endpoints or end products of cellular processes (as opposed to those gained from genomics or transcriptomics) at the time when sampling was made that they are useful in understanding how plants respond to environmental conditions (e.g., biotic or abiotic stress) at specific point in time (Jorge et al., 2016; Hall, 2006; Majumdar and Keller, 2020). As metabolites are influenced by both genetic and environmental factors, changes in metabolite composition provides insights on plants' physiological or phenotypic responses to NPs exposure. Metabolomics analysis can also be employed for rapid assessment of the nutritional or commercial value (e.g., quality features such as flavor, aroma, shelf life, mineral and antioxidant contents) of food crops, which could be compromised from widespread applications of engineered NP as nanofertilizers or nanopesticides in agriculture. Likewise, mechanistic understanding of how nanobased technologies could make a more efficient and sustainable agriculture can be greatly advanced by metabolomics approaches. In this way, metabolomics can help to predict the suitability of nanobased agricultural implements to improve yields and nutritional value or enhance resistance against pest and diseases of crops.

This chapter discusses published reports that measured metabolite changes in NP-treated plants using advanced metabolomics techniques (GC- or LC-MS/MS). Several reviews have thoroughly analyzed photosynthetic activity, nutrient and antioxidant levels, membrane damage, and enzyme activity modulations in NP-exposed plants using nonmetabolomics technologies (Rico et al., 2011; Gardea-Torresdey et al., 2014). Currently, literature is sparse on metabolomics analysis of plants exposed to NPs (Table 4.2). In the last five years (2016–21), there have been only around thirty metabolomics studies that measured the impacts of NPs (TiO_2, CeO_2, Fe_3O_4, SiO_2, $Cu(OH)_2$, CdO, CdS, Ag, Cu, graphene oxide, and multiwalled carbon nanotubes) on food crops such as rice (*Oryza sativa*), wheat (*Triticum aestivum*), barley (*Hordeum vulgare*), corn (*Zea mays*), cucumber (*Cucumis sativus*), spinach (*Spinacia oleracea*), lettuce (*Lactuca sativa*), soybean (*Glycine max*), kidney beans (*Phaseolus vulgaris*), tomato (*Lycopersicon esculentum*), and fenugreek (*Trigonella foenum*), and other plants such as *Pistia stratiotes* and *Arabidopsis thaliana*. Most data were obtained using leaf or root samples from juvenile stage of plants (one to four weeks old) grown in hydroponics solution or soil. There were two reports on long-term exposure (28 days) in cucumber and *Arabidopsis thaliana*. There was one publication each on metabolite profiling of rice grain treated with TiO_2 NPs and tomato fruit exposed to MWCNT, and another report on wheat grains generationally exposed to CeO_2 NPs. Majority of the studies employed GC-QTOF-MS and GC-MSD, but LC-QTOF-MS, LC-PDA, or LC-MS-Orbitrap were also widely used (Table 4.2). There were two publications from the same research group that reported the use of NMR technique (Zhao et al., 2016a; Zhao

TABLE 4.2 Plants exposed to different types of engineered nanoparticles and major metabolites that were significantly altered.

Plant	Nanoparticles (size; concentration)	Mode and duration of exposure	Plant parts used	Instrumentations used	Major observations	References
Barley (*Hordeum vulgare*)	CdO (7–60 nm); $2.03 \pm 0.45 \times 10^5$ particles/cm^3	Air (3 weeks)	Leaves Roots	HPLC-LTQ Orbitrap XL HRMS (Thermo Fisher)	• Affected primary (amino acids, saccharides) but not secondary (phenolic compounds, Krebs cycle acids, fatty acids) metabolites.	Vecerova et al. (2016)
Fenugreek (*Trigonella foenum*)	Ag (200 μL of 1 μg mL^{-1})	Hydroponics (5 days)	Seedling	HPLC (Thermo Fisher)	• Increased plant height and production of phytochemical diosgenin.	Jasim et al. (2017)
Arabidopsis thaliana	Ag (citrate / polyvinylpyrolidinone coated, 1 mg L^{-1})	Hydroponics (24 h)	Leaves	LC-QTOF-MS (Agilent Technologies)	• Altered metabolome; accumulation of phytosphingosines and N-acylethanolamines.	Chavez Soria et al. (2017)
	Ag (citrate capped; 10, 40, 100 nm; 0.5–5.0 ppm)	Hydroponics (25 days)	Seedling	UPLC-PDA (Thermo Fisher)	• Increased biosynthesis of phytoalexin compounds (i.e., camalexin, hydroxycamalexin O-hexoside, hydroxycamalexin malonyl-hexoside).	Kruszka et al. (2020)
	Ag (citrate capped; 10 nm; 1.0 and 2.5 mg L^{-1} in hydroponics; 12.5 mg kg^{-1} in soil)	Soil (30 days) and hydroponics (3 days)	Leaves	GC-MSD (Agilent Technologies)	• There were Ag NPs specific effects on shikimate-phenylpropanoid biosynthesis, and tryptophan and galactose metabolism.	Ke et al. (2018)

Broad beans (*Vicia fab* L.)	CdS	Soil (28 days)	Leaves Roots	GC-MSD (Agilent Technologies)	• Increased production of antioxidative metabolites (N-acetyl-5-hydroxytryptamine, 2-hydroxybutanoic acid, putrescine, flavone). • Modulated more metabolites (purine metabolism, alanine, aspartate, and glutamate metabolism; β-alanine, pantothenate and CoA biosynthesis and carbon fixation) in leaves than roots.	Tian et al. (2020)
Phaseolus vulgaris L.	CeO_2 (10–30 nm; 0, 250, 500, 1000, 2000 mg L^{-1})	Foliar, soil grown (14 days)	Leaves	LC-QTOF-MS (Agilent Technologies)	• Altered 154 and 63 metabolites in foliar and soil applications, respectively.	Salehi et al. (2018)
Soybean (*Glycine max*)	CdS quantum dots (8 nm; 200 μg L^{-1})	Virmiculite (14 days)	Shoots Roots	LC-MS-TQ-(Agilent Technologies)	• Perturbed metabolic pathways including glutathione metabolism, TCA cycle, glycolysis, fatty acid oxidation, and biosynthesis of phenylpropanoid and amino acids.	Majumdar et al. (2019)

(*Continued*)

TABLE 4.2 (Continued)

Plant	Nanoparticles (size; concentration)	Mode and duration of exposure	Plant parts used	Instrumentations used	Major observations	References
Maize (*Zea mays* L.)	SiO_2, TiO_2, Fe_3O_4 (20, 5–10, and 30 nm, respectively; 100 mg kg^{-1})	Soil (28 days)	Leaves Roots	GC-MSD (Agilent Technologies)	• Altered amino acid pool in leaf tissues. • Increased levoglucosan, linolenic acid, 4-hydroxycinnamic acid, and allo-inositol.	Zhao et al. (2019b)
	$Cu(OH)_2$ (50–1000 nm; 0, 10, 100 mg $plant^{-1}$)	Foliar exposure, soil grown (7 days)	Leaves	GC-QTOF-MS (Agilent Technologies)	• Highest treatment decreased leaf chlorophyll content, but increased precursors of phenolic acids, 4-hydroxycinnamic acid, total phenolic content.	Zhao et al. (2017b)
	$Cu(OH)_2$ (50–1000 nm; 0, 10, 100 mg $plant^{-1}$)	Foliar exposure, soil grown (7 days)	Leaves	GC-QTOF-MS (Agilent Technologies)	• Activated metabolites of the glycolysis, tricarboxylic acid cycle, and shikimate-phenylpropanoid pathways.	Zhao et al. (2018)
	Fe and Fe_3O_4 (50 and 10–20 nm respectively; 50 mg L^{-1})	Foliar exposure, soil grown (10 days)	Leaves	GC-MS (Agilent Technologies)	• Reduction in antioxidant metabolites (4-hydroxycinammic acid, ferulic acid, and vanillic acid). • Fe and Fe_3O_4 perturbed 8 and 7 biological pathways, respectively. There were 4 shared perturbed pathways. • Increased leaf chlorophyll content.	Li et al. (2020)

	Fe_3O_4 (30 nm; 0, 50 500 mg kg^{-1})	Soil (28 days)	Roots	GC-MS (Agilent Technologies)	• Nitrogen metabolism decreased. • Inhibited the metabolic pathways related to antioxidant defense.	Yan et al. (2020)
Cucumber (*Cucumis sativus*)	Cu (40 nm; 0, 10, 20 mg L^{-1})	Hydroponics (7 days)		GC-QTOF-MS (Agilent Technologies)	• Up-regulated amino acids, ascorbic acid, phenolic compounds; down-regulated citric acid.	Zhao et al. (2016b)
	Cu (40 nm, 0, 200, 400, 800 mg kg^{-1})	Soil (87 days)	Fruit	GC-QTOF-MS (Agilent Technologies)	• Proline, glycine, valine, pelargonic acid, arachidic acid, xylose, benzoic acid were upregulated. • Citric acid, myo-inositol, ornithine, 1-kestose were downregulated. • C and N related pathways were perturbed (galactose metabolism and TCA cycle)	Zhao et al. (2016c)
	Cu (40 nm; 0, 400, 800 mg kg^{-1})	Soil (60 days)	Leaves	LC-MS/MS (Agilent Technologies)	• Benzoic acid, gallic acid hydrate, and p-coumaric acid were upregulated. • 10 out of 13 targeted antioxidants were consumed due to overproduction of ROS.	Huang et al. (2019)

(*Continued*)

TABLE 4.2 (Continued)

Plant	Nanoparticles (size; concentration)	Mode and duration of exposure	Plant parts used	Instrumentations used	Major observations	References
	Cu (40 nm; 0, 200, 400, 800 mg kg^{-1})	Soil (60 days)	Leaves	HILIC-LC-MS/MS	• 13 out of 23 targeted amino acids were upregulated.	Huang et al. (2018)
	Cu (40 nm; 0, 200, 400, 800 mg kg^{-1})	Soil (68 days)	Fruit	800 MHz NMR (Bruker Corp.) and GC-QTOF-MS (Agilent Technologies)	• Twenty metabolites were identified as discriminating metabolites. • The upregulation of carbohydrates (xylose, fructose), amino acids (glycine, proline), fatty acids (linolenic acid, caprylic acid) show potential fruit benefits.	Zhao et al. (2017a)
	$Cu(OH)_2$ (50–1000 nm; 0, 10, 100 mg $plant^{-1}$)	Foliar, soil grown (7 days)	Leaves	GC-QTOF-MS (Agilent Technologies)	• Altered the arginine and proline metabolic pathways, and the levels of fatty acid and polysaccharides.	Zhao et al. (2017b)
	Ag (20 nm; 4, 40 mg $plant^{-1}$)	Foliar, soil grown (7 days)	Leaves	GC-MSD (Agilent Technologies)	• Upregulated phenolic compounds, TCA intermediates, pentadecanoic and arachidonic acids. • Downregulated glycine/serine ratio, glutamine and asparagine.	Zhang et al. (2018)

	C60 fullerol and CuSO5 (0, 1, 2 mg $plant^{-1}$, and 5 mg $plant^{-1}$, respectively. Combined 1 + 5 and 2 + 5 mg $plant^{-1}$)	Foliar, soil grown (2 days)	Leaves	GC-MS (Agilent Technologies)	• Upregulated antioxidant metabolites (3-hydroxyflavone, 1,2,4-benzenetriol, and methyl *trans*-cinnamate). • Downregulated cell membrane metabolites (linolenic and palmitoleic acid).	Zhao et al. (2019a)
Lettuce (*Lactuca sativa*)	$Cu(OH)_2$ (50–1000 nm; 8.75, 12.9, 17.5 mg pot^{-1})	Foliar, soil grown (24 days)	Leaves Roots	GC-MSD (Agilent Technologies)	• Disturbed TCA cycle and amino acid-related biological pathways. • Decreased cis-caffeic acid, chlorogenic acid, 3,4-dihydroxycinnamic acid, dehydroascorbic acid. • Increased nicotianamine.	Zhao et al. (2016d)
	$Cu(OH)_2$ (50–1000 nm; 0, 1050, 1555 mg L^{-1})	Foliar, soil grown (30 days)	Leaves	GC-QTOF-MS (Agilent Technologies)	• Decreased dehydroascorbic acid, cis-caffeic acid, 4-hydroxybutyric acid, and total antioxidant capacity.	Zhao et al. (2016e)
Spinach (*Spinacia oleracea*)	CeO_2 (0.3, 3 mg $plant^{-1}$)	Foliar, soil grown (21 days)	Leaves Roots	GC-MSD (Agilent Technologies)	• Low dose (0.3 $mgplant^{-1}$) affected metabolic profile in leaves, but high dose (3 $mgplant^{-1}$) decreased amino acids (threonine, trypthophan, cysteine, methionine, cycloleucine, aspartic acid, asparagine, tyrosine, and glutamic acid) in roots.	Zhang et al. (2019)

(*Continued*)

TABLE 4.2 (Continued)

Plant	Nanoparticles (size; concentration)	Mode and duration of exposure	Plant parts used	Instrumentations used	Major observations	References
	$Cu(OH)_2$ (50–1000 nm; 0.18, 18 mg $plant^{-1}$)	Foliar, hydroponics grown (7 days)	Leaves Roots	GC-QTOF-MS (Agilent Technologies)	• Perturbed nitrogen metabolism, and ascorbate and aldarate metabolic pathways. • Reductions in antioxidant metabolites (ascorbic acid, α-tocopherol, ferulic acid, phenolics, threonic acid, β-sitosterol, 4-hydroxybutyric acid.)	Zhao et al. (2017c)
Pistia stratiotes L. (Araceae)	Cu and Zn (<100 nm; 0.75 and 0.89 mg L^{-1})	Hydroponics (3 days)	Leaves Shoots Roots	Tandem MS	• Both NPs decreased ascorbic acid, total amino acid contents, and ten acylcarnitines metabolites.	Olkhovych et al. (2016)
Tomato (*Solanum esculentum*)	Multiwalled carbon nanotubes (50 mg L^{-1})	Hydroponics (10 weeks)	Fruit	LC-MS	• There were 93 up-regulate and 70 down-regulated metabolites associated with secondary metabolites including flavone and flavonol biosynthetic pathways, terpenoid and alkaloid biosynthetic pathways.	McGehee et al. (2017)

Rice (*Oryza sativa* L.)	TiO_2 (0, 100, 250, 500 mg L^{-1})	Hydroponics (14 days)	Leaves	GC-MSD (Agilent Technologies)	• Altered the metabolite profile of rice seedlings; increased glucose-6-phosphate, glucose-1-phosphate, succinic and isocitric acid, but decreased sucrose, isomaltulose, and glyoxylic acid.	Wu et al. (2017)
	TiO_2 (20 nm; 0, 25, 50, 150, 250, 500, 750 mg kg^{-1})	Soil (P deficient)	Grains	GC-MSD (Agilent Technologies)	• Increased levels of amino acids, palmitic acid, and glycerol content in rice grains.	Zahra et al. (2017)
	Graphene oxide (0.01, 0.1, 1.0 mg L^{-1})	Hydroponics (7 days)	Roots	GC-MS (Agilent Technologies)	• Down-regulated 39 metabolites	Li et al. (2018)
Wheat (*Triticum aestivum* L.)	CeO_2 (8 nm; 0, 500 mg kg^{-1})	Soil (full life cycle)	Grains	GC-QTOF-MS (Agilent Technologies)	• Altered DNA/RNA (thymidine, uracil, guanosine, deoxyguanosine, adenosine monophosphate) in wheat grains. • Decreased nicotianamine production in grains.	Rico et al. (2020b)
Eggplant (*Solanum melongena)*	NiO, CuO, ZnO (10–20, 25–55, and 18 nm respectively; 100, 250, 500, 1000 mg L^{-1})	Hydroponics (15 days)	Seedling	UV-Vis	• Decreased anthocyanin at highest concentration. Increased anthocyanin at lower concentrations of NPs. • Flavonoids and phenolics were altered at different NP concentrations.	Baskar et al. (2018)

(*Continued*)

TABLE 4.2 (Continued)

Plant	Nanoparticles (size; concentration)	Mode and duration of exposure	Plant parts used	Instrumentations used	Major observations	References
Hypericum perforatum	Perlite, TiO_2/Perlite (14.51–23.34, and 15.50–24.61 nm respectively; 0, 25, 50, 100, 150, 200 mg L^{-1})	*In vitro*; Solid Murashige and Skoog (MS) medium (30 days)	Calli	GC-MS	• *In vitro* grown plants exposed to TiO_2/perlite and perlite NPs produced more volatile compounds than field grown plants. • Both NPs induced the production of hypericin and pseudohypericin.	Ebadollahi et al. (2019)
Satureja khuzestanica	Multiwalled carbon nanotubes (0, 25, 50, 100, 250, and 500 μg mL^{-1})	*In vitro*; B5 medium (30 days)	Calli	HPLC and UV-Vis (PG Instruments)	• Downregulation of phenolics observed at highest concentration. • Upregulation of phenolics and flavonoids in 100 and 250 μg mL^{-1}	Ghorbanpour and Hadian (2015)

et al., 2017a). Also, most of the published reports performed the data analysis in MetaboAnalyst (http://www.metaboanalys.ca). Recently, Majumdar and Keller (2020) and Li et al. (2019) have reviewed metabolomics in NP-exposed plants covering about 15 publications also examined here. The readers are referred to these reviews for detailed discussions on how various NPs altered metabolite compositions in plants. Majumdar and Keller (2020) discussed metabolomics changes per type of NPs while Li et al. (2019) showed alterations per group of metabolites.

4.4.1 Using metabolomics to investigate biochemical modes of actions of NPs in plants

The literature reports on metabolomics of plants exposed to NPs focused on understanding physiological responses for relating metabolite composition to phenotypic changes. Majority of these studies exposed plants for a short duration to chronic or acute doses of different NPs (Table 4.2). Majumdar and Keller (2020) provided a thorough discussion of the metabolic perturbations in NP-treated plants; and analysis of more recent reports also revealed significant metabolic reprogramming in plants exposed to NPs. The findings are showing clear species-dependent metabolic perturbations in plants similar to trends found in nonmetabolomics studies discussed previously (Rico et al., 2011).

Zhao et al. conducted a series of metabolomics analysis on foliage of food crops (corn, cucumber, spinach, and lettuce) exposed to commercial biocide Kocide 3000 that contains $Cu(OH)_2$ NPs (50–1000 nm nanosheets) as active antifungal component (Zhao et al., 2017b,c; Zhao et al., 2018; Zhao et al., 2016d). Most of these studies involved plants cultivated in hydroponics growth media for a short period of time (7–28 days), but the researchers employed relevant doses by following the recommended dosage of the nanopesticide. These experiments provide rare perspectives on understanding physiological responses in plants exposed to commercially available nanobased agricultural input that is being used in real-life agricultural production. The authors found no visible toxic phenotypic responses in foliar treated plants, and there were no visible leaf damage or growth retardation observed while some plants even grew bigger and produced more biomass. However, metabolomics revealed alterations in the abundances of over 100 metabolites that are implicated in important metabolic pathways as summarized in Fig. 4.3. The metabolites could be categorized into different biological pathways (i.e., glycolytic pathway, TCA cycle, shikimate-phenylpropanoid biosynthesis, ascorbate and aldarate metabolism, and sugar, organic acid, lipid/fatty acid, and nucleotide metabolisms); however, only glucose-6-phosphate was consistently up-regulated in all crops tested (corn, cucumber, spinach, and lettuce) while the organic acids saccharic, gluconic, galactonic, mannoic, and isohexonic were consistently promoted in corn only (Fig. 4.3). Some

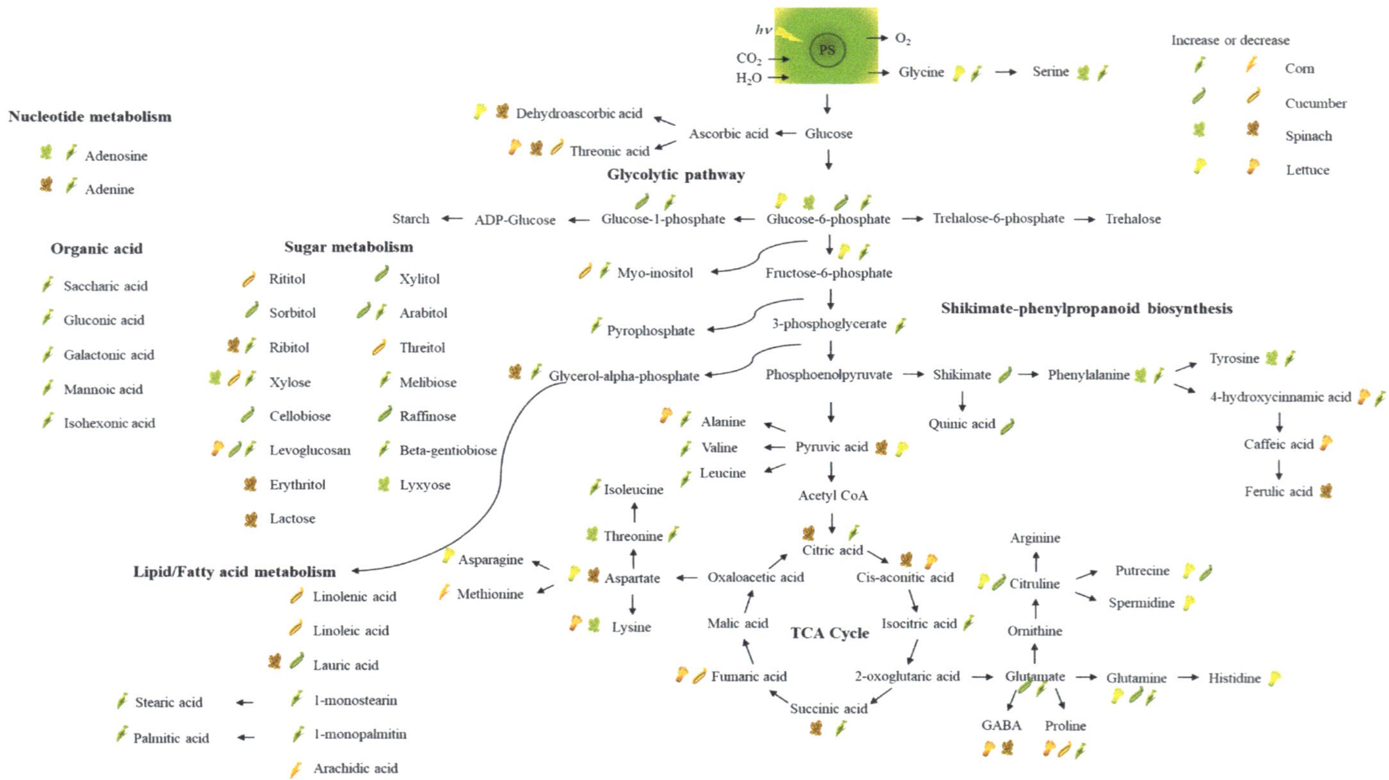

FIGURE 4.3 Changes in metabolite concentrations in foliage of corn, cucumber, spinach, and lettuce sprayed with Koicide 3000 having $Cu(OH)_2$ NPs as active component (Zhao et al., 2017b; Zhao et al., 2018; Zhao et al., 2016d; Zhao et al., 2016e). The shade of green or brown color indicates an increase or decrease in metabolite concentrations in plants exposed to the NPs.

metabolites had contrasting responses in different crops; for example aspartate level was up-regulated in lettuce but down-regulated in spinach, and succinic acid level decreased in spinach but increased in corn (Fig. 4.3). The researchers concluded that the metabolic reprogramming, some of which was NPs-specific, was involved mostly in promoting antioxidant levels and defense response that serve as a mechanism for plants to increase tolerance to $Cu(OH)_2$ NPs.

Table 4.2 presents other short studies that investigated metabolic responses in plants exposed to CdO, CdS, TiO_2, CeO_2, SiO_2, Fe_2O_3, Y_2O_3, Cu, Zn, Ag, C60 fullerols, graphene oxide, and carbon nanotubes/nanodots. Fig. 4.4 also depicts the trend in metabolic patterns induced by the metal and metal oxide NPs. Ag NPs down-regulated several metabolites (glycine, sucrose, pyruvic acid, glutamin, and linolenic acids), but up-regulated more metabolites including the known nonenzymatic antioxidants (ferulic acid, salicylic acid, glutamate, ornithine) (Zhang et al., 2018). It is interesting to note that CeO_2 NPs down-regulated metabolites (tyrosine, asparagine, methionine, threonine, and aspartate) in the roots, but up-regulated metabolites (dehydroascorbic acid, glucose-1-phosphate, sucrose, cysteine, hydroxycinnamic acid, glutamine, methionine, aspartate, asparagine, and phenylalanine) in the leaves (Zhang et al., 2019). Similarly, CdS mostly up-regulated metabolites (1-ketose, norleucine, cycloleucine, alanine, glycine, malate, fumaric acid, isocitric acid, glutamate, putrecine, allantoate, hydroxybenzoate, hydroxycinnamic acid, flavone, and hydroxytryptamine) in the leaves (Majumdar et al., 2019). Data from other studies also showed mixed trends that it remains a challenge to isolate specific metabolites to use as biomarkers or bioindicators of NPs impacts in plants. For example, findings in cucumber exposed to Cu NPs upregulated levels of few antioxidant and amino acids to boost the plants defense system against stress (Huang et al., 2019; Huang et al., 2018).

These studies performed in chronic exposure toxicity conditions found that it is difficult to infer on long-term toxicity implications from NPs exposure. In fact, most studies found no toxic responses from plants; instead growth promotion, biomass production, photosynthetic activity, chlorophyll content, and metabolic reprogramming happened to enhance tolerance to stress brought about by NPs (Li et al., 2020; Zhang et al., 2018; Zhang et al., 2019; Huang et al., 2019). Plants reach full maturity and produce comparable yield biomass even at doses of NPs higher than realistic environmental concentrations. Zhao et al. (2016c) reported that cucumber exposed to Cu NPs at 200, 400, and 800 mg kg^{-1} soil had comparable fruit yields with untreated control. Similarly, previous reports found that rice, wheat, and barley produced grains (Rico et al., 2020a; Rico et al., 2015; Rico et al., 2017; Rico et al., 2014; Rico et al., 2013a; Rico et al., 2020b) even at high concentrations of 250 or 500 mg CeO_2 NPs per kg soil, which suggests that plants overcome stress experienced at juvenile stage making it more challenging to extrapolate metabolic perturbations at young stage to long-term effects.

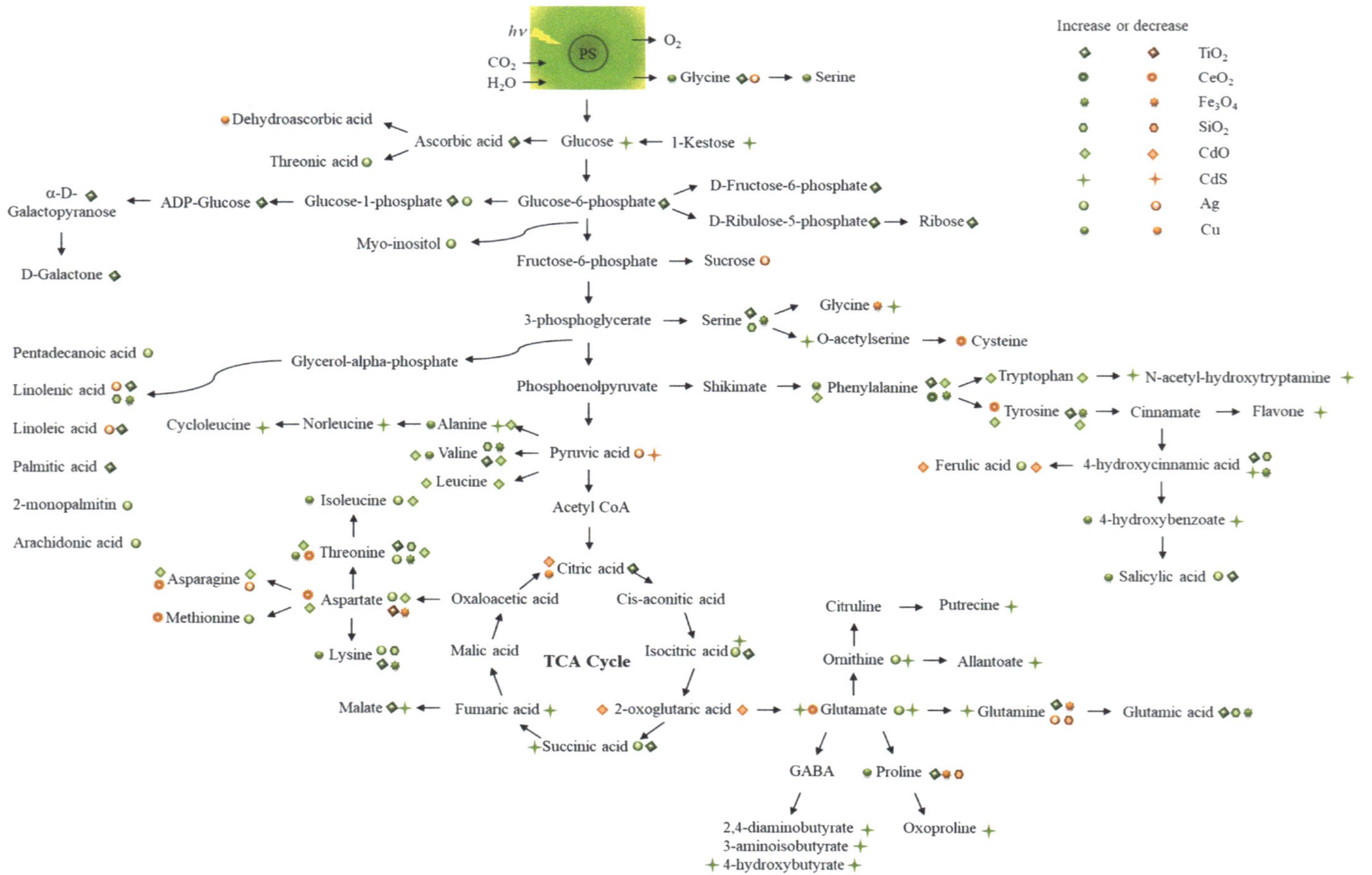

FIGURE 4.4 Changes in metabolite concentrations in root (left side) or leaves (right side) of plants exposed to different NPs (Zhao et al., 2017b; Zhao et al., 2018; Zhao et al., 2016d; Zhao et al., 2016e). The shade of green or brown color indicates an increase or decrease in metabolite concentrations in plants exposed to the NPs.

Reports are also showing that Calvin cycle is not affected in plants exposed to NPs. The chlorophyll pigments (*a* and *b*) in spinach, kidney beans, or *A. thaliana* did not change when exposed to carbon nanodots and $Cu(OH)_2$ NPs (Zhao et al., 2017c; Tian et al., 2020; Chen et al., 2018), but decreased in corn and beans at high concentrations of CeO_2 and $Cu(OH)_2$ NPs (Zhao et al., 2017b; Salehi et al. 2018). Nevertheless, glucose production in leaves was mostly not affected wherein only three studies reported modifications in glucose response: glucose increased in leaves of kidney beans and soybean grown in vermiculite or soil amended with CdS quantum dots for 14–28 days, but decreased in corn exposed to Fe_3O_4 (Majumdar et al., 2019; Tian et al., 2020; Li et al., 2020). Similarly, the glycolytic pathway did not show adverse response (Figs. 4.3 & 4.4). Glucose-6-phosphate was upregulated in leaves of corn, cucumber, spinach, lettuce, and rice exposed to TiO_2 and $Cu(OH)_2$ NPs, and glucose-6-phosphate, fructose-6-phosphate and 3 phosphoglycerate were also upregulated in leaves exposed to $Cu(OH)_2$ NPs (Zhao et al., 2016e,d; Zhao et al., 2018; Zhao et al., 2016a; Zhao et al., 2017c; Wu et al., 2017). Likewise, generational exposures to CeO_2 NPs increased glucose-6-phosphate in wheat grains (Rico et al., 2020b). These are interesting results because chloroplasts synthesis of NADPH and ATP (which are needed in the Calvin cycle to fix CO_2 to glucose) and glycolysis were not compromised. Together, these data support the findings that biomass production is not affected in fully mature plants exposed to low or environmentally-relevant doses of NPs.

4.4.2 Using metabolomics to measure impacts of nanoparticles on quality of food crops

Metabolomics provide an excellent platform to detect changes in metabolite composition of plants and appraise nutritional quality of food crops. Previous reviews emphasized the usefulness of metabolomics in providing detailed biochemical composition that define quality aspects of plants (taste, flavor, aroma). Likewise, a few studies have employed metabolomics to investigate the quality changes in food crops exposed to NPs. For example, spinach leaves exposed to $Cu(OH)_2$ nanopesticide exhibited decreases in levels of beneficial antioxidants (ascorbic acid, α-tocopherol, threonic acid, β-sitosterol, 4-hydroxybutyric acid, ferulic acid, and total phenolics), which indicate a reduction in its nutritional value (Zhao et al., 2017c). The same reduction in nutritional quality was recorded in lettuce exposed to $Cu(OH)_2$ nanopesticide wherein important antioxidants such as dehydroascorbic acid, caffeic acid, 4-hydroxybutyric acid, and total antioxidant capacity decreased significantly in the leaves of $Cu(OH)_2$-exposed plants (Zhao et al., 2016e). In a more long-term exposure scenario in soil, Cu NPs modified the nutritional content in cucumber fruits. Some essential amino acids (lysine and methionine) and other metabolites (myo-inositol, citric acid, ornithine, and citrulline) were down-regulated, whereas some sugars (xylose and fructose), amino acids (glycine and proline), and fatty acids (linolenic acid and caprylic acid) were up-regulated (Zhao et al., 2017a).

Zahra et al. (2017) investigated the changes in grain quality of rice cultivated in phosphorus-deficient soil amended with TiO_2 NPs. They found improved grain quality of rice; there were increased levels of amino acids (isoleucine, proline, glycine, aspartic acid, lysine), palmitic acid, glutamic acid, glycerol-3-phosphate, and phosphorus in grains harvested from plants exposed to TiO_2 NPs (Zahra et al., 2017). In another study, tomato plants were exposed to multiwalled carbon nanotubes (MWCNT) in hydroponics system until fruit production. Results showed that MWCNT improved yield and fruit biomass, and that very small amount of MWCNT (up to 0.137 μg per mg DW fruit) accumulated in the fruits (McGehee et al., 2017). However, the fruit metabolome was significantly impacted showing up-regulation in biochemical pathways involved in secondary metabolisms (e.g., terpenoid and alkaloid). Other studies also reported that Ag NPs enhanced the production of secondary metabolites and major phytochemicals (artemisinin, diosgenin, phytoalexin compounds) (Jasim et al., 2017; Zhang et al., 2013; Kruszka et al., 2020). Marslin et al. (2017) noted that the impacts of NPs on the secondary metabolite production in plants is still a field largely unexplored, although these researchers noted that the induction of reactive oxygen species could be linked to regulation of secondary metabolism in plants following exposure to NPs.

There are several other studies demonstrating the impacts on metabolic pools in plant organs (leaves and roots), but the effects of these changes in metabolite compositions in parts directly consumed for source of nutrition in human have not been established. For example, primary and secondary metabolites were impacted in the leaves and roots of plants exposed to NPs; nevertheless, the implications of those metabolome changes on the nutritional quality were not fully assessed because the experiments were performed at short exposure scenarios (Vecerova et al., 2016; Huang et al., 2019; Kruszka et al., 2020). However, current reports show that these changes may have significant impacts on nutritional value of food crops; therefore assessment of NPs-exposed plant parts for human consumption is an urgent priority (McGehee et al., 2017; Marslin et al., 2017).

4.4.3 Discriminating plant responses between exposure scenarios

One major research question in the field of nanophytotoxicity that has far-reaching regulatory implications is whether NPs induce unique toxicity responses in plants compared to their bulk equivalents or whether surface coatings affect the toxicity of NPs. These hypotheses have generated several publications, but the question on differential impacts between NPs and bulk particles (BPs) still continues to persist in the literature. Metabolomics has been employed to provide more robust data in understanding metabolite composition and biosynthetic pathways that could potentially discriminate impacts between NP and bulk counterparts. Studies suggest that metal ions released from NPs or BPs are responsible for metabolic response in plants. For example, Kruszka et al. (2020) exposed *A. thaliana* to either Ag NPs or bulk $AgNO_3$ and found that both particles released silver ions, which induced the biosynthesis of phytoalexin compounds (camalexin, hyroxycamalexin O-hexoside and hydroxycamalexin malonyl-hexoside) in plants. Similarly, Zhao et al. (2017c) reported that both $Cu(OH)_2$ NPs and $CuSO_4$ BPs released Cu^{2+} ions that induced reduction in antioxidant levels and perturbed the metabolisms of ascorbate and aldarate in spinach. In another study, plants exhibited contrasting reprogramming in metabolic patterns due to the rapid availability of metal ions in BPs compared to surface-coated NPs. This was observed in *A. thaliana*, which showed greatest metabolic changes in $AgNO_3$ exposure compared to Ag NPs (coated with citrate or PVP) due to the Ag^+ ions readily available for plant uptake in BPs than in NPs (Chavez Soria et al., 2017). The same study found differences in metabolic reprogramming between surface-coated Ag NPs (i.e., six metabolites accumulated in citrate-coated Ag NPs versus three metabolites in PVP-coated Ag NPs), which was due to coating affecting the interactions of NPs with plants or modifying the release and availability of Ag^+ ions for plant uptake (Chavez Soria et al., 2017). Majumdar et al. (2019) also reported influence of coating on toxicity modulation of CdS quantum dots to soybean plants. Published reports also revealed NP-specific effects in plants. For example, $Cu(OH)_2$ NPs altered nitrogen metabolism in spinach whereas $CuSO_4$ BPs did not, and Ag NPs impacted a few metabolites (carbazole, lactulose, raffinose, citraconic acid, lactamide, acetanilide, and *p*-benzoquinone) in cucumber differently than $AgNO_3$ (Zhao et al., 2017c; Zhang et al., 2018). Likewise, there were distinct changes in the accumulations of metabolites in soybean roots and shoots that were recorded in CdS quantum dots but not in the Cd^{2+} ions or CdS BPs (Majumdar et al., 2019). These findings suggest that the molecular mechanisms of plant responses to NPs exposure are different than those of BPs or metal ions (Ruotolo et al., 2018).

The high sensitivity capability of metabolomics also made possible the discrimination between exposure scenarios that are not yet well explored in literature. This includes differentiating impacts in foliar versus soil treatments, high versus low soil N availability, and transgenerational exposure of plants. Salehi et al. (2018) reported differential metabolic responses in beans exposed to CeO_2 NPs for 15 days via foliar or soil exposure wherein there were 154 metabolites altered in plants under spray application than 63 in soil exposure. The researchers concluded that aerial deposition of CeO_2 NPs could pose greater concern compared to soil contamination. In another study, metabolomics profiling distinguished responses between plants exposed to CeO_2 NPs in low versus high soil N. Results revealed that CeO_2 NPs affected 18 metabolites in wheat grown at low soil N compared to only 11 metabolites impacted at high soil N (Rico et al., 2020b). The same study reported for the first time the metabolic reprogramming in wheat grains exposed to CeO_2 NPs for three generations. Data showed that DNA/RNA metabolic pools (thymine, uracil, guanosine, deoxyguanosine, and adenosine monophosphate) were significantly impacted, signifying that continued exposures of parent plants to CeO_2 NPs affect the integrity of progeny plants (Rico et al., 2020b). In addition, nicotianamine, a nonprotenaceous endogenous chelator of metals (Fe, Zn, Cu) in plants, was severely affected in generationally-exposed plants. Nicotianamine is an important metabolite involved in the translocation and accumulation of elemental nutrients (Fe, Zn, Cu) in shoots and grains of wheat, rice, and barley (Takahashi et al., 2003; Zheng et al., 2010; Singh et al., 2017). Fig. 4.5 demonstrates the decreased nicotianamine level, which coincided with reduced accumulation of Fe in wheat grains harvested from plants exposed to CeO_2 NPs for three generations (Rico et al., 2020b). In related studies in NPs-plant interactions, nicotianamine level was up-regulated in lettuce exposed to $Cu(OH)_2$ NPs as

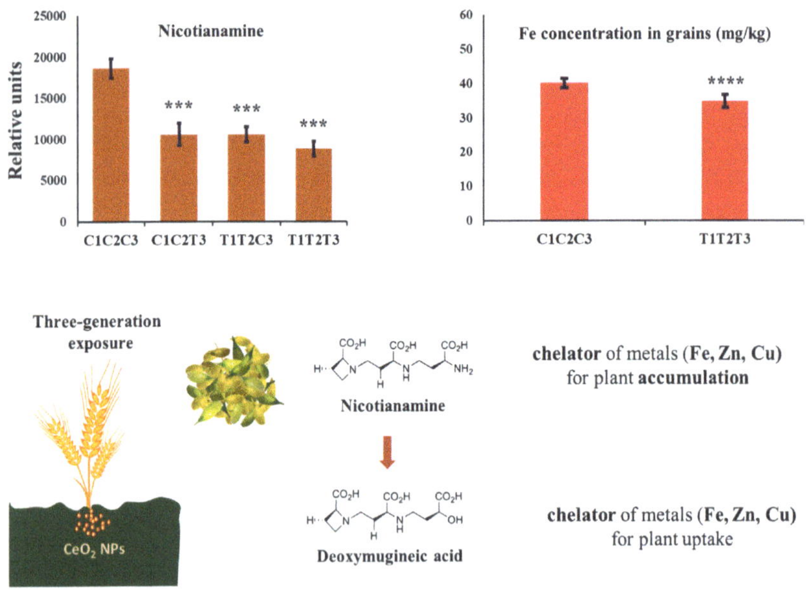

FIGURE 4.5 Changes in nicotianamine abundance and Fe concentration in grains of wheat exposed to CeO_2 NPs for three generations. C1C2C3 = wheat grains harvested from control treatment (0 mg kg^{-1}) for three generations; T1T2T3 = wheat grains harvested from CeO_2 NPs treatment (500 mg kg^{-1}) for three generations. Values are mean ± SE ($n = 6$); ***, **** indicate $p < 0.01, 0.001$. Data taken from Rico et al. (2020b).

a means to increase tolerance to Cu^{2+} ions similar to the observed tolerance to nickel in tobacco plants (Kim et al., 2005; Zhao et al., 2016d). Nicotianamine is also the precursor of deoxymugineic acid, which is a metal chelator in roots (Fig. 4.5). Interestingly, metabolomics showed that decreases in nicotianamine occurred only in high N soil.

4.5 Conclusion

Metabolomics is expected to be an indispensable technique in assessing the impacts of environmental contaminants on organisms, particularly plants. Literature reports are already demonstrating that metabolomics provides significant information on biological processes sensitive to NPs exposure, which are much more helpful in understanding the impacts of NPs in plants compared to traditional single toxicological assays. The literature is saturated with reports focusing on metabolomics changes as a measure of toxicological response at short exposure regimes; therefore it is difficult to make predictions on long-term implications of NPs in plants. Other important research questions that should perhaps be considered in future studies include long-term exposure (i.e., generational exposure), mixed exposure of two or more different plant species, and wild plant vegetation. In addition, the concerns raised by experts on factors affecting the integrity of data should be carefully considered and observed during plant cultivation, sample extraction, and metabolite analysis.

References

Baskar, V., Nayeem, S., Kuppuraj, S.P., Muthu, T., Ramalingam, S., 2018. Assessment of the effects of metal oxide nanoparticles on the growth, physiology and metabolic responses in in vitro grown eggplant (Solanum melongena). 3 Biotech. 8.

Chavez Soria, N.G., Montes, A., Bisson, M.A., Atilla-Gokcumen, G.E., Aga, D.S., 2017. Mass spectrometry-based metabolomics to assess uptake of silver nanoparticles by Arabidopsis thaliana. Environ. Science: Nano 4, 1944–1953.

Chen, J., Liu, B., Yang, Z., Qu, J., Xun, H., Dou, R., et al., 2018. Phenotypic, transcriptional, physiological and metabolic responses to carbon nanodot exposure in Arabidopsis thaliana (L.). Environ. Science: Nano 5, 2672–2685.

Creydt, M., Arndt, M., Hudzik, D., Fischer, M., 2018. Plant Metabolomics: Evaluation of Different Extraction Parameters for Nontargeted UPLC-ESI-QTOF-Mass Spectrometry at the Example of White Asparagus officinalis. J. Agric. Food Chemistry 66, 12876–12887.

Dunn, W.B., Ellis, D.I., 2005. Metabolomics: Current analytical platforms and methodologies. TrAC. Trends Anal. Chemistry 24, 285–294.

Ebadollahi, R., Jafarirad, S., Kosari-Nasab, M., Mahjouri, S., 2019. Effect of explant source, perlite nanoparticles and TiO2/perlite nanocomposites on phytochemical composition of metabolites in callus cultures of Hypericum perforatum. Sci. Reports 9.

Gardea-Torresdey, J.L., Rico, C.M., White, J.C., 2014. Trophic Transfer, Transformation, and Impact of Engineered Nanomaterials in Terrestrial Environments. Environ. Sci. & Technology 48, 2526–2540.

Gauthier, L., Tison-Rosebery, J., Morin, S., Mazzella, N., 2019. Metabolome response to anthropogenic contamination on microalgae: a review. Metabolomics 16, 8.
Ghorbanpour, M., Hadian, J., 2015. Multi-walled carbon nanotubes stimulate callus induction, secondary metabolites biosynthesis and antioxidant capacity in medicinal plant Satureja khuzestanica grown in vitro. Carbon 94, 749–759.
Hall, R.D., 2006. Plant metabolomics: from holistic hope, to hype, to hot topic. N. Phytologist 169, 453–468.
Hatami, M., Kariman, K., Ghorbanpour, M., 2016. Engineered nanomaterial-mediated changes in the metabolism of terrestrial plants. Sci. Total. Environment 571, 275–291.
Hong, J., Wang, C., Wagner, D.C., Gardea-Torresdey, J.L., He, F., Rico, C.M., 2021. Foliar application of nanoparticles: mechanisms of absorption, transfer, and multiple impacts. Environ. Science: Nano .
Huang, Y., Li, W., Minakova, A.S., Anumol, T., Keller, A.A., 2018. Quantitative analysis of changes in amino acids levels for cucumber (Cucumis sativus) exposed to nano copper. NanoImpact 12, 9–17.
Huang, Y., Adeleye, A.S., Zhao, L., Minakova, A.S., Anumol, T., Keller, A.A., 2019. Antioxidant response of cucumber (Cucumis sativus) exposed to nano copper pesticide: Quantitative determination via LC-MS/MS. Food Chem. 270, 47–52.
Jasim, B., Thomas, R., Mathew, J., Radhakrishnan, E.K., 2017. Plant growth and diosgenin enhancement effect of silver nanoparticles in Fenugreek (Trigonella foenum-graecum L.). Saudi Pharm. J. 25, 443–447.
Jorge, T.F., Mata, A.T., António, C., 2016. Mass spectrometry as a quantitative tool in plant metabolomics. Philos. Trans. R. Soc. A: Mathematical, Phys. Eng. Sci. 374, 20150370.
Ke, M., Qu, Q., Peijnenburg, W.J.G.M., Li, X., Zhang, M., Zhang, Z., et al., 2018. Phytotoxic effects of silver nanoparticles and silver ions to Arabidopsis thaliana as revealed by analysis of molecular responses and of metabolic pathways. Sci. Total. Environment 644, 1070–1079.
Kim, S., Takahashi, M., Higuchi, K., Tsunoda, K., Nakanishi, H., Yoshimura, E., et al., 2005. Increased Nicotianamine Biosynthesis Confers Enhanced Tolerance of High Levels of Metals, in Particular Nickel, to Plants. Plant. Cell Physiology 46, 1809–1818.
Kole, C., Kole, P., Randunu, K.M., Choudhary, P., Podila, R., Ke, P.C., et al., 2013. Nanobiotechnology can boost crop production and quality: first evidence from increased plant biomass, fruit yield and phytomedicine content in bitter melon (Momordica charantia). BMC Biotechnology 13, 37.
Kruszka, D., Sawikowska, A., Kamalabai Selvakesavan, R., Krajewski, P., Kachlicki, P., Franklin, G., 2020. Silver nanoparticles affect phenolic and phytoalexin composition of Arabidopsis thaliana. Sci. Total. Environment 716, 135361.
Lankadurai, B.P., Nagato, E.G., Simpson, M.J., 2013. Environmental metabolomics: an emerging approach to study organism responses to environmental stressors. Environ. Reviews 21, 180–205.
Li, X., Mu, L., Hu, X., 2018. Integrating proteomics, metabolomics and typical analysis to investigate the uptake and oxidative stress of graphene oxide and polycyclic aromatic hydrocarbons. Environ. Science: Nano 5, 115–129.
Li, X., Peng, T., Mu, L., Hu, X., 2019. Phytotoxicity induced by engineered nanomaterials as explored by metabolomics: Perspectives and challenges. Ecotoxicol. Environ. Safety 184, 109602.
Li, P., Wang, A., Du, W., Mao, L., Wei, Z., Wang, S., et al., 2020. Insight into the interaction between Fe-based nanomaterials and maize (*Zea mays*) plants at metabolic level. Sci. Total. Environ 738, 139795.

Long, S.M., Tull, D.L., De Souza, D.P., Kouremenos, K.A., Dayalan, S., Mcconville, M.J., et al., 2020. Metabolomics Provide Sensitive Insights into the Impacts of Low Level Environmental Contamination on Fish Health-A Pilot Study. Metabolites 10, 24.

Ma, C., White, J.C., Zhao, J., Zhao, Q., Xing, B., 2018. Uptake of engineered nanoparticles by food crops:characterization, mechanisms, and implications. Annu. Rev. Food Sci. Technol. 9, 129–153.

Majumdar, S., Keller, A.A., 2020. Omics to address the opportunities and challenges of nanotechnology in agriculture. Crit. Rev. Environ. Sci. Technol. 1–42.

Majumdar, S., Pagano, L., Wohlschlegel, J.A., Villani, M., Zappettini, A., White, J.C., et al., 2019. Proteomic, gene and metabolite characterization reveal the uptake and toxicity mechanisms of cadmium sulfide quantum dots in soybean plants. Environ. Science: Nano 6, 3010–3026.

Marslin, G., Sheeba, C.J., Franklin, G., 2017. Nanoparticles Alter Secondary Metabolism in Plants via ROS Burst. Front. Plant. Sci. 8.

Matich, E.K., Soria, N.G.C., Aga, D.S., Atilla-Gokcumen, E., 2019. Applications of metabolomics in assessing ecological effects of emerging contaminants and pollutants on plants. J. Hazard. Materials 373, 527–535.

McGehee, D.L., Lahiani, M.H., Irin, F., Green, M.J., Khodakovskaya, M.V., 2017. Multiwalled Carbon Nanotubes Dramatically Affect the Fruit Metabolome of Exposed Tomato Plants. ACS Appl. Mater. & Interfaces 9, 32430–32435.

Miralles, P., Church, T.L., Harris, A.T., 2012. Toxicity, Uptake, and Translocation of Engineered Nanomaterials in Vascular plants. Environ. Sci. & Technology 46, 9224–9239.

Olkhovych, O., Volkogon, M., Taran, N., Batsmanova, L., Kravchenko, I., 2016. Effect of Copper And Zinc Nanoparticles on the Growth Parameters, Contents of Ascorbic Acid, and Qualitative Composition of Amino Acids and Acylcarnitines in *Pistia stratiotes* L. (Araceae). Nanoscale Res. Letters 11, 218.

Paramo, L.A., Feregrino-Pérez, A.A., Guevara, R., Mendoza, S., Esquivel, K., 2020. Nanoparticles in Agroindustry: Applications, Toxicity, Challenges, and Trends. Nanomaterials 10, 1654.

Piasecka, A., Kachlicki, P., Stobiecki, M., 2019. Analytical Methods for Detection of Plant Metabolomes Changes in Response to Biotic and Abiotic Stresses. Int. J. Mol. Sci 20.

Pošćić, F., Mattiello, A., Fellet, G., Miceli, F., Marchiol, L., 2016. Effects of Cerium and Titanium Oxide Nanoparticles in Soil on the Nutrient Composition of Barley (Hordeum vulgare L.) Kernels. Int. J. Environ. Res. Public. Health 13, 577.

Rico, C.M., Majumdar, S., Duarte-Gardea, M., Peralta-Videa, J.R., Gardea-Torresdey, J.L., 2011. Interaction of Nanoparticles with Edible Plants and Their Possible Implications in the Food Chain. J. Agric. Food Chemistry 59, 3485–3498.

Rico, C.M., Morales, M.I., Barrios, A.C., Mccreary, R., Hong, J., Lee, W.-Y., et al., 2013a. Effect of Cerium Oxide Nanoparticles on the Quality of Rice (Oryza sativa L.) Grains. J. Agric. Food Chemistry 61, 11278–11285.

Rico, C.M., Morales, M.I., Mccreary, R., Castillo-Michel, H., Barrios, A.C., Hong, J., et al., 2013b. Cerium oxide nanoparticles modify the antioxidative stress enzyme activities and macromolecule composition in rice seedlings. Env. Sci. Technol 47, 14110–14118.

Rico, C.M., Lee, S.C., Rubenecia, R., Mukherjee, A., Hong, J., Peralta-Videa, J.R., et al., 2014. Cerium oxide nanoparticles impact yield and modify nutritional parameters in wheat (*Triticum aestivum* L.). J. Agric. Food Chem 62, 9669–9675.

Rico, C.M., Barrios, A.C., Tan, W., Rubenecia, R., Lee, S.C., Varela-Ramirez, A., et al., 2015. Physiological and biochemical response of soil-grown barley (*Hordeum vulgare* L.) to cerium oxide nanoparticles. Env. Sci. Pollut. Res 22, 10551–10558.

Rico, C.M., Johnson, M.G., Marcus, M.A., Andersen, C.P., 2017. Intergenerational responses of wheat (*Triticum aestivum* L.) to cerium oxide nanoparticles exposure. Environ. Science: Nano 4, 700–711.

Rico, C.M., Abolade, O.M., Wagner, D., Lottes, B., Rodriguez, J., Biagioni, R., et al., 2020a. Wheat exposure to cerium oxide nanoparticles over three generations reveals transmissible changes in nutrition, biochemical pools, and response to soil N. J. Hazard. Materials 384, 121364.

Rico, C.M., Wagner, D., Abolade, O., Lottes, B., Coates, K., 2020b. Metabolomics of wheat grains generationally-exposed to cerium oxide nanoparticles. Sci. Total. Environment 712, 136487.

Ruotolo, R., Maestri, E., Pagano, L., Marmiroli, M., White, J.C., Marmiroli, N., 2018. Plant Response to Metal-Containing Engineered Nanomaterials: An Omics-Based Perspective. Environ. Sci. & Technology 52, 2451–2467.

Salehi, H., Chehregani, A., Lucini, L., Majd, A., Gholami, M., 2018. Morphological, proteomic and metabolomic insight into the effect of cerium dioxide nanoparticles to *Phaseolus vulgaris L.* under soil or foliar application. Sci. Total. Environment 616–617, 1540–1551.

Singh, S.P., Keller, B., Gruissem, W., Bhullar, N.K., 2017. Rice NICOTIANAMINE SYNTHASE 2 expression improves dietary iron and zinc levels in wheat. Theor. Appl. Genetics 130, 283–292.

Takahashi, M., Terada, Y., Nakai, I., Nakanishi, H., Yoshimura, E., Mori, S., et al., 2003. Role of nicotianamine in the intracellular delivery of metals and plant reproductive development. Plant. Cell 15, 1263–1280.

Tang, J., 2011. Microbial metabolomics. Curr. genomics 12, 391–403.

Tian, L., Zhang, H., Zhao, X., Gu, X., White, J.C., Zhao, L., et al., 2020. CdS nanoparticles in soil induce metabolic reprogramming in broad bean (*Vicia faba L.*) roots and leaves. Environ. Science: Nano 7, 93–104.

Vecerova, K., Vecera, Z., Docekal, B., Oravec, M., Pompeiano, A., Triska, J., et al., 2016. Changes of primary and secondary metabolites in barley plants exposed to CdO nanoparticles. Env. Pollut 218, 207–218.

White, J.C., Gardea-Torresdey, J., 2018. Achieving food security through the very small. Nat. Nanotechnology 13, 627–629.

White, W.H.E.J.C., 2016. The use of metallic oxide nanoparticles to enhance growth of tomatoes and eggplants in disease infested soil or soilless medium. Environ. Science: Nano 3, 1072–1079.

Wu, B., Zhu, L., Le, X.C., 2017. Metabolomics analysis of TiO2 nanoparticles induced toxicological effects on rice (Oryza sativa L.). Env. Pollut. 230, 302–310.

Yan, L., Li, P., Zhao, X., Ji, R., Zhao, L., 2020. Physiological and metabolic responses of maize (Zea mays) plants to Fe3O4 nanoparticles. Sci. Total. Environment 718, 137400.

Yuan, J., He, A., Huang, S., Hua, J., Sheng, G.D., 2016. Internalization and Phytotoxic Effects of CuO Nanoparticles in Arabidopsis thaliana as Revealed by Fatty Acid Profiles. Environ. Sci. & Technol. 50, 10437–10447.

Zahra, Z., Waseem, N., Zahra, R., Lee, H., Badshah, M.A., Mehmood, A., et al., 2017. Growth and Metabolic Responses of Rice (Oryza sativa L.) Cultivated in Phosphorus-Deficient Soil Amended with TiO2 Nanoparticles. J. Agric. Food Chemistry 65, 5598–5606.

Zhang, B.Z., Zheng, L.P.;, Li, W.Y.;, Wang, J.W., 2013. Stimulation of Artemisinin Production in Artemisia annua Hairy Roots by Ag-SiO2 Core-shell Nanoparticles. Curr. Nanoscience 9, 363–370.

Zhang, H., Du, W., Peralta-Videa, J.R., Gardea-Torresdey, J.L., White, J.C., Keller, A., et al., 2018. Metabolomics Reveals How Cucumber (*Cucumis sativus*) Reprograms Metabolites To Cope with Silver Ions and Silver Nanoparticle-Induced Oxidative Stress. Env. Sci. Technol 52, 8016–8026.

Zhang, H., Lu, L., Zhao, X., Zhao, S., Gu, X., Du, W., et al., 2019. Metabolomics Reveals the "Invisible" Responses of Spinach Plants Exposed to CeO2 Nanoparticles. Env. Sci. Technol. 53, 6007–6017.

Zhao, L., Huang, Y., Hu, J., Zhou, H., Adeleye, A.S., Keller, A.A., 2016a. (1)H NMR and GC-MS Based Metabolomics Reveal Defense and Detoxification Mechanism of Cucumber Plant under Nano-Cu Stress. Env. Sci. Technol 50, 2000–2010.

Zhao, L., Huang, Y., Hu, J., Zhou, H., Adeleye, A.S., Keller, A.A., 2016b. 1H NMR and GC-MS Based Metabolomics Reveal Defense and Detoxification Mechanism of Cucumber Plant under Nano-Cu Stress. Environ. Sci. & Technol. 50, 2000–2010.

Zhao, L., Huang, Y., Zhou, H., Adeleye, A.S., Wang, H., Ortiz, C., et al., 2016c. GC-TOF-MS based metabolomics and ICP-MS based metallomics of cucumber (Cucumis sativus) fruits reveal alteration of metabolites profile and biological pathway disruption induced by nano copper. Environ. Science: Nano 3, 1114–1123.

Zhao, L., Ortiz, C., Adeleye, A.S., Hu, Q., Zhou, H., Huang, Y., et al., 2016d. Metabolomics to Detect Response of Lettuce (Lactuca sativa) to Cu(OH)2 Nanopesticides: Oxidative Stress Response and Detoxification Mechanisms. Env. Sci. Technol 50, 9697–9707.

Zhao, L.H., Huang, Y., Hannah-Bick, C., Fulton, A.N., Keller, A.A., 2016e. Application of metabolomics to assess the impact of $Cu(OH)_2$ nanopesticide on the nutritional value of lettuce (*Lactuca sativa*): Enhanced Cu intake and reduced antioxidants. NanoImpact 3–4, 58–66.

Zhao, L., Hu, J., Huang, Y., Wang, H., Adeleye, A., Ortiz, C., et al., 2017a. 1H NMR and GC–MS based metabolomics reveal nano-Cu altered cucumber (Cucumis sativus) fruit nutritional supply. Plant. Physiol. Biochem. 110, 138–146.

Zhao, L., Hu, Q., Huang, Y., Keller, A.A., 2017b. Response at Genetic, Metabolic, and Physiological Levels of Maize (Zea mays) Exposed to a Cu(OH)2 Nanopesticide. ACS Sustain. Chem. & Eng. 5, 8294–8301.

Zhao, L., Huang, Y., Adeleye, A.S., Keller, A.A., 2017c. Metabolomics Reveals Cu(OH)2 Nanopesticide-Activated Anti-oxidative Pathways and Decreased Beneficial Antioxidants in Spinach Leaves. Env. Sci. Technol 51, 10184–10194.

Zhao, L., Huang, Y., Keller, A.A., 2018. Comparative Metabolic Response between Cucumber (Cucumis sativus) and Corn (Zea mays) to a Cu(OH)2 Nanopesticide. J. Agric. Food Chem. 66, 6628–6636.

Zhao, L., Zhang, H., Wang, J., Tian, L., Li, F., Liu, S., et al., 2019a. C60 Fullerols Enhance Copper Toxicity and Alter the Leaf Metabolite and Protein Profile in Cucumber. Env. Sci. Technol 53, 2171–2180.

Zhao, L., Zhang, H., White, J.C., Chen, X., Li, H., Qu, X., et al., 2019b. Metabolomics reveals that engineered nanomaterial exposure in soil alters both soil rhizosphere metabolite profiles and maize metabolic pathways. Environ. Science: Nano 6, 1716–1727.

Zheng, L., Cheng, Z., Ai, C., Jiang, X., Bei, X., Zheng, Y., et al., 2010. Nicotianamine, a Novel Enhancer of Rice Iron Bioavailability to Humans. PLOS ONE 5, e10190.

Chapter 5

The importance of system complexity in understanding plant responses to engineered nanoparticles: direct versus indirect effects

Christian P. Andersen[1] and Cyren M. Rico[2]
[1]*Consulting Plant Physiologist, Corvallis, OR, United States,* [2]*Department of Chemistry and Biochemistry, Missouri State University, Springfield, MO, United States*

Chapter Outline

5.1 Introduction

Engineered nanoparticles (ENPs) are widespread and can reach terrestrial ecosystems through a multitude of pathways. ENPs are incorporated into numerous products used both commercially and by consumers directly. ENPs are used in surface coatings including paints, packaging, cosmetic products, antimicrobial agents, and electronics (Ju-Nam and Lead, 2008; Klaine et al., 2008). Soils are exposed to ENPs both intentionally and through weathering/breakdown of products, and bioavailability is affected by a range

Plant Exposure to Engineered Nanoparticles. DOI: https://doi.org/10.1016/B978-0-323-85032-2.00008-7

of factors (Cornelis et al., 2014; Tripathi et al., 2017). Routes include biosolid application, landfills, wastewater, as well as aerial deposition and other routes (Cornelis et al., 2014; Gottschalk and Nowack, 2011).

There is significant interest in identifying the mechanisms underlying the effects of ENPs on plants and animals following release to the environment. Much of the research employs a toxicological approach, whereby organisms are exposed to specific ENPs under highly controlled conditions, and molecular, biochemical and physiological responses are measured. There is a substantial body of literature that focuses on the mechanisms of uptake, and how various particle attributes may affect underlying plant responses (e.g., Tripathi et al., 2017). Considerably less attention has been paid to the role of biotic factors, particularly external to the plant, in understanding differential responses to ENPs.

Recently, several approaches have used the adverse outcome pathway (AOP) framework to identify molecular initiating events (MIEs) following exposure to a toxicant that leads to higher-level responses at scales relevant in a risk assessment (Villeneuve et al., 2014). The AOP approach is attractive because it provides a mechanistic basis for observed responses, and because downstream responses can be predicted whenever a specific MIE is triggered. The AOP framework requires good understanding of the molecular perturbation that occurs following ENP uptake, and the key events (KEs) that link each response to higher order responses within an organism.

The AOP framework applies especially well for human health exposures, particularly when exposure routes and receptor sites (e.g., cellular uptake, etc.) have been identified. For human health applications, the AOP focuses on a single target organism (humans) even though animal analogs may be used experimentally to identify MIEs and KEs. Unfortunately, the AOP framework can be difficult to apply to ecological risk assessment since assessments rarely include only one organism of interest, exposure routes can be complex, chemical transformation prior to uptake may lead to more than one potential toxicant, and effects on a target organism may include indirect responses unrelated to toxicant uptake. Specifically, identifying MIEs and KEs in a target plant can be challenging as a result of the various modes through which ENPs may be affecting plant growth and development, and the numerous feedback effects that may occur among species. It is likely that this complexity has led to a wide array of observed responses to ENPs, even when using similar ENPs.

One way to organize the components of this complexity may be through an aggregate exposure pathway (AEP) framework, which allows consideration of contaminant pathways from the source, through the environment, to the site of action in target organisms (Hines et al., 2018; Teeguarden et al., 2016). Using an AEP approach coupled with AOPs would allow consideration of response prior to or in addition to uptake that also may lead to responses in the target organism. In the case of external factors, we would consider these responses indirect with respect to the target organism, compared to a direct effect resulting from ENP uptake.

It is clear that effects of ENPs on plants may result even when the particle is not taken up, complicating risk assessment using AOP models based on uptake and a molecular-level perturbation in the target organism. As primary producers, plants are part of ecological systems that are comprised of thousands of species organized in a multi-level trophic structure. As a result, ENPs may elicit a response through a direct uptake by a particular plant species, but also may elicit a plant response indirectly by altering trophic interactions, for example, through competitive interactions, or by disrupting energy transfer through the food chain. A better understanding of these complex interactions, particularly below ground, is needed to understand the longer-term consequences of ENP exposure. Without a better understanding of direct and indirect responses following ENP exposure, it will be difficult to predict higher level responses at scales relevant in risk assessment.

In this chapter, we will highlight the importance of distinguishing between direct and indirect effects when attempting to identify mechanisms of response to ENPs in plants and ecological systems. The discussion will initially focus on the belowground ecosystem to illustrate how interactions external to the plant may lead to differential plant responses to ENPs. We will then discuss several higher-level responses to illustrate the difficulty of understanding longer-term effects on ecological systems without a better understanding of the underlying mechanisms. The goals are to bring awareness to these important areas, and to promote the use of experimental protocols that recognize the possible interactions and feedbacks that are occurring in natural systems to better identify the longer-term risks of ENPs.

5.2 Distinguishing between direct and indirect ENP effects

When evaluating the effects of ENPs on plant physiology and growth, it is important to understand whether the particle is acting directly on cellular metabolism of the target organism or indirectly through altered interactions external to the plant. The discussion will focus on belowground complexity to illustrate some of the factors that can complicate the interpretation of plant responses following ENP exposure.

Fig. 5.1 shows a conceptual pathway for how ENPs applied to soil may influence plant growth and development directly or indirectly, leading to longer-term ecological shifts. When applied to soil, ENPs first interact with soil and rhizosphere microbial communities, which may lead to chemical transformation of the ENP. ENPs or their transformed chemical byproducts may then be taken up by roots, leading directly to plant effects. Uptake and translocation are complex, and may lead to changes both in the root and the shoot (Ali et al., 2021). Any direct changes in root growth and/or exudation may lead to changes in additional uptake and also may lead to shifts in

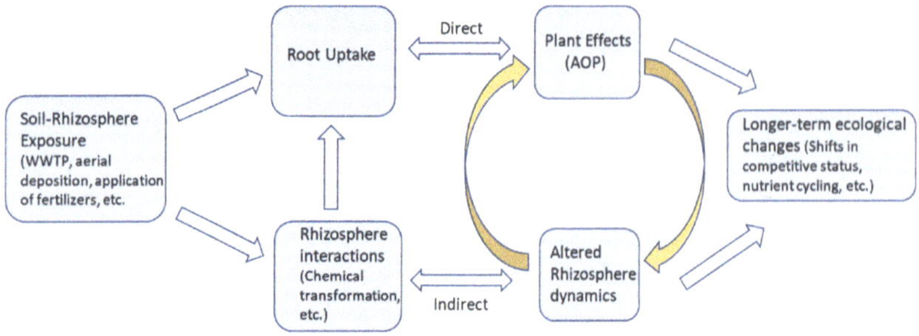

FIGURE 5.1 Conceptual framework on how soil-applied engineered nanoparticles (ENPs) may directly or indirectly influence plant growth and development.

microbial communities; changes in microbial communities in turn may affect uptake. When translocated to the shoot, ENPs have been shown to affect photosynthesis and carbohydrate dynamics (Ali et al., 2021), which may subsequently lead to changes in root dynamics (i.e., growth, exudation) and rhizosphere diversity and function. Therefore mechanistic changes underlying shifts in plant growth and development may be the result of direct uptake into the plant, but may include indirect feedbacks affecting roots and root-associated microbial communities. Regardless of the mechanism, any changes in growth and development are likely to lead to longer-term effects on plant competitive status and ultimately longer-terms shifts in the system.

ENPs also may be excluded, either in particle form or following particle transformation into other chemical forms (Fig. 5.1). Importantly, ENMs may alter soil microbial diversity and complexity in the rhizosphere, leading indirectly to shifts in plant growth and development. For example, an effect on microbial species diversity in the rhizosphere may result in altered availability of specific nutrients, which could influence plant growth and development. The potential for indirect effects is high even if particles are excluded from uptake since interactions in the rhizosphere can affect formation of beneficial symbioses and susceptibility to pathogens. ENP toxicity to root-associated symbionts such as mycorrhizal fungi is particularly important since they may influence nutrient acquisition of immobile nutrients such as P, leading to changes in growth. These effects are expected to be ongoing, and are likely to be plant species and soil-specific. All of these effects in the rhizosphere have the potential to alter plant physiology and growth, even when the ENP is not taken up into the plant cells.

It is likely that both direct and indirect effects are occurring following ENP exposure, with a high likelihood of feedbacks (Fig. 5.1). From an ecosystem standpoint, distinguishing between direct and indirect effects may not seem important, but mechanistically it is important, particularly when trying to develop frameworks that rely on a mechanistic understanding to predict longer-term responses (Boyes et al., 2017). For example, indirect effects on the rhizosphere leading to changes in nutrient availability may lead to reduced root growth and subsequent uptake of additional ENP (Fig. 5.1). These changes can also lead to competitive shifts among species. Similarly, a direct effect on root growth and branching may lead to altered exudation, leading to further rhizosphere changes affecting both nutrient availability and additional ENP uptake/exclusion. These potential feedback loops are complex in space and time.

Regardless of whether the particle is taken up, resulting in direct molecular perturbation, or whether the ENP is excluded, leading to altered root-rhizosphere interactions, exposure can lead to longer-term effects that are difficult to quantify in ecological systems (Fig. 5.1). Changes in competitive status both within and among species could lead to structural and functional changes in ecosystems over time. Changes in soil and rhizosphere organisms may alter nutrient cycling and subsequent changes in ecosystem productivity. Recent research has also shown that changes resulting from ENP exposure can be carried from one generation to the next through an unknown mechanism. Below we will focus on soil and rhizosphere interactions, as well as responses observed at larger scales, to illustrate the importance of incorporating ecological complexity into experimental protocols.

5.3 Importance of understanding the effects of ENPs on below-ground ecosystems

Soil organisms are dependent on plant-derived carbon and other compounds, and processing of these compounds through food-webs is an important property of a functional below-ground ecosystem (Wolters, 1998). One gram of soil can contain tens of thousands of species of bacteria, fungi, and invertebrates (Dindal, 1990; Torsvik and Øvreås, 2002). Plants exude a high concentration of root exudates which promote the development of a rhizosphere zone around roots, and a rhizoplane zone on root surfaces which have higher microbial and fungal density than the bulk soil (Marschner, 2012; Tinker and Nye, 2000). Roots also form important symbiotic relationships with arbuscular mycorrhizae, ectomycorrhizae, rhizobia, and actinorhizal species that are important in nutrient uptake and can provide protection from pests and pathogens (Marschner, 2012).

Despite the higher density of bacteria and fungi in the rhizosphere and rhizoplane compared to bulk soil, the diversity of species in these zones is often reduced (Tinker and Nye, 2000). This may be due to the coupling between the types of root exudates and specific bacterial groups. Organisms in the rhizoplane may be imbedded in root surface mucigel, but often consist of similar species as in the rhizosphere (Campbell and Greaves, 1990). Bacteria and fungi are also found in the endorhizosphere in between cortical cells and intercellular spaces (Patriquin et al., 1983). All of these communities interact with and are affected by ENPs prior to and during uptake.

Interactions in the rhizosphere/rhizoplane among microorganisms and roots are complex. Bacterial species can form communication networks termed quorum sensing (Sharma et al., 2003). These networks are able to influence cell density through changes in gene expression, and in some cases inhibit development of pathogens. Nodule formation on roots with beneficial bacteria associated with N-fixation can be promoted by quorum sensing. Several growth promoting bacteria such as pseudomonas can influence quorum sensing communication systems and influence plant growth and development. These complex interactions are bacterial and plant species-specific, and also vary by soil type, adding to the difficulty of understanding how stresses such as ENPs may influence them.

Rhizosphere organisms can alter root morphology and therefore can influence plant growth and development. For example, Martín et al. (1989) found that application of *Azosirillum brasilense* to young wheat plants increased total root length, the number of lateral roots, and number of root hairs. It is thought that this species produces IAA and may be responsible for the increased growth observed. Changes in root growth through such a mechanism has the potential to influence nutrient and water uptake, further affecting plant growth and development.

Root exudation allows roots to regulate microbial interactions in the rhizosphere in such a way as to benefit the plant, including the establishment of communication networks. Roots may help protect plants from herbivores, promote the development of symbiotic relationships, inhibit root growth by other competing species, and alter soil physical characteristics beneficial to the plant (Walker et al., 2003). For example, exudation of plant flavonoids of some legumes can activate *Rhizobium sp.* genes that promote the nodulation process (Peters et al., 1986). Exudation of various proteins are also thought to serve in a protective role against pathogenic fungi (Flores et al., 1999). This comes at a cost to the plant, however, with estimates of C lost to the rhizosphere ranging from 5%–21% of photosynthetically-fixed carbon (Marschner, 1995). Although extremely difficult to study, these exudates likely establish communication networks between roots of competing species, between roots and microbial communities, and between roots and insect communities (Walker et al., 2003). These communication networks serve both to attract beneficial organisms as well as protect against predators.

Root exudates in the rhizosphere can affect the availability of certain nutrients. Several metals including Fe and other trace metal oxides and hydroxides that have relatively low solubility in soils can be more readily absorbed following rhizosphere acidification (Tinker and Nye, 2000). Another mechanism involves the exudation of ligand siderophores that can bind metals such as Fe and Cu, making them more available for plant uptake (Tinker, 1986). The degree to which these interactions occur varies with species and soils, but are important in allowing plants to access key nutrients. It is critical to understand how ENP's influence these complex interactions to better understand both the direct and indirect effects on plants.

5.4 Effects of ENPs on soil and rhizosphere organisms

ENPs affect soil and rhizosphere microbial communities (Simonin and Richaume, 2015). For example, Dai et al. (2020) found that CeO_2-NPs supplied to strawberry ramet plants ended up as $CePO_4$ in the rhizosphere, which was associated with decreased rhizosphere microbial diversity but increased abundance of plant growth promoting rhizobacteria. Similarly, Ge et al. (2014) found that CeO_2-NPs affected the soil bacterial community in soils containing soybean plants, but did not influence the bacterial community in the soil without plants. This shows the importance of plant-soil feedbacks that influence rhizosphere responses to ENPs.

Although CeO_2-NPs are relatively insoluble (Walser et al., 2012), studies with plants have shown that it can transform in the rhizosphere of some species (Dai et al., 2020; Rico et al., 2018; Zhang et al., 2012). Studies found that a majority of the CeO_2-NPs in the rhizosphere of barley was reduced from Ce^{4+} to Ce^{3+} but no such transformation was found in wheat (Rico et al., 2017, 2018) (Fig. 5.2). Dai et al. (2020) also found Ce^{3+} in the form of $CePO_4$ in the rhizospheres of strawberry exposed to CeO_2-NPs. These various forms may be more readily taken up by some plant species, and in some cases the Ce taken up may be transformed back to Ce^{4+} within the plant (Rico et al., 2018). The importance of these rhizosphere interactions is critical not only for understanding differential uptake of CeO_2-NPs among species, but also for understanding shifts in rhizosphere chemistry and biology that may secondarily influence plant growth and development.

Results from some studies with metal ENPs suggest a role for rhizosphere microflora in affecting plant nutrition. Dimkpa et al. (2018) exposed wheat to nano-, ionic or bulk Mn via the root or the shoot, and followed plant growth and nutrient uptake. Both shoot- and soil-applied Mn reduced N accumulation in wheat shoots; however, there were marked differences in the uptake of other nutrients. Soil applied Mn-NPs reduced shoot Mn accumulation, reduced shoot P and K uptake, and increased residual nitrate-N in soil, responses that were not observed for soil applied ionic or bulk Mn, or for foliar-applied Mn-NPs. Compared to soil application, foliar application of Mn-NPs showed higher shoot and grain Mn contents, and lower soil nitrate-N, and higher soil and shoot P. These effects lead the authors to caution against the use of soil applied Mn-NPs fertilizers and instead consider the use of foliar Mn-NPs. Although the authors ascribe most of the differences to higher reactivity and slower solubility of Mn-NPs in soil, the higher levels of residual nitrate-N in soil applied with Mn-NPs led them to conclude that Mn-NPs may have interfered with denitrification in the wheat rhizosphere.

The interactions occurring below ground provide a good illustration of the complexity of interpreting plant responses to ENPs. Above-ground interactions are also complex in space and time as species compete for light and are

FIGURE 5.2 Growth and productivity of wheat [top panel (A), (B)] and barley [bottom panel (C), (D)] exposed to cerium oxide nanoparticles. The line graphs in (A) and (C) show the trend in delayed or decreased grain maturity from control to treated. The insets (A1) and (C1) show spikes at the highest treatment. *Data and images taken from Rico et al. (2014, 2015, 2017, 2018).*

exposed to other stresses. Future research protocols need to examine these interactions to understand initial mechanisms of response, and how these changes translate to changes at large ecological scales. We discuss examples of higher-level responses following ENP exposure in the following sections.

5.5 Higher-level responses: importance of understanding mechanisms

The concern that ENPs adversely affect terrestrial plants has triggered a wide variety of research on ENP-plant interactions and how ENPs may be affecting ecosystems. There is now a large body of literature regarding the impacts of ENPs on plant performance (e.g., germination, fitness, reproductive output, molecular changes). But in the vast majority of these studies, the underlying mechanisms leading to these responses remain poorly understood. A good example is the impacts of maternal plant exposure on fitness and performance of the succeeding generations, which likely has longer-term consequences for ecosystems, but a mechanism has not been documented. Also, many of the current studies were performed using individual plants in the absence of competition from another individual of the same or different species (intra-specific or inter-specific interactions). From a mechanistic standpoint it is important to understand the impacts of ENPs on plant interactions and what role does mixed-culture play on ENPs toxicity in single-generation or multi-generational exposures. Currently, there are few publications that address generational exposure, and there are no reports assessing mixed cultivation of plants to ENPs.

It is important to understand the underlying mechanisms (i.e., direct vs indirect) associated with initial plant response leading to subsequent intergenerational changes to help understand likely ecological consequences. The literature that does exist clearly shows the potential for intergenerational effects, but are limited to only a few studies on a limited number of species and particles (Rico et al., 2017, 2020a,b). Servin and White (2016) summarized research areas (e.g., trophic transfer and trans-generational studies, nutritional quality, rhizosphere biota, cocontaminant uptake, and sensitive endpoints involved in low dose exposures) peculiar to understanding ENPs risks and benefits in terrestrial environment. The role that mixed cultivation of plants plays in regulating the impact of ENPs is poorly understood.

5.5.1 Effects of ENPs on intergenerational shifts

Plant exposure to other environmental stresses such as drought or nutrient limitation has revealed parental, environmental, and genetic influence on transgenerational alterations of phenotypes to promote offspring fitness and function (He et al., 2014; Herman et al., 2012; Kochanek et al., 2011; Mondoni et al., 2014). In general, transgenerational effects commonly occur and may have considerable impact on the way plants respond to environmental stress (Duan et al., 2020; Groot et al., 2016). Parent plants that encounter stress produce larger seeds and improve seed-derived quality to increase fitness and success of offspring. For example, *Avena barbata* and *Hordeum vulgare* that were stressed in the first generation promoted seed provisioning via increased seed mass or seed mineral content to ensure offspring lifetime fitness (Germain et al., 2013; Nosalewicz et al., 2016). On the other hand, there are also reports showing that transgenerational phenotypic changes are not always positive (i.e., promoting fitness to the same stress the parents encountered) (Groot et al., 2016; Rasmann et al., 2011; Sandner et al., 2018). This has been observed in *Silene vulgaris* wherein some phenotypic adjustments were recorded in parental plants but those changes were not effective in improving performance and fitness in the succeeding filial generations (Sandner et al., 2018).

A thorough discussion on generational exposures of plants to engineered nanomaterials is provided in another chapter. A few cases showed that the altered physiological characteristics and elemental/nutrient profile of seeds harvested from maternal plants following exposure to ENPs induced changes in plant responses of succeeding generations (e.g., daughter plants). In two separate studies, *Arabidopsis thaliana* or tomato (*Solanum lycopersicum* L.) exposed to CuO-NPs or CeO_2-NPs produced daughter plants that exhibited growth retardation and higher accumulation rate of nanosize CuO or CeO_2 (Wang et al., 2013, 2016b). Similarly, sequential exposure to TiO_2-NPs affected the photosynthetic activity of basil (*Ocimum basilicum*) daughter plants (Tan et al., 2018), whereas continuous exposure to CeO_2-NPs for three generations decreased germination, growth, and productivity of *Brassica rapa* and *A. thaliana* (Geisler-Lee et al., 2014; Ma et al., 2016). On the other hand, Medina-Velo et al. (2018) found only minimal effect of generational exposures of bean (*Phaseolus vulgaris*) to ZnO-NPs. Among these studies only *B. rapa* exposure to CeO_2-NPs did not measure uptake in plants and only *Arabidopsis* exposure to CuO-NPs showed nanoparticle accumulation in first generation seeds. In contrast, basil exposure to TiO_2-NPs only showed increase in titanium concentration in roots but not in shoots.

Rico et al. (2017) has also reported that second generation wheat from parents exposed to CeO_2-NPs behaved differently when cultivated to full maturity in either clean or CeO_2-NPs amended soil (treatments: C1C2, C1T2, T1C2, T1T2, C = control/clean soil, T = CeO_2-NPs treated soil, 1 = first generation, 2 = second generation) (Fig. 5.3). The data revealed that compared to C1, T1 plants had accelerated growth and grain production regardless of second generation treatments (e.g., either C2 or T2). The roots accumulation of Al, Fe, and Mn was significantly altered: at $T = 125$ mg CeO_2-NPs per kg soil, the trend for Al, Fe, and Mn concentration in roots was C1C2 >> T1C2, whereas at $T = 500$ mg CeO_2-NPs per kg soil the trend was C1C2 >> C1T2, T1T2 >> T1C2. The mineral content of the grains was likewise affected at $T = 125$ mg CeO_2-NPs per kg soil; the trend C1T2 >> T1T2 was recorded for Ca, K, Mg, and P concentration in grains. Nitrogen content was also affected wherein T1 plants had decreased N and ^{15}N uptake compared to C1 plants.

The results of Rico et al. (2017) suggest that the mechanisms involved changes in rhizosphere chemistry and microbial activity because there was no uptake and accumulation of CeO_2-NPs in the aerial parts of wheat (i.e., shoots and grains). Analysis of the roots also showed no changes in speciation, that is CeO_2-NPs remained in nanoscale particles. There were clear changes on the biochemical composition and physiological responses despite the lack of transformation and uptake of CeO_2-NPs in wheat. Since exposure to CeO_2-NPs occurred only in the roots, these generational responses were likely due to mechanisms involving indirect effects at the root-rhizosphere interface (Fig. 5.1). The findings of Tan et al. (2018) on basil exposed to TiO_2-NPs could also indicate indirect effects on plants since titanium was not found accumulating in the aerial parts of the plants.

There is mounting evidence that other stresses can result in changes in gene expression and DNA methylation, and such epigenetic changes may

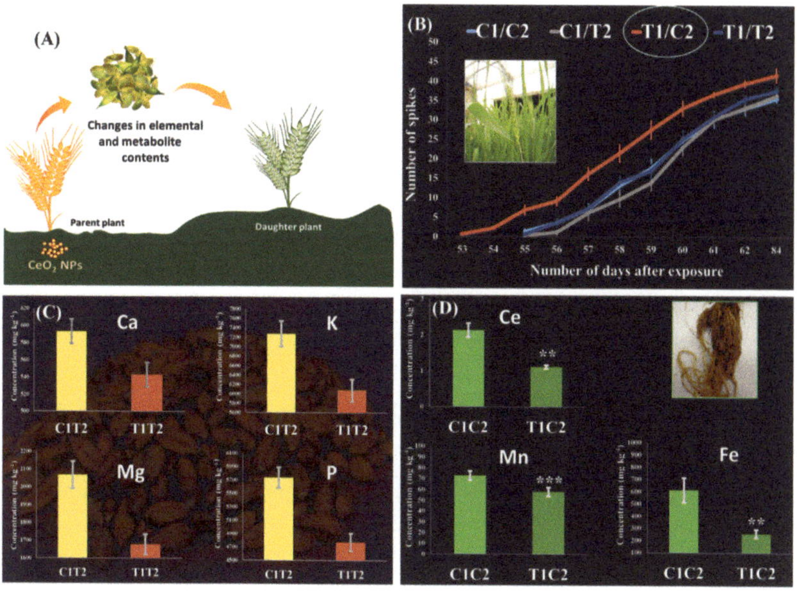

FIGURE 5.3 Second generation exposure of wheat to cerium oxide nanoparticles. (A) Carton depicting the 2nd generation treatments: C1C2, C1T2, T1C2, and T1T2; C = control/clean soil, T = CeO_2-NPs treated soil, 1 = first generation, 2 = second generation. (B) Plants previously exposed to cerium oxide nanoparticles exhibited enhanced and earlier spike formation in. (C) Repeated exposure for two generations decreased macroelements in grains. (D) Plants previously exposed to cerium oxide nanoparticles decreased Ce, Mn, and Fe concentrations in the roots. *Data derived from Rico et al. (2017).*

provide a possible mechanism explaining plant response to ENPs (Duan et al., 2020). Environmental stresses such as water, temperature, pathogen, salt, and heavy metals have been shown to change levels of DNA methylation in plants (Al-Lawati et al., 2016; Cicatelli et al., 2014; Duan et al., 2020; Khan and Zinta, 2016). Changes in DNA methylation could result in aberrant growth or phenotypes in plants. Studies have found that decreased DNA methylation in *Arabidopsis* caused the plant to grow shorter with smaller leaves and reduced maturity (Jeddeloh et al., 1998), whereas increased methylation caused delayed fruit ripening and fruit discoloration in tomato (*S. lycopersicum*) (Manning et al., 2006). Genome-wide epigenetic variations in natural and experimental populations of *A. thaliana*, soybean (*Glycine max*), maize (*Zea mays*), and wheat (*Triticum aestivum*) have also been documented (Becker et al., 2011; Duan et al., 2020; Eichten et al., 2013; Kooke et al., 2015).

Although the mechanisms underlying generational responses are poorly understood, epigenetic changes resulting from ENP exposure should be examined as a possible way that effects are carried from one generation to the next. There are limited studies on epigenetic effects of ENPs in organisms. Changes in DNA methylation in zebrafish (*Danio rerio*) exposed to single- or multi-walled carbon nanotubes and Ag-NPs, and changes in microRNA expression in nematodes exposed to multi-walled carbon nanotubes have been reported (Gorrochategui et al., 2016; Xu et al., 2018). Changes in microRNA expression in *Nicotiana tabacum* exposed to Al_2O_3-NPs have also been found (Stoccoro et al., 2013). Recently, epigenetics effects of CuO-NPs on *Enchtraeus cryptus* (Oligochaeta: Enchytraeidae) have also been reported (Bicho et al., 2020). However, epigenetic modifications in plants due to ENPs exposure are not well explored and therefore it remains a challenge to understand the potential role these changes may play in affecting plant health and performance.

5.5.2 Importance of including multiple species when evaluating the effects of ENPs

In the absence of ENPs, studies have found that plant performance and productivity are very different in mixed-culture than monoculture, and that individual or isolated species studies are not strong predictors of plant growth and reproductive output in real mixed culture environment. Studies have shown that mixed cultures could induce competition or facilitation in plants as responses to environmental stress (Pierik et al., 2013). For example, plants grown in mixtures have shown changes in gene regulation as a result of changes in use of water, light, nutrients, root exudates and volatile organic compounds (Khashi u Rahman et al., 2019; Pierik et al., 2013; Venturelli et al., 2015). Kotowska et al. (2010) found higher plant productivity in nine genotypes of *A. thaliana* that were grown in mixtures compared to those grown in mono-cultures. Several research groups using elevated levels of CO_2 have also demonstrated that the data derived from isolated plants are poor predictors of plants responses to stress compared to data obtained from mixed culture (Poorter et al., 2012). This was observed in annual plants exposed to elevated amount of CO_2 which responded differently in growth and nitrogen uptake when grown in community in comparison to those grown individually (Berntson and Wayne, 2000; Wayne and Bazzaz, 1997). Similarly, genotypic diversity conferred differences in productivity and nitrogen demand between two *Salix* genotypes that were grown in mono-culture than those cultivated in mixture (Hoeber et al., 2017). Another study found that N utilization strategy of neighboring plants could induce N competitiveness in dominant species and disturb plant community dynamics (Wang et al., 2016a).

There is also evidence that plants growing in mixed-culture respond differently to heavy metal stress than those grown individually. Zhao et al. (2017) reported that when duckweeds (*Lemna aequinoctialis* and *Spirodela polyrhiza*) were co-exposed to heavy metals (Ni, Cu, Zn), they exhibited enhanced fitness and survival (i.e., increased plant growth and decreased metal accumulation) compared to when they were grown in mono-culture. Subsequent experiments by these researchers further revealed that mixed culture of duckweed altered the bacterial communities and utilization of carbon type which protected the plants from damages to metal exposure (Zhao et al., 2017, 2018). Koelbener et al. (2008) also found that plants modified the impacts of Ni and Zn in mixed cultures of *Carex flava, Centaurea angustifolia*, and *Salix caprea* by increasing metal uptake of *S. caprea* that reduced the harmful effects on neighboring species *C. flava*. Similar effects of mixed cultures have been documented in biotic stress such as herbivore and pest, and in abiotic factors such as light environment (Baker et al., 2019; Brooker, 2006).

Mixed-species studies have also revealed that plant traits (cultivar or genotype) and environmental factors affect the biochemical pathways in which maternal plant provisions its seeds (Marinoni et al., 2018; Violle et al., 2009). Seed provisioning (i.e., plant allocations of seed nutrients or reserves) depends on the biotic and abiotic environment (temperature, light, moisture, soil nutrient) in which the parents were exposed and the modulation of these environmental factors, especially nutrient resources usage by the interacting neighbors. For example, Violle et al. (2009) performed studies of phytometer plants (i.e., indicator transplants that provide information on site conditions based on plant survival, growth, and reproduction (Strobl et al., 2018)) and several garden species at different soil nitrogen gradients and found that alterations in seed quality (mass, nitrogen concentration, germinability, timing of germination) of the phytometer depended on the neighbor identity.

Based on these studies, it is probable that performance of mixed-species systems consisting of two or more plant species would respond differently to ENP exposure than single-plant systems. In addition, differences belowground between single and mixed species systems likely alter the dynamics of ENP uptake/exclusion by plants, affecting whether responses are direct or indirect (Fig. 5.1). However, there are no studies that document the effects of ENPs on mixed-species systems. Also, studies on mixed cultures of generational plants and their influence on the succeeding generations are not readily available in literature.

Preliminary studies on co-exposures of wheat and barley to CeO_2-NPs (0 and 500 mg kg^{-1}) have been performed in our lab. Interestingly, the data showed that barley was extremely delayed in grain formation compared to wheat that there were no grains obtained in barley in both 0 and 500 mg kg^{-1} CeO_2-NPs treatments at harvest. Wheat showed a higher grain yield in CeO_2-NPs treated soil than in control, a finding consistent with our previous study (Rico et al., 2014). Our previous study found that 500 mg kg^{-1} CeO_2-NPs treatment was detrimental to barley grain production (Rico et al., 2015), which possibly explains the lack of grain formation in barley co-exposed with wheat. However, the lack of grain formation in barley in the control clean soil was unexpected considering that barley in control soil in our previous study produced grains (Rico et al., 2015).

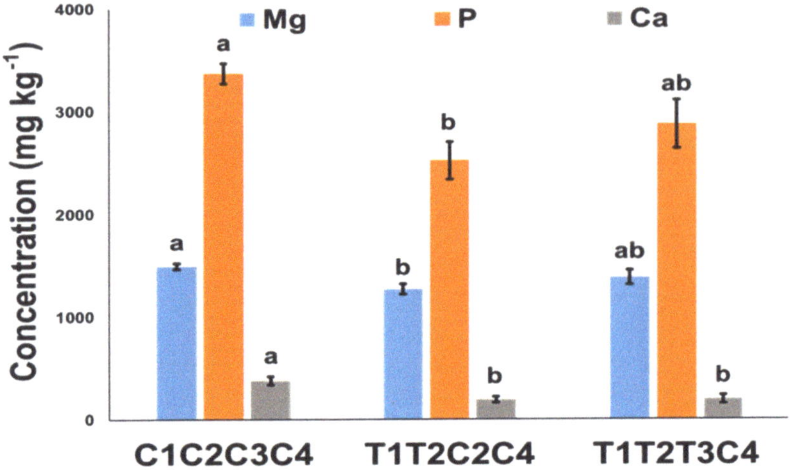

FIGURE 5.4 Elemental concentrations in grains of generational wheat cultivated in mixed culture at 4th generation in clean soil. Values are means ± SE ($n = 8$). Different letters indicate statistical differences between treatments ($p \leq 0.05$).

Preliminary data from our laboratory has also been obtained regarding mixed cultivation of generational wheat (C1C2C3, T1T2C3, T1T2T3) in clean untreated soil and in CeO_2-NPs treated soil to produce 4th generation plants (C1C2C3C4, T1T2C3C4, T1T2T3C4). Ancestral exposure had significant influence on the elemental accumulation of progeny seeds. Prior exposure to CeO_2-NPs reduced Mg, P, and Ca concentrations in grains of daughter plants in clean soil (Fig. 5.4), which occurred despite the uptake of these elements (except P which showed decreasing trend as observed in grains) was not affected in shoots (data not shown). Surprisingly results from the CeO_2-NPs treated soil showed that grain elemental accumulation was not affected (data not shown). These changes occurred even without Ce accumulation in aboveground parts of the plants, a finding consistently observed in wheat (Rico et al., 2014, 2017, 2020a) (Fig. 5.2). Together, these findings strongly signify influence of plant interactions on modulating impacts of ENMs in coexposed plants that were not observed in mono-cultured environment. Moreover, plants can exhibit indirect effects from ENP exposures few generations away from first exposure.

5.6 Summary and future research

ENPs can alter plant nutrition, physiology, growth and development, and even long-term generational performance. What is often poorly understood is the mechanistic basis for these changes, and whether they result from direct uptake and perturbation of a molecular event, or from indirect effects owing to the complexity of ecological interactions occurring external to the plant.

One major area of future research needs is to include mesocosms with more realistic soil microbial communities to allow for rhizosphere interactions that are known to affect both plant growth and uptake of ENPs. Without a better understanding of the interactions occurring in the root rhizosphere, it will be difficult to determine whether observed effects are direct or indirect, and what the possible effects are at scales relevant to ecological risk assessment. It is likely that even when the particle is not taken up, there can be changes at larger scales that affect ecological systems. It is also important to consider that even when the particle is taken up, leading to direct plant effects, it is likely that indirect effects are also occurring both above and belowground, influencing ecological effects.

It is also important to bring species complexity into our laboratory/mesocosm systems designed to characterize the effects of ENPs. Future research needs to include a range of soils from different ecosystems that match native plant mixtures. In addition, plants and plant systems should be exposed to multiple ENPs and other abiotic stresses simultaneously to quantify the potential interactive effects of co-occurring stresses. Finally, it is important to follow the effects of ENPs on plants and plant systems over long time frames. This will allow establishment of belowground microbial communities, and also the ability to follow more subtle effects that may be amplified over multiple growing seasons or from one generation to the next.

References

Ali, S., Mehmood, A. & Khan, N., 2021. Uptake, translocation, and consequences of nanomaterials on plant growth and stress adaptation. J. Nanomater., 6677616.

Al-Lawati, A., Al-Bahry, S., Victor, R., Al-Lawati, A.H., Yaish, M.W., 2016. Salt stress alters DNA methylation levels in alfalfa (Medicago spp). Genet. Mol. Res. 15.

Baker, B.H., Sultan, S.E., Lopez-Ichikawa, M., Waterman, R., 2019. Transgenerational effects of parental light environment on progeny competitive performance and lifetime fitness. Philos. Trans. R. Soc. B Biol. Sci. 374, 20180182.

Becker, C., Hagmann, J., Müller, J., Koenig, D., Stegle, O., Borgwardt, K., et al., 2011. Spontaneous epigenetic variation in the Arabidopsis thaliana methylome. Nature 480, 245–249.

Berntson, G.M., Wayne, P.M., 2000. Characterizing the size dependence of resource acquisition within crowded plant populations. Ecology 81, 1072–1085.

Bicho, R.C., Roelofs, D., Mariën, J., Scott-Fordsmand, J.J., Amorim, M.J.B., 2020. Epigenetic effects of (nano)materials in environmental species—Cu case study in Enchytraeus crypticus. Environ. Int. 136105447.

Boyes, W.K., Thornton, B.L.M., Al-Abed, S.R., Andersen, C.P., Bouchard, D.C., Burgess, R.M., et al., 2017. A comprehensive framework for evaluating the environmental health and safety implications of engineered nanomaterials. Crit. Rev. Toxicol. 47, 771–814.

Brooker, R.W., 2006. Plant–plant interactions and environmental change. New Phytologist 171, 271–284.

Campbell, R., Greaves, M.P., 1990. Anatomy and Community Structure of the Rhizosphere. Wiley and Sons, Chichester.

Cicatelli, A., Todeschini, V., Lingua, G., Biondi, S., Torrigiani, P., Castiglione, S., 2014. Epigenetic control of heavy metal stress response in mycorrhizal vs non-mycorrhizal poplar plants. Environ. Sci. Pollut. Res. 21, 1723–1737.

Cornelis, G., Hund-Rinke, K., Kuhlbusch, T., Van Den Brink, N., Nickel, C., 2014. Fate and bioavailability of engineered nanoparticles in soils: a review. Crit. Rev. Environ. Sci. Technol. 44, 2720–2764.

Dai, Y., Chen, F., Yue, L., Li, T., Jiang, Z., Xu, Z., et al., 2020. Uptake, transport, and transformation of CeO_2 nanoparticles by strawberry and their impact on the rhizosphere bacterial community. ACS Sustain. Chem. Eng. 8, 4792–4800.

Dimkpa, C.O., Singh, U., Adisa, I.O., Bindraban, P.S., Elmer, W.H., Gardea-Torresdey, J.L., et al., 2018. Effects of manganese nanoparticle exposure on nutrient acquisition in wheat (Triticum aestivum L.). Agronomy 8, 158.

Dindal, D.L., 1990. Soil Biology Guide. John Wiley & Sons, NY.

Duan, H., Li, J., Zhu, Y., Jia, W., Wang, H., Jiang, L., et al., 2020. Responsive changes of DNA methylation in wheat (Triticum aestivum) under water deficit. Sci. Rep. 10, 7938.

Eichten, S.R., Briskine, R., Song, J., Li, Q., Swanson-Wagner, R., Hermanson, P.J., et al., 2013. Epigenetic and genetic influences on DNA methylation variation in maize populations, Plant Cell, 25, pp. 2783–2797.

Flores, H.E., Vivanco, J.M., Loyola-Vargas, V.M., 1999. Radicle biochemistry: the biology of root-specific metabolism. Trends Plant Sci. 4, 220–226.

Ge, Y., Priester, J.H., Van De Werfhorst, L.C., Walker, S.L., Nisbet, R.M., An, Y.-J., et al., 2014. Soybean plants modify metal oxide nanoparticle effects on soil bacterial communities. Environ. Sci. Technol. 48, 13489–13496.

Geisler-Lee, J., Brooks, M., Gerfen, J.R., Wang, Q., Fotis, C., Sparer, A., et al., 2014. Reproductive toxicity and life history study of silver nanoparticle: effect, uptake and transport in Arabidopsis thaliana, Nanomaterials (Basel), 4, pp. 301–318.

Germain, R.M., Caruso, C.M., Maherali, H., 2013. Mechanisms and consequences of water stress–induced parental effects in an invasive annual grass. Int. J. Plant Sci. 174, 886–895.

Gorrochategui, E., Li, J., Fullwood, N.J., Ying, G.-G., Tian, M., Cui, L., et al., 2016. Diet-sourced carbon-based nanoparticles induce lipid alterations in tissues of zebrafish (Danio rerio) with genomic hypermethylation changes in brain. Mutagenesis 32, 91–103.

Gottschalk, F., Nowack, B., 2011. The release of engineered nanomaterials to the environment. J. Environ. Monit. 13, 1145–1155.

Groot, M.P., Kooke, R., Knoben, N., Vergeer, P., Keurentjes, J.J.B., Ouborg, N.J., et al., 2016. Effects of multi-generational stress exposure and offspring environment on the expression and persistence of transgenerational effects in Arabidopsis thaliana. PLoS One 11, e0151566.

He, H., De Souza Vidigal, D., Snoek, L.B., Schnabel, S., Nijveen, H., Hilhorst, H., et al., 2014. Interaction between parental environment and genotype affects plant and seed performance in Arabidopsis. J. Exp. Bot. 65, 6603–6615.

Herman, J.J., Sultan, S.E., Horgan-Kobelski, T., Riggs, C., 2012. Adaptive transgenerational plasticity in an annual plant: grandparental and parental drought stress enhance performance of seedlings in dry soil. Integr. Comp. Biol. 52, 77–88.

Hines, D.E., Edwards, S.W., Conolly, R.B., Jarabek, A.M., 2018. A case study application of the aggregate exposure pathway (AEP) and adverse outcome pathway (AOP) frameworks to facilitate the integration of human health and ecological end points for cumulative risk assessment (CRA). Environ. Sci. Technol. 52, 839–849.

Hoeber, S., Fransson, P., Prieto-Ruiz, I., Manzoni, S., Weih, M., 2017. Two salix genotypes differ in productivity and nitrogen economy when grown in monoculture and mixture. Front. Plant Sci. 8, 231. -231.

Jeddeloh, J.A., Bender, J., Richards, E.J., 1998. The DNA methylation locus DDM1 is required for maintenance of gene silencing in Arabidopsis. Genes Dev. 12, 1714–1725.

Ju-Nam, Y., Lead, J.R., 2008. Manufactured nanoparticles: an overview of their chemistry, interactions and potential environmental implications. Sci. Total Environ. 400, 396–414.

Khan, A., Zinta, G., 2016. Drought stress and chromatin: an epigenetic perspective. In: Hossain, M.A., Wani, S.H., Bhattacharjee, S., Burritt, D.J., Tran, L.-S.P. (Eds.), Drought Stress Tolerance in Plants, Vol 2: Molecular and Genetic Perspectives. Springer International Publishing, Cham.

Khashi u Rahman, M., Zhou, X., Wu, F., 2019. The role of root exudates, CMNs, and VOCs in plant–plant interaction. J. Plant Interact. 14, 630–636.

Klaine, S.J., Alvarez, P.J., Batley, G.E., Fernandes, T.F., Handy, R.D., Lyon, D.Y., et al., 2008. Nanomaterials in the environment: behavior, fate, bioavailability, and effects. Environ. Toxicol. Chem. 27, 1825–1851.

Kochanek, J., Steadman, K.J., Probert, R.J., Adkins, S.W., 2011. Parental effects modulate seed longevity: exploring parental and offspring phenotypes to elucidate pre-zygotic environmental influences. New Phytol. 191, 223–233.

Koelbener, A., Ramseier, D., Suter, M., 2008. Competition alters plant species response to nickel and zinc. Plant Soil 303, 241–251.

Kooke, R., Johannes, F., Wardenaar, R., Becker, F., Etcheverry, M., Colot, V., et al., 2015. Epigenetic basis of morphological variation and phenotypic plasticity in Arabidopsis thaliana. Plant Cell 27, 337–348.

Kotowska, A.M., Cahill Jr., J.F., Keddie, B.A., 2010. Plant genetic diversity yields increased plant productivity and herbivore performance. J. Ecol. 98, 237–245.

Ma, X., Wang, Q., Rossi, L., Ebbs, S.D., White, J.C., 2016. Multigenerational exposure to cerium oxide nanoparticles: physiological and biochemical analysis reveals transmissible changes in rapid cycling Brassica rapa. NanoImpact 1, 46–54.

Manning, K., Tör, M., Poole, M., Hong, Y., Thompson, A.J., King, G.J., et al., 2006. A naturally occurring epigenetic mutation in a gene encoding an SBP-box transcription factor inhibits tomato fruit ripening, Nat. Genet., 38, pp. 948–952.

Marinoni, L., Zabala, J.M., Parra-Quijano, M., Fernández, R.J., Pensiero, J.F., 2018. Genetic and environmental variation of seed weight in Trichloris species (Chloridoideae, Poaceae) and its association with seedling stress tolerance. Plant Ecol. Div. 11, 173–184.

Marschner, H., 1995. Mineral Nutrition of Higher Plants. Academic Press, London.

Marschner, P., 2012. Rhizosphere biology. In: Marscher, H. (Ed.), Mineral Nutrition of Higher Plants, third ed.

Martín, P., Glatzle, A.F., Kolb, W., Omay, H.S., Schmidt, W., 1989. N2 fixing bacteria in the rhizosphere: quantification and hormonal effects on root development. J. Plant Nutrit. Soil Sci. 152, 237–245.

Medina-Velo, I.A., Zuverza-Mena, N., Tamez, C., Ye, Y., Hernandez-Viezcas, J.A., White, J.C., et al., 2018. Minimal transgenerational effect of ZnO nanomaterials on the physiology and nutrient profile of Phaseolus vulgaris. ACS Sustain. Chem. Eng. 6, 7924–7930.

Mondoni, A., Orsenigo, S., Donà, M., Balestrazzi, A., Probert, R.J., Hay, F.R., et al., 2014. Environmentally induced transgenerational changes in seed longevity: maternal and genetic influence. Ann. Bot. (Lond.) 113, 1257–1263.

Nosalewicz, A., Siecińska, J., Śmiech, M., Nosalewicz, M., Wiącek, D., Pecio, A., et al., 2016. Transgenerational effects of temporal drought stress on spring barley morphology and functioning. Environ. Exp. Bot. 131, 120–127.

Patriquin, D.G., Döbereiner, J., Jain, D.K., 1983. Sites and processes of association between diazotrophs and grasses. Can. J. Microbiol. 29, 900–915.

Peters, N.K., Frost, J.W., Long, S.R., 1986. A plant flavone, luteolin, induces expression of Rhizobium meliloti nodulation genes. Science 233, 977–980.

Pierik, R., Mommer, L., Voesenek, L.A., 2013. Molecular mechanisms of plant competition: neighbour detection and response strategies. Funct. Ecol. 27, 841–853.

Poorter, H., Bühler, J., Van Dusschoten, D., Climent, J., Postma, J.A., 2012. Pot size matters: a *meta*-analysis of the effects of rooting volume on plant growth. Funct. Plant Biol. 39, 839–850.

Rasmann, S., De Vos, M., Casteel, C.L., Tian, D., Halitschke, R., Sun, J.Y., et al., 2011. Herbivory in the previous generation primes plants for enhanced insect resistance. Plant Physiol. 158, 854–863.

Rico, C.M., Lee, S.C., Rubenecia, R., Mukherjee, A., Hong, J., Peralta-Videa, J.R., et al., 2014. Cerium oxide nanoparticles impact yield and modify nutritional parameters in wheat (Triticum aestivum L.). J. Agric. Food Chem. 62, 9669–9675.

Rico, C.M., Barrios, A.C., Tan, W., Rubenecia, R., Lee, S.C., Varela-Ramirez, A., et al., 2015. Physiological and biochemical response of soil-grown barley (Hordeum vulgare L.) to cerium oxide nanoparticles. Environ. Sci. Pollut. Res. 22, 10551–10558.

Rico, C.M., Johnson, M.G., Marcus, M.A., Andersen, C.P., 2017. Intergenerational responses of wheat (Triticum aestivum L.) to cerium oxide nanoparticles exposure. Environ. Sci. Nano 4, 700–711.

Rico, C.M., Johnson, M.G., Marcus, M.A., 2018. Cerium oxide nanoparticles transformation at the root–soil interface of barley (Hordeum vulgare L.). Environ. Sci. Nano 5, 1807–1812.

Rico, C.M., Abolade, O.M., Wagner, D., Lottes, B., Rodriguez, J., Biagioni, R., et al., 2020a. Wheat exposure to cerium oxide nanoparticles over three generations reveals transmissible changes in nutrition, biochemical pools, and response to soil N. J. Hazard. Mater. 384121364.

Rico, C.M., Wagner, D., Abolade, O., Lottes, B., Coates, K., 2020b. Metabolomics of wheat grains generationally-exposed to cerium oxide nanoparticles. Sci. Total Environ. 712136487.

Sandner, T.M., Van Braak, J.L., Matthies, D., 2018. Transgenerational plasticity in Silene vulgaris in response to three types of stress. Plant Biol. (Stuttgart, Germany) 20, 751–758.

Servin, A.D., White, J.C., 2016. Nanotechnology in agriculture: next steps for understanding engineered nanoparticle exposure and risk. NanoImpact 1, 9–12.

Sharma, A., Sahgal, M., Johri, B.N., 2003. Microbial communication in the rhizosphere: operation of quorum sensing. Curr. Sci. 85, 1164–1172.

Simonin, M., Richaume, A., 2015. Impact of engineered nanoparticles on the activity, abundance, and diversity of soil microbial communities: a review. Environ. Sci. Pollut. Res. Int. 22, 13710–13723.

Stoccoro, A., Karlsson, H.L., Coppedè, F., Migliore, L., 2013. Epigenetic effects of nano-sized materials. Toxicology 313, 3–14.

Strobl, K., Schmidt, C., Kollmann, J., 2018. Selecting plant species and traits for phytometer experiments. The case of peatland restoration. Ecol. Indic. 88, 263–273.

Tan, W., Du, W., Darrouzet-Nardi, A.J., Hernandez-Viezcas, J.A., Ye, Y., Peralta-Videa, J.R., et al., 2018. Effects of the exposure of TiO_2 nanoparticles on basil (Ocimum basilicum) for two generations. Sci. Total Environ. 636, 240–248.

Teeguarden, J.G., Tan, Y.-M., Edwards, S.W., Leonard, J.A., Anderson, K.A., Corley, R.A., et al., 2016. Completing the link between exposure science and toxicology for improved environmental health decision making: the aggregate exposure pathway framework, Environ. Sci. Technol., 50, pp. 4579–4586.

Tinker, P.B., 1986. Trace elements in arable agriculture. Hills bequest lecture of the Royal Agricultural Society of England. J. Soil Sci. 37, 587–601.

Tinker, P.B., Nye, P.H., 2000. Solute Movement in the Rhizosphere. Oxford University Press, Inc, New York.

Torsvik, V., Øvreås, L., 2002. Microbial diversity and function in soil: from genes to ecosystems. Curr. Opin. Microbiol. 5, 240–245.

Tripathi, D.K., Shweta, Singh, S., Singh, S., Pandey, R., Singh, V.P., et al., 2017. An overview on manufactured nanoparticles in plants: uptake, translocation, accumulation and phytotoxicity. Plant Physiol. Biochem. 110, 2–12.

Venturelli, S., Belz, R.G., Kämper, A., Berger, A., von Horn, K., Wegner, A., et al., 2015. Plants release precursors of histone deacetylase inhibitors to suppress growth of competitors. Plant Cell 27, 3175–3189.

Villeneuve, D.L., Crump, D., Garcia-Reyero, N., Hecker, M., Hutchinson, T.H., Lalone, C.A., et al., 2014. Adverse outcome pathway (AOP) development I: strategies and principles. Toxicol. Sci. 142, 312–320.

Violle, C., Castro, H., Richarte, J., Navas, M.-L., 2009. Intraspecific seed trait variations and competition: passive or adaptive response? Funct. Ecol. 23, 612–620.

Walker, T.S., Bais, H.P., Grotewold, E., Vivanco, J.M., 2003. Root exudation and rhizosphere biology. Plant Physiol. 132, 44–51.

Walser, T., Limbach, L.K., Brogioli, R., Erismann, E., Flamigni, L., Hattendorf, B., et al., 2012. Persistence of engineered nanoparticles in a municipal solid-waste incineration plant. Nat. Nanotechnol. 7, 520–524.

Wang, Q., Ebbs, S.D., Chen, Y., Ma, X., 2013. Trans-generational impact of cerium oxide nanoparticles on tomato plants. Metallomics 5, 753–759.

Wang, J., Luo, P., Yang, H., Mou, C., Mo, L., 2016a. Different responses of alpine plants to nitrogen addition: effects on plant-plant interactions. Sci. Rep. 6, 38320.

Wang, Z., Xu, L., Zhao, J., Wang, X., White, J.C., Xing, B., 2016b. CuO nanoparticle interaction with Arabidopsis thaliana: toxicity, parent-progeny transfer, and gene expression. Environ. Sci. Technol. 50, 6008–6016.

Wayne, P.M., Bazzaz, F.A., 1997. Light acquisition and growth by competing individuals in CO_2-enriched atmospheres: consequences for size structure in regenerating birch stands. J. Ecol. 85, 29–42.

Wolters, V., 1998. Functional aspects of animal diversity in soil—introduction and overview. Appl. Soil Ecol. 10, 185–190.

Xu, Q.-H., Guan, P., Zhang, T., Lu, C., Li, G., Liu, J.-X., 2018. Silver nanoparticles impair zebrafish skeletal and cardiac myofibrillogenesis and sarcomere formation. Aquat. Toxicol. 200, 102–113.

Zhang, P., Ma, Y., Zhang, Z., He, X., Zhang, J., Guo, Z., et al., 2012. Biotransformation of ceria nanoparticles in cucumber plants. ACS Nano 6, 9943–9950.

Zhao, Z., Shi, H., Kang, X., Liu, C., Chen, L., Liang, X., et al., 2017. Inter- and intra-specific competition of duckweed under multiple heavy metal contaminated water. Aquat. Toxicol. 192, 216–223.

Zhao, Z., Shi, H., Liu, C., Kang, X., Chen, L., Liang, X., et al., 2018. Duckweed diversity decreases heavy metal toxicity by altering the metabolic function of associated microbial communities. Chemosphere 203, 76–82.

Chapter 6

Physiological and molecular responses of plants exposed to engineered nanomaterials

Keni Cota-Ruiz[1], Carolina Valdes[2], Kenneth Flores[2], Ye Yuqing[2], Jesus Cantu[2] and Jorge Gardea-Torresdey[2,3,4,5]

[1]*MSU DOE-Plant Research Laboratory, Michigan State University, East Lansing, MI, United States,* [2]*Department of Chemistry and Biochemistry, The University of Texas at El Paso, El Paso, TX, United States,* [3]*Environmental Science and Engineering Ph.D. Program, The University of Texas at El Paso, El Paso, TX, United States,* [4]*UC Center for Environmental Implications of Nanotechnology (UC CEIN), The University of Texas at El Paso, El Paso, TX, United States,* [5]*NSF-ERC Nanotechnology-Enabled Water Treatment Center (NEWT), Rice University, Houston, TX, United States*

Chapter Outline

Plant Exposure to Engineered Nanoparticles. DOI: https://doi.org/10.1016/B978-0-323-85032-2.00005-1

6.1 Introduction

Engineered nanomaterials (ENMs) comprise particles with at least one dimension below 100 nm. They are classified as organic and inorganic nanomaterials (NMs) based on their chemical nature. Examples of the former group include fullerenes and carbon nanotubes, while metal oxides, metals, and quantum dots (QDs) belong to inorganic ENMs (Fig. 6.1) (Peralta-Videa et al., 2011). Many ENMs are manufactured to fulfill the multiple needs in electronics, biomedicine, agriculture, automotive industries, cosmetics, and many more (Cota-Ruiz et al., 2018).

The production of ENMs has been increasing in the past years (Ma et al., 2015). Among the main types of fabricated and consumed ENMs are the nanoforms of TiO_2, ZnO, SiO_2, FeO_x, AlO_x, CeO_x CNT, and Ag. TiO_2 and SiO_2 nanoparticles (NPs) are at the top of the list with production/utilization median quantities above 3000 t/year (Piccinno et al., 2012). By 2030, the use of SiO_2 NPs will be approximately 30,000 t/year (Giese et al., 2018). The tremendous utilization and disposal of ENMs have created societal concerns as they could impact the biological systems. It has been anticipated that the soils will become the major destination site of the ENMs and their byproducts (Gardea-Torresdey et al., 2014), which will directly impact the terrestrial plants. Nowadays, the studies dealing with ENMs' physiological, biochemical, agronomical, and toxicological effects are increasing but still limited (Zhao et al., 2020).

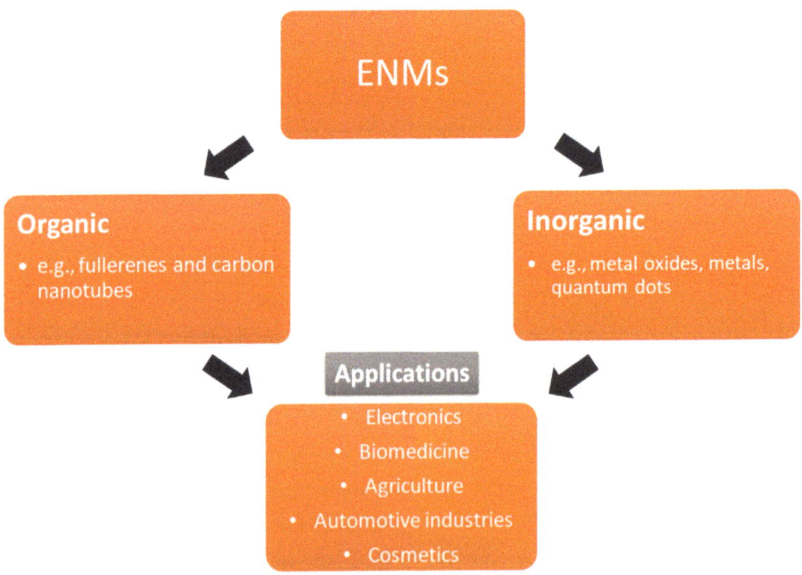

FIGURE 6.1 The engineered nanomaterials can be conventionally classified into organic and inorganic groups. They have at least one dimension no greater than 100 nm.

As global climate conditions are changing, it is expected that pathogen activity will increase, resulting in higher losses in crop yields. Conventional practices used to combat plant disease involve the direct application of agrochemicals; however, average losses of 10%–75% resulting from leaching, degradation, and immobilization make these efforts ineffective (White and Gardea-Torresdey, 2018). Recently, the application of nanotechnology in agriculture has gained attention as a way to overcome inefficiencies in current practices. The appeal for nanotechnology relies on several factors, including precise delivery and the slow release of active components (Xu et al., 2020). Extensive research is currently focused on evaluating the effects of ENMs in plant systems involving those pathogen-infected plants. The ENMs of interest for plant disease suppression include metalloid, metallic oxide, nonmetal, and carbon-based materials (Elmer et al., 2018).

Studying the mechanistic responses of plants and their associated microorganisms exposed to ENM is fundamental to understand plants' biological performance. Gene expression analyses give ample information to understand it as they provide evidence on how genes are expressed at transcript and protein levels. Although several crops such as maize, wheat, alfalfa, and rice, among others, have been explored, many significant contributions have been made using the higher plant *Arabidopsis thaliana.*

Studies on *A. thaliana* plants have helped us understand many biological functions in plants and other eukaryotic organisms. It has been considered a model organism since the 80s due to its information availability and the relative ease of being subjected to experimentation (Hayashi and Nishimura, 2006). More recently, more comprehensive information has accumulated due to high throughput technologies such as genomic and proteomic approaches. In this chapter, part of our discussion regarding the direct impact of ENMs on plants will focus on *A. thaliana*. We also discussed other novel studies about ENMs impacts on essential crop species. The analyses include evaluating the molecular responses of plants exposed to ENMs and tracking the speciation and visualization of the ENMs. Finally, we explore the effects and benefits of using ENMs in the agrosystem, specifically when applied to combat plant diseases.

6.2 Arabidopsis as a model organism to assess engineered nanomaterials-plant interactions

Among more than 300,000 plant species, researchers have preferred *A. thaliana* to study cell and molecular responses in flowering plants. Arabidopsis plants are suitable for experimentation as they have a relatively short life cycle and produce many seeds (Koornneef and Meinke, 2010). Furthermore, it has become the first plant whose genome sequence was obtained (Arabidopsis Genome Initiative, 2000). Recent physiological, cellular, and molecular investigations in Arabidopsis have resulted in a deeper understanding of the responses of plants to EMs exposure (Chen et al., 2018; Montes et al., 2017; Yang et al., 2018). In this section, these mechanistic responses are discussed.

6.2.1 Effect of carbonaceous nanomaterials and quantum dots on the physiology of Arabidopsis

Some studies have described the physiological impacts of the carbonaceous NMs (CNMs) or QD in the plants, while others have assessed the mechanistic responses by using cell lines or protoplast as models. A recent and comprehensive study investigated the effect of carbon nanodots (C-dots) at 62.5–1000 mg L^{-1} on the physiological and molecular responses of Arabidopsis seedlings (Chen et al., 2018). The authors reported that C-dots were translocated from the roots to the leaves, accumulating into the apoplastic spaces. Also, the root elongation was reduced when the seedlings were treated with higher concentrations of the C-dots. This study also demonstrated that the most affected genes (in terms of a downregulation) after C-dots exposure were those involved in the chloroplast function. In another research, the use of CNMs (500 mg L^{-1}) achieved from graphite reduced the root elongation of 8-day old Arabidopsis seedlings (Kumar et al., 2018). Also, no effects in the antioxidant responses, root elongation, and gene expression involved in the germination and development of the seedlings were found in *A. thaliana* sprouts subjected up to 1 mg L^{-1} of graphene oxide (Zhao et al., 2015). These results suggest that the physiological effects at the seedling level greatly depend on the type of CNMs form and concentration exposure.

In the 4-week-old *A. thaliana* plants, the foliar application of ZnSe QDs (100 and 250 μM) increased the expression of antioxidant responsive genes and promoted the production of antioxidant metabolites (Kolackova et al., 2019). In another study, CdSe/ZnS QDs were applied at 5 mg L^{-1} of Cd^{2+} in 3–4-weeks old *A. thaliana* seedlings and incubated for 1–7-days period (Navarro et al., 2012). The results showed that Cd and Se from QDs were only translocated to the root systems and not to aerial tissues. Furthermore, as demonstrated by confocal fluorescence microscopy, the QDs were not internalized in the root cells, but rather accumulated at the root surface. When the QDs were incubated with humic acids, the oxidized glutathione in plants increased, probably due to oxidative stress triggered by the QDs. Meanwhile, the interaction of multiwall carbon nanotubes (MWCNT) with other environmental compounds such as soil xenobiotic contaminants has been poorly investigated. Accordingly, a recent hydroponics investigation showed that when MWCNT interacted with the herbicide methyl viologen (MV), less deleterious physiological effects were registered than in the plants incubated only with MV (Fan et al., 2018).

On the other hand, other studies have demonstrated that CNTs negatively affected the functioning of Arabidopsis cells. For instance, the incubation of protoplasts isolated from 20–25 days-old plants with MWCNT led to increased oxidative stress, chromatin condensation, and programmed cell death (Shen et al., 2010). Similarly, the MWCNTs had adverse effects on the viability of *Arabidopsis* T87 suspension cells in a dose-dependent manner (10–600 mg L^{-1}) (Lin et al., 2009). In the previous study, the toxicity effect was more pronounced when MWCNTs agglomerated in smaller sizes. In the same T87cell line (Columbia ecotype), it was also demonstrated that graphene at 40–80 mg L^{-1} caused mitochondrial dysfunction, increased the production of ROS, and initiated cell necrosis (Begum and Fugetsu, 2013).

6.3 Use of engineered nanomaterials to suppress plant disease: gene activation at transcript and protein levels

Applications of ENMs in infected plants often involve the use of metal-based NMs. Considering that certain metals such as Cu, Mn, and Zn are essential for plant metabolic processes, they are often evaluated to determine their effects on disease suppression. Most importantly, these essential micronutrients are vital cofactors for activating defense-related enzymes. Infection-induced reaction cascades involving these enzymes can also generate secondary metabolites to further enhance disease resistance (Servin et al., 2015). Furthermore, metals including Cu are already present in commercially available fungicides and bactericides such as Kocide® and CuPro® manufactured by Dupont and Sepro, respectively. The evaluation of ENMs on fungal-infected plants extends from model systems to fruiting plants such as tomato and watermelon. Although phenotypic evidence can elucidate the effectiveness of ENMs for disease suppression, evaluation of key gene expressions can be used to determine the efficiency of ENMs before the onset of visible signs, thus giving a better understanding of the mechanistic pathway elicited by these ENMs.

Literature on ENMs and fungal-infected plants suggests that rather than combating plant disease itself, ENMs can be used to boost the defense system of plants to suppress disease progression (Fig. 6.2) (Elmer and White, 2016; Lopez-Lima et al., 2021). In an earlier study, Elmer and White (2016) evaluated the interaction of several ENMs, including CuO on *Fusarium oxysporum*-infected tomatoes. While CuO application up to 1000 ppm did not affect *F. oxysporum* growth, CuO NPs foliarly-applied to *F. oxysporum*-infected tomato at a concentration of 1 ppm reduced disease progression by 31% as measured by the area-under disease-progress-curve (AUDPC) method (Elmer and White, 2016). Conventional methods use visual estimation of disease symptoms and AUDPC to calculate disease progression, which involves frequent measurements of plant size and weight to produce the AUDPC. Although these techniques are still in practice, there has been a

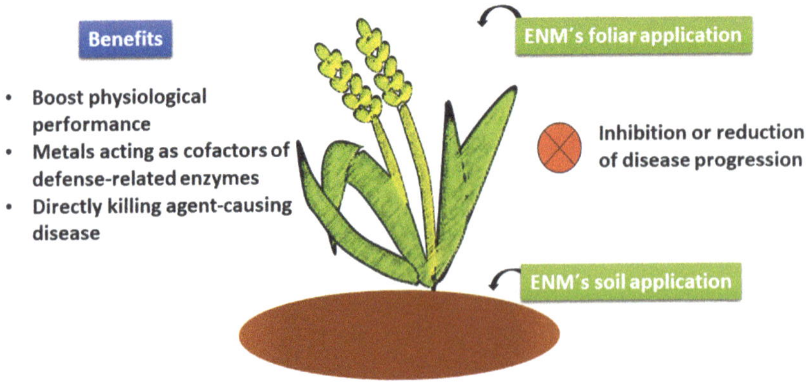

FIGURE 6.2 The role of engineered nanomaterials (ENMs) in alleviating plant diseases. Once soil or foliarly-applied, ENMs can aid in blocking or reducing disease progression directly (e.g., killing agents causing diseases) or indirectly (e.g., improving plant physiological performance).

shift in evaluating disease suppression by monitoring specific nutrients, genetic expression, and metabolic by-products.

When assessing the potential of foliar-applied commercial CuO-NPs vs in-house synthesized $Cu_3(PO4)_2$ nanosheets (NS) on *F. oxysporum* infected tomato, Ma et al. (2019) found that both ENMs significantly reduced disease presence by an average of 31%. However, in comparison to CuO-NPs and control, $Cu_3(PO_4)_2$ specifically increased the expressions of plant defense genes such as pathogenesis-related transcription activator (PTI5) genes, polyphenol oxidase (PPO) and plant resistance protein 1A1, before phenotypic evidence of disease was observed (Ma et al., 2019). Seed priming is another technique to introduce ENMs to plant systems: when evaluating chitosan-coated (CTS-MSN) and uncoated (MSN) mesoporous silica NPs at 500 mg L^{-1} to *F. oxysporum*-infected watermelon, it was found that CT-MSN and MSN significantly reduced AUDPC by 27% and 40%, respectively. Furthermore, under all CTS-MSN and MSN treatments, there was a reduced expression of PPO, polyamine oxidase 1 (PAO1) and heavy metal exporting ATPase (RAN1), suggesting a reduced disease burden on the plant (Buchman et al., 2019).

Other means to determine ENMs effectiveness in disease suppression involve evaluating metabolic by-products such as phytohormone and fatty acids. To determine the effects of several ENMs such as MWCNT, reduced graphene oxide (rGO), Cu-NPs and TiO_2-NPs, Hao et al. (2019) evaluated these materials on *Podosphaera pannosa* fungal-infected rose leaves. It was determined that while all ENMs inhibited fungal growth at 200 mg L^{-1}, CuO-NPs minimized fungal growth at a lower concentration of 50 mg L^{-1}. CuO-NPs applications induced an increase in the phytohormone zeatin riboside (ZR) content. This phytohormone is an endogenous cytokinin in leaves that can delay senescence, accelerate plant cell division and promote stress resistance (Hao et al., 2019). Furthermore, Shang et al. (2020) evaluated

CuS-NPs at 1:1 and 1:4 ratio of Cu and S on *Gibberella fujikuroi* infected rice. In this study, CuS-NPs significantly decreased infection incidence by 35.1% and 45.9%, respectively. The CuS-NPs at 1:4 ratio modulated salicylic acid (SA) and jasmonic acid (JA) phytohormones, enhancing plant defense mechanism against *G. fujikuroi* infection (Shang et al., 2020). Salicylic acid and its conjugates are known to coincide with increases in the genetic expression of pathogenesis-related proteins and the activation of disease resistance (Shah, 2003). Cross talk between phytohormones such as SA and other signaling pathways is essential for fine-tuning plant defense (Shah, 2003). Meanwhile, jasmonic acid is a phytohormone that strongly regulates the expression of defense genes in plants (Stintzi et al., 2001). Furthermore, each hormone can activate a specific metabolic pathway, which can act individually, synergistically or antagonistically to each other (Kachroo and Kachroo, 2007). Fatty acid (FA) profiles have also been evaluated to determine the effectiveness of ENM applications on infected plants. Following the application of $Cu_3(PO_4)_2$-NS, CuO-NS and CuO-NPs on *Fusarium virguliforme*-infected soybean, it was determined that foliar applications of CuO-NS at 250 mg L^{-1} significantly increased total FA content by 60.6% in infected plants. This CuO-NS treatment restored FA levels in infected soybean to those similar to the healthy plant (Ma et al., 2020). As the primary components of cell membranes and cutin wax, FAs are fundamental in forming structural barriers against pathogens and minimizing disease progression.

The evaluation of disease suppression induced by ENMs has also been extended to full plant life cycles, with an emphasis on fruit quality. Adisa et al. (2018) assessed the effects of CeO_2-NPs applied to leaves or via the roots on *F. oxysporum*-infested tomatoes. Overall, CeO_2-NPs effectively suppressed disease severity by 53% and 57% when applied through the roots or leaves, respectively (Adisa et al., 2018). Most importantly, fruit quality was determined by measuring dry weight, lycopene content and sugar levels of the tomato fruit. It was reported that foliar applications of CeO_2-NPs at 250 mg L^{-1} increased fruit dry weight by 67%, lycopene content by 9%, while root applications at 50 and 250 mg kg^{-1} decreased sugar content by 63% and 54%, respectively (Adisa et al., 2020). Overall, ENMs can have positive disease suppressing effects on fruiting plants with variations in fruit quality depending on the mode of application.

The use of ENMs has also been applied to bacterial-infected crops. However, in the case of Cu bactericides, an excess in these products can result in Cu-tolerant strains, making these materials ineffective (Liao et al., 2019a). Thus the use of hybrid Cu ENMs such as Cu/Zn and MoS_2-Cu are being explored, in addition to Cu core-shell composites, multivalent Cu and other metal oxides such as MgO (Carvalho et al., 2019; Li et al., 2020b; Liao et al., 2019b; Strayer-Scherer et al., 2018).

Although applications of ENMs on infected plant systems have demonstrated to be promising toward combating plant disease, it is relevant to note that many factors are involved in the efficacy of ENMs in plant defense. More investigations are needed as varying effects can be obtained between greenhouse and field studies, mode of application, morphology and dissolution profile in the overall performance of ENMs in infected plant systems (Adisa et al., 2020; Elmer et al., 2021; Ma et al., 2020).

6.4 Modulations of the enzymatic processes in plants treated with engineered nanomaterials

UV-Vis has been used to investigate the changes in plant total proteins, enzyme activity, and substrate changes with respect to the presence of NPs. Several enzymes are involved in ROS regulation, including superoxide dismutase (SOD), catalase (CAT), and peroxidases. These enzymatic processes directly scavenge and convert superoxide radicals and hydrogen peroxide into less reactive molecules. SOD converts superoxide radicals into hydrogen peroxide (H_2O_2) (Scandalios, 2005). Plants have different classes of SOD, including Cu/Zn SOD, Fe SOD, and Mn SOD. Both Cu/Zn SOD and Fe SOD can be found in chloroplasts, while Mn SOD can be found in the mitochondria. When H_2O_2 is present in the cells, CAT starts its degradation and converts it to H_2O. Additionally, CAT decomposes H_2O_2 at a rapid rate without consuming cellular energy. Moreover, peroxidases catalyze the reaction between the substrate and either hydrogen peroxide or an organic peroxide. A few examples of peroxidases include ascorbate peroxidase (APX), an iron heme protein, and glutathione peroxidase (GPX), enzyme-containing selenium. Similar to CAT, APX reduces H_2O_2 to water; however, for this reaction to occur would require ascorbic acid to act as an electron donor. GPX reduces peroxides into water and alcohols by utilizing reduced glutathione (GSH) and oxidizing it into its disulfide form (GSSH). In order to analyze the activity of these enzymes, they first are extracted from the tissue. The enzymes mentioned can all be extracted together in a single step. For the extraction, the fresh or lyophilized tissue is homogenized with phosphate buffer and polyvinnylpyrrolidone followed by centrifugation at 4°C (Gallego et al., 1996). Subsequent to centrifugation, the supernatant is transferred and kept for quantification of the respective enzymes.

The determination of total proteins in plant tissues can be conducted by reacting the tissue extract with Coomassie (Bradford) protein assay reagent and measuring its absorbance at 595 nm (Bradford, 1976). Valdes et al. (2020) found that the total protein content in roots and shoots of corn seedlings were unaffected by the exposure to 10 mg L^{-1} of either Cu-NPs, bulk Cu, or $CuSO_4$. However, López-Vargas et al. (2018) showed that exposure to Cu-NPs between 50–250 mg L^{-1} caused an increase in total protein content in tomato fruits, but at a concentration of 500 mg L^{-1} Cu-NPs, the total protein content of the fruit was similar with the control.

SOD activity evaluation can be conducted using a SOD test kit. The absorbance for this assay is measured at 450 nm. The study conducted by López-Vargas et al. (2018) showed that the addition of Cu-NPs to tomatoes increased the SOD activity. However, when tomato plants were exposed to either chitosan encapsulated Cu-NPs or bare Cu-NPs, the SOD activity in the fruit and leaves were unaffected. The application of CuO-NPs on green onions had no effect on the SOD activity in the roots and leaves at any concentration investigated (Wang et al., 2020).

The quantification of the CAT activity can be determined by measuring the absorbance of the degradation of H_2O_2 at 240 nm. For the quantification of the CAT reaction, kinetics can be conducted following various spectrophotometric methods (Aebi, 1974; Cakmak and Horst, 1991; Dhindsa et al., 1981). One method by Dhindsa et al. (1981) measures the degradation of H_2O_2 at two different times (time 0 and time 1). In this method, H_2SO_4 can be used to inhibit the degradation of H_2O_2 at any point (Dhindsa et al., 1981). In a different method by Aebi et al. (1974), the degradation of H_2O_2 is measured at specific intervals over a set time. Several studies have investigated the effects of NPs on the enzyme activity in plants (Dimkpa et al., 2012; Jalali et al., 2017; Juárez-Maldonado et al., 2016; López-Vargas et al., 2018; Ochoa et al., 2017; Tamez et al., 2019; Tombuloglu et al., 2019; Valdes et al., 2020). For instance, the addition of CuO to green pea plants via soil application decreased the CAT activity in the roots of plants (Ochoa et al., 2017). Additionally, the application of Cu-NPs to cucumber plants caused a decrease in CAT activity in the roots but not in the leaves (Tamez et al., 2019). However, when cucumber plant was exposed to CuO-NPs, the enzyme activity was not affected (Tamez et al., 2019). A similar effect was observed when Fe_3O_4-NPs were applied to maize plants, where the application of the NPs decreased the CAT activity (Jalali et al., 2017). Conversely, in two different studies where bare Cu-NPs were applied to tomato plants, the CAT activity increased drastically by over 150% compared to the control (Juárez-Maldonado et al., 2016; López-Vargas et al., 2018). Chitosan-encapsulated Cu-NPs also increased the enzyme activity, however not to the same extent as the bare Cu-NPs (Juárez-Maldonado et al., 2016). Moreover, the addition of Fe_3O_4 onto barley plants showed mixed results for the CAT activity. For instance, in the roots, the enzyme activity increased in the presence of 250 and 500 mg L^{-1} of Fe_3O_4-NPs while in the leaves, the CAT activity increased with 125 mg L^{-1} but decreased when the plant was in contact with 500 and 1000 mg L^{-1} of Fe_3O_4 (Tombuloglu et al., 2019). The catalase activity directly correlates to the amount of hydrogen peroxide in the tissue. Therefore an increase in catalase is indicative of oxidative stress. Previous studies have stated that some NPs can be harmful to plants and biota by increasing the amount of reactive oxygen species such as Cu and Fe partaking in Fenton and/or Haber-Weis reactions (Anjum et al., 2015).

APX activity can be quantified by measuring the absorbance for the ascorbate oxidation rate at 266 nm. For quantification, the extract is mixed with ascorbate and H_2O_2, and the absorbance is measured over time. While for GPX, quantification of the enzyme activity can be conducted by measuring the absorbance at 412 nm (Flohé and Günzler, 1984; Xue et al., 2001). Juárez-Maldonado et al. (2016) found no effect on the APX and GPX content in tomato leaves and fruit exposed to chitosan-encapsulated Cu-NPs and bare Cu-NPs. However, López-Vargas et al. (2018) found a decrease in both APX and GPX in tomato fruits exposed to Cu-NPs. Similarly, Tamez et al. (2019) also found a decrease in APX activity in root tissues exposed to CuO-NPs and bulk CuO. Interestingly, the APX activity in the roots of scallions was unaffected by the exposure of CuO-NPs; however, the activity in the leaves was increased at concentrations greater than 100 mg kg^{-1} (Wang et al., 2020).

The use of UV-Vis has been crucial for understanding the effects of NPs on the enzyme activities of different plant tissues. Studies have shown that the exposure of different plants to metal and metal oxide NPs is dependent on a complex interplay between the plant, exposure route, concentration, and the physicochemical properties of the NPs. Therefore, the effects can not be generalized, and further investigations are required.

6.5 Evaluation of infested-plant protein changes using spectroscopic approaches

For evaluating the effect of ENM in agriculture (e.g., fungal and disease suppression), utilizing spectroscopy-based techniques can be an extremely versatile tool. Monitoring changes in protein content and phytohormones (absorbance), reactive oxygen species (fluorescence), disease progression (fluorescence), and changes in cellular content of fungal and bacterial cells (Raman and IR) can be descriptive when investigating the effect ENMs on infected plants. Examples of these spectroscopic techniques and their impacts in investigating disease progression in plants will be discussed in detail, followed by new novel approaches for utilizing spectroscopy in this domain.

UV-Vis spectroscopy has been widely utilized as a tool for measuring protein and phytohormone content in infected plants that may be utilized for comparisons against control samples (infected plants without the addition of ENPs). Li et al. (2020b) found that foliar application of MoS_2-Cu-NPs in rice infected with *Xanthomonas oryzae* pv. *Oryzae* led to increased levels of superoxide dismutase, peroxidase, and catalase, indicating the activation of the enzymatic antioxidant system. It was also found that malondialdehyde was reduced compared to the control in samples treated with the MoS_2-Cu-NPs, indicating a reduction in oxidative stress. All antioxidant enzymes were quantified by absorbance spectroscopy (Li et al., 2020b). Absorbance can also be utilized in tandem with enzyme-linked immunosorbent assay (ELISA) to quantify plant phytohormones and protein coats. For tobacco plants infected with Turnip mosaic virus, it was found that foliar application of MWCNT, C_{60} fullerenes, and metal NP (Fe_2O_3 and TiO_2) all led to an increase in both brassinolide (BR) and Zeatin riboside (ZR) phytohormones. BR is a steroid hormone that regulates plant growth and abiotic stress, while ZR accelerates plant cell division. The same type of spectroscopy (adsorption) was used with TAS-ELISA to quantify the amount of viral protein accumulation in newly emerged tobacco leaves. Compared to the control, a reduction of approximately 15%–60% in the amount of Turnip Mosaic virus was present in the leaf tissue (Hao et al., 2018). Absorbance spectroscopy has also been utilized in determining the state of NP in various media. Dimkpa et al. (2013) utilized absorbance spectroscopy to evaluate the state of ZnO-NPs in mung bean broth, applied for the suppression of *Fusarium graminearum*. The characteristic absorbance peak of these particles (374–375 nm) was significantly reduced, signaling the agglomeration of ZnO particles directly affecting the particle speciation in this medium (Dimkpa et al., 2013). This technique has been utilized for analyzing ZnO-NPs in other various mediums (Baalousha et al., 2016; Fang et al., 2013).

Fluorescence spectrometry has proven to be a valuable tool for evaluating ROS production in plants when used in tandem with microscopy. In tomato plants, it was reported that the application of Cu-NPs ($0.5\ mg\ mL^{-1}$) to growth medium acted as a strong inhibitor (67.3%) for *F. oxysporum* f. sp. *lycopersici*. Researchers believe that Cu-NPs produced ROS that caused degradation to the membrane lipids and eventually led to the death of the fungal cells (Lopez-Lima et al., 2021). Evaluation of ROS production was accomplished by utilizing 2′, 7′-dichlorodihydrofluorescein that becomes oxidized in the presence of ROS creating a fluorescent probe (2′, 7′-dichlorofloures-cein). The product of this reaction is easily detectable under microscopes equipped with fluorescent detectors and appropriate excitation source (Chen et al., 2010; Lopez-Lima et al., 2021). A similar analytical approach using a different fluorescent probe was employed to observe the spread of the Turnip mosaic virus (TuMV) in tobacco plants. A genetically modified version of the virus, expressing a green fluorescent protein, was utilized to observe the spread of the TuMV virus in new leaves produced after inoculation with the virus. Tobacco plants treated with Fe_2O_3-NPs, TiO_2-NPs, and multiwalled carbon nanotubes (50 ppm) saw a significant decrease in viral proliferation in comparison to the control (Hao et al., 2018).

Vibrational spectroscopy has been explored to investigate changes in bacterial and fungal cells following interaction with ENM. He et al. (2011) utilized raman spectroscopy to investigate changes to the fungi *Botrytis cinerea* following interaction with ZnO-NPs. Characteristic bands for nucleic acids (743, 772, 938 and 1082 cm^{-1}), carbohydrates (848, 868, 895, 965 and 1048 cm^{-1}), proteins (1005, 1254 and 1668 cm^{-1}), and lipids (1416 & 1460 cm^{-1}) were monitored to better understand the interaction of ZnO-NPs with these fungal cells. There was a noticeable increase in nucleic acid and carbohydrates, with no change to proteins and lipids, when the fungus was treated with 12 $mmol\ L^{-1}$ of a ZnO-NPs suspension. This accumulation of proteins and lipids was thought to eventually deform the structure of hyphal cells inhibiting the growth of the fungus strain. The same approach was taken for investigating the interaction of ZnO-NPs with *Penicillium expansum*. Complete removal of all characteristic bands revealed high sensitivity of this fungus to ZnO-NPs. The authors believed this signaled complete fungal inhibition/destruction after interacting with the ZnO-NPs (He et al., 2011). In a related approach, researchers utilized infrared spectroscopy to investigate the antibacterial activity of MgO-NPs on bacterial cells (*Bacillus subtilis* and *Staphylococcus aereus*). FT-IR analysis of *B. subtilis* found a dramatic increase in peaks situated around 1078 cm^{-1}. This peak was representative of the C-O stretching, which the authors interpreted as a change to the proteins in the cell wall of the bacteria. In the presence of dissolved oxygen, the reactive species superoxide (O_2^-) can be easily generated by single-electron reduction, and will stay stable in a basic environment ($pH > 7$) (Sawai et al., 2000). These ions were predicted to attack the carbonyl group in the peptide linkage, leading to the degradation of these proteins in the bacterial cell wall. The same pathway for the destruction of the peptide linkage was also observed for *S. aereus* following interaction with MgO-NPs (Huang et al., 2005).

Monitoring the migration and accumulation of NMs throughout plant systems can give researchers a better understanding of the overall effect of ENMs and their role in combating infection. By tracing the accumulation of ENMs in various plant components, researchers can compare differences in genetic expression and nutrient content between areas of high and low accumulation. Khodakovskaya et al. (2011) developed a novel nanobubble amplified photothermal/photoacoustic spectroscopy technique to detect carbon nanotubes in tomato plants, down to single nanoparticle levels. Upon irradiation, internalized carbon nanotubes would absorb this energy and convert it to detectable heat and acoustic signals. It was found that nanobubbles generated around the overheated CNTs amplified these photoacoustic (PA) and photothermal (PT) signals. Utilizing the aforementioned detection method, in tandem with microscopy, it was found that CNTs accumulated in leaves and roots of the tomato plants leading researchers to further investigate the genetic expression in these areas of high accumulation. Compared to the control specimen there were elevated levels of LeAqp2 and TDR3 proteins in roots, and heat shock protein 90 within the leaves of the tomato plants exposed to CNTs (Khodakovskaya et al., 2011). LeApq2 is an aquaporin protein essential to germination and plant growth by controlling water permeability of cells during growth, development, and stress responses (Maurel, 1997). Plant aquaporins have also been reported to play a major role in immunity against plant infection, via gating and trafficking of H_2O_2 transport in *A. thaliana* (Zhang et al., 2019). Heat shock protein 90 has also been reported to play an essential role in various stress responses, as well as growth and development, in a large variety of plant species (Li et al., 2020a). Recently, single particle ICP-MS has become a powerful tool for the rapid detection and quantification of ENMs in plant tissues. When used with enzymatic digestions, this analysis can quickly count and measure the size of a large variety of metal NP (e.g., Ag, Ce, and Cu) in numerous plant species. The enzymatic digestion leaves the particles unaltered while dissolving the organic plant matter, thus, liberating particles for detection by mass spectroscopy (Bao et al., 2016; Dan et al., 2016). Single-particle ICP-MS relies on rapid analysis windows (as short as 50 μs), allowing for the detection of ion pulses arising from the ionization of single particles. The frequency of pulses in the analysis is related to the concentration of particles in solution, while the pulse intensity is related to particle size, or mass (Pace et al., 2011). By utilizing such a technique, investigators can quickly identify areas of high accumulation of ENMs, highlighting what portions of the plants should be isolated for further genetic and nutrient testing.

6.6 Microscopy studies to evaluate impacts in plants exposed to engineered nanomaterials

Many analytical methods have been developed to study the plant responses toward ENMs exposure; they are discussed in this chapter. However, the methods mentioned earlier mainly provide indirect measurements of nano-plant interaction at different levels. Lately, microscopic-based techniques, including spectral/electron imaging with qualitative analysis capabilities, have been widely used in evaluating cellular interaction between organisms and ENMs (Graham et al., 2020). In electron microscopy, the signals generated by electron-specimen interaction can be captured by different detectors, thus forming high-resolution images (Carroni and Saibil, 2016). In spectral microscopy, on the other hand, detectors acquire a single or broad spectrum of light characteristic of an object/objects, providing spectral information in addition to 2D spatial images (Armin Schneider and Hubertus Feussner, 2017). These techniques allow the direct analysis of ENMs in plant cells, facilitating understanding ENMs' biological effects and promoting safe designs for nanotechnology. This section focuses on recent developments in microscopic applications for the investigation of nanoplant interactions at the cellular level; and the section elaborates the major aims of microscopic-based techniques in this field, including the localization of ENMs in plant cells, identification of organelle-specific ENMs accumulations, and prediction of ENMs potential transportation mechanism within the plant.

6.6.1 Translocation and accumulation of engineered nanomaterials in plant cells

In nanoagriculture studies, one of the challenges is to localize ENMs within plants. It is because they are extremely small and there is a lack of well-developed methods. Current studies have adopted electron microscopy and its coupled energy-dispersive X-ray spectroscopy (EDS) to achieve qualitative and quantitative microanalyses. In multiple plant systems, the studies were able to localize ENMs such as CuO-NPs (Xiong et al., 2017), Pt-NPs (Kranjc et al., 2018), and Mn_2O_3-NPs (Ye et al., 2020) in different plant tissues as in their nanoforms. Due to the high resolution of electron micrographs, plant structures such as leaf stomata (Kranjc et al., 2018; Xiong et al., 2017) and seed interior skin (Ye et al., 2020) were revealed with the presence of accumulated ENMs. These studies suggested potential ways for ENMs to enter the plant system and initiate further positive and negative bioeffects. In order to acquire more cellular information, studies have also shown ENMs accumulations in different cellular compartments (Deng et al., 2017; Leonardo et al., 2016).

Under electron microscopy, TiO_2-NPs were found extensively on the epidermal surface of root of rice and tended to agglomerate near plasma membranes (Deng et al., 2017). Deng et al. (2017) further conducted elemental analysis by inductively coupled plasma mass spectroscopy and confirmed TiO_2-NPs accumulation. Similarly, monodispersed Ag-NPs were found mainly accumulated in mitochondria and chloroplast of microalgae, but appeared in almost all cellular compartments when intracellular Ag concentration increased (Leonardo et al., 2016). In this study, Leonardo et al. (2016) also used other techniques such as X-ray absorption spectroscopy and X-ray diffraction revealing that accumulated Ag existed as Ag^0 with a face-centered cubic structure. The high amount of Ag inside microalgae indicated a potential candidate for heavy metal contamination water remediation.

A recent study explored the impact of the CuO-NPs in hydroponically-grown *A. thaliana*, and showed that Cu (CuO-NPs at 100 mg L^{-1} and Cu^{2+} ions at 1.2 mg L^{-1}) was accumulated in higher amount in plants exposed to the nanoforms (Yuan et al., 2016). Moreover, different spectroscopic techniques such as High-Resolution Transmission Electron Microscopy (HRTEM) and the Fast Fourier transform (FFT) techniques gave evidence that the root cells accumulated CuO-NPs in their vacuoles, while the aerial parts translocated Cu mainly as Cu^{2+}. In another microscopic experiment using Scanning Electron Microscopy (SEM) and EDS, the acquisition of Ag-NPs by Arabidopsis plants was explored on a time-frame period (Geisler-Lee et al., 2014). The authors found that the Ag-NPs can be translocated from root to shoot as a function of time. Moreover, they also reported that the bioaccumulation of silver affected the germination rate of third generation plants.

6.6.2 Factor-dependent internalization of engineered nanomaterials into plants

In addition to the studies in ENMs translocation in plants, researchers used electron microscopy to demonstrate how ENMs charge and size can significantly affect their internalization and delivery efficiency to leaf cells and organelles (Hu et al., 2020; Kranjc et al., 2018). Kranjc et al. (2018) reported that in the experiment of arugula (*Eruca vesicaria* Mill.) and escarole (*Cichorium endivia* L.) leaves exposed to Pt-NPs, the leaf with lower surface free energy (SFE) are more likely to adhere and retain ENMs than those with high SFE, thus resulting in the uptake and translocation of Pt-NPs from leaf to root. The electron micrography provided evidence of a higher degree of Pt-NPs aggregation in arugula leaf due to its higher SFE than escarole. On the other hand, Hu et al. (2020) reported that the patterns of ENMs delivered to different cellular compartments are highly related to ENMs hydrodynamic size and zeta potential. In this study, confocal fluorescence microscopy videos indicated real-time movements of different ENMs, including carbon dots, cerium oxide, and silica ENMs, from the leaf surface into cellular compartments. Both Kranjc et al. and Hu et al.'s studies include relevant factors that should be increasingly taken into account and will lead to the prediction of nanoplant interaction.

6.6.3 Evaluation of the engineered nanomaterials transport pathways in plants

Microscopic techniques have recently been used to localize different NMs and their elements on plant tissues. One of the well-designed studies utilized a spilt-root system to cultivate hydroponic cucumber plants. In this study (He et al., 2017), half-plant root were exposed to 200 or 2000 mg L^{-1} of CeO_2-NPs suspensions while the other half remained untreated. He et al. (2017) used transmission electron microscope to successfully observe CeO_2-NPs in nontreated roots, thus confirming the transport of CeO_2-NPs throughout the plant through the vascular system. Moreover, X-ray absorption near-edge structure was adopted to characterize the speciation of Ce, indicating that Ce was transported as Ce^{4+} and Ce^{3+} within the plant. This study was first to report the root-to-shoot-to-root redistribution pattern of CeO_2-NPs in plants, which has significant implications for environmental safety and human health. In addition to electron microscopy to reveal the presence of ENMs inside plants, confocal microscopy has also been used alone in ENMs transport studies. For example, Karny et al. (2018) found that NP-liposomes could penetrate the leaf and distribute to other leaves and to the roots. Confocal microscopy provided the evidence of the moving NP-liposomes, which contains a fluorescence tracer that appears green on the image. The images showed the movement was in a bidirectional manner, which supports the implication of nanotechnology for micronutrient delivery, especially for agricultural crops to increase yield. Similarly, Zhao et al. (2012) investigated ZnO-NPs transport in sandy loam soil and uptake by corn plants. In this study, confocal microscopy confirmed the penetration of ZnO-NPs aggregates in the cortex and root epidermis via apoplastic pathway.

All the examples have taken advantage of tracking ENMs within plant tissues to evaluate the transport pathways.

Overall, microscopy techniques can achieve in situ visualization of ENMs in plants, thus providing direct measurements without potentially being interfered by experimental factors associated with standard assays. So far, microscopy techniques have revealed that ENMs can enter the plant, get transported in plant tissues, and even accumulate in different cellular compartments due to their distinct physiochemical properties. However, there are aspects that microscopy techniques certainly fall behind. For example, the sampling can be inherently biased for any microscopy techniques and the sample preparation can distort the sample tissues, thus causing inaccuracy. Furthermore, imaging of some ENMs, particularly the carbon-based ENMs, can be challenging due to the background interference of carbon in plants (Zhang et al., 2020). Therefore practices such as lab and multitechnique cross-validations are needed in future studies.

6.7 Risk assessment and toxicity of engineered nanomaterials

Althoug ENMs can reach the environment due to occasional spills, most of the release occurs during manufacture, transport, and disposal of ENM-containing products, or intentional use in agriculture (Giese et al., 2018). Government and civil organizations worldwide have leaned toward the regulation of the safe disposal of ENMs. To develop accurate assessment methodologies that evaluate the risk associated with ENMs, it is important to understand their life cycle, their interactions with abiotic and biotic factors, and their accumulation in trophic levels (Caballero-Guzman and Nowack, 2016; Navarro et al., 2008).

The main causes of ENP toxicity to living cells can be attributed to the following: (1) chemical composition of released ions, (2) minuscule size that allows their uptake, and (3) specific conformational structures (Bystrzejewska-Piotrowska et al., 2009). Evidence demonstrates that various metallic nanomaterials are toxic to animal cell cultures and animal models (Cox et al., 2016). Additionally, the toxicity of biotransformed products from ENMs is not well understood (Montes et al., 2017; Servin et al., 2017). Pradhan and Mailapalli (2017) have discussed some criteria to evaluate ENMs previous to commercialization. These plans include: (1) documenting physicochemical properties, (2) evaluating the biotransformation, (3) defining environmental fates, (4) describing and clarifying effects in humans and ecosystems, and (5) characterizing the shelf life and transformations (Pradhan and Mailapalli, 2017).

Probabilistic models have been created to evaluate the concentration of ENMs in environmental samples. Gottschalk et al. (2013) summarized the literature on this topic. A modeling study in the San Francisco Bay area suggests that, within a few years, the concentration of ENMs will pose a threat to the organisms inhabiting the Bay (Garner et al., 2017). Thus more research focused on understanding the relationships between ENMs and biotic/abiotic systems, especially those related to the phytoremediation potential, are needed.

6.8 Conclusion

ENM differentially impact the coordinated biochemical responses of plants. Depending on the type and mode of ENM application, several signaling cascades are activated, resulting in the activation of specific genes. Their expression can promote plant fitness or coordinate defense responses. The techniques discussed in this chapter, such as those looking into the gene activation at transcript and protein levels and spectroscopic and microscopic studies to monitor plant protein changes, bring insight into the cellular alteration of plants exposed to ENMs. Combining each technique can assist researchers in drawing more comprehensive conclusions on the impacts of nanotechnology. Meanwhile, the sophistication of several methods and the development of high throughput approaches contribute to building mechanistic models, including gene network regulation pathways, which help understand the overall molecular responses of plants grown in ENM-laden environments.

Acknowledgments

The authors acknowledge the USDA grant 2016–67021–24985 and the NSF Grant CHE-0840525. Partial funding was also provided by the NSF ERC-1449500 grant on Nanotechnology-Enabled Water Treatment. This work was also supported by the National Science Foundation and the Environmental Protection Agency under Cooperative Agreement Number DBI-1266377. This work has not been subjected to EPA review, and no official endorsement should be inferred. The Grant 2G12MD007592 from the National Institutes on Minority Health and Health Disparities (NIMHD), a component of the National Institutes of Health (NIH), also supported this work. J. L. Gardea-Torresdey (J-GT) acknowledges the Dudley family for the Endowed Research Professorship. J-GT acknowledges the LERR and STARs Retention Award (2018) of the University of Texas System.

References

Adisa, I.O., et al., 2018. Role of cerium compounds in fusarium wilt suppression and growth enhancement in tomato (*Solanum lycopersicum*). J. Agric. Food Chem. 66 (24), 5959–5970. Available from: https://doi.org/10.1021/acs.jafc.8b01345. American Chemical Society.

Adisa, I.O., et al., 2020. Nutritional status of tomato (Solanum lycopersicum) fruit grown in fusarium-infested soil: impact of cerium oxide nanoparticles. J. Agric. Food Chem. 68 (7), 1986–1997. Available from: https://doi.org/10.1021/acs.jafc.9b06840.

Aebi, H., 1974. lrsquo; Methods of Enzymatic Analysis', pp. 673–684.

Anjum, N.A., et al., 2015. Nanoscale copper in the soil-plant system - toxicity and underlying potential mechanisms. Environ. Res. 138, 306–325. Available from: https://doi.org/10.1016/j.envres.2015.02.019. Elsevier.

Arabidopsis Genome Initiative, 2000. Analysis of the genome sequence of the flowering plant Arabidopsis thaliana. Nature 408 (December), 796–815. Available from: https://doi.org/10.1038/35048692.

Armin Schneider and Hubertus Feussner, 2017. Diagnostic procedures. Biomedical Engineering in Gastrointestinal Surgery. Academic Press, pp. 87–220. Available from: https://doi.org/10.1016/B978-0-12-803230-5.00005-1.

Baalousha, M., et al., 2016. Outdoor urban nanomaterials: the emergence of a new, integrated, and critical field of study. Sci. Total Environ. 557–558, 740–753. Available from: https://doi.org/10.1016/j.scitotenv.2016.03.132. Elsevier B.V.

Bao, D., Oh, Z.G., Chen, Z., 2016. Characterization of silver nanoparticles internalized by arabidopsis plants using single particle ICP-MS analysis. Front. Plant Sci. 7 (FEB2016), 1–8. Available from: https://doi.org/10.3389/fpls.2016.00032.

Begum, P., Fugetsu, B., 2013. Induction of cell death by graphene in Arabidopsis thaliana (Columbia ecotype) T87 cell suspensions. J. Hazard. Mater. 260, 1032–1041. Available from: https://doi.org/10.1016/j.jhazmat.2013.06.063. Elsevier B.V.

Bradford, M.M., 1976. A rapid and sensitive method for the quantitation of microgram quantities of protein utilizing the principle of protein-dye binding. Anal. Biochem. 72 (1–2), 248–254. Available from: https://doi.org/10.1016/0003-2697(76)90527-3.

Buchman, J.T., et al., 2019. Chitosan-coated mesoporous silica nanoparticle treatment of Citrullus lanatus (watermelon): enhanced fungal disease suppression and modulated expression of stress-related genes. ACS Sustain. Chem. Eng. 7 (24), 19649–19659. Available from: https://doi.org/10.1021/acssuschemeng.9b04800.

Bystrzejewska-Piotrowska, G., Golimowski, J., Urban, P.L., 2009. Nanoparticles: their potential toxicity, waste and environmental management. Waste Manag. 29 (9), 2587–2595. Available from: https://doi.org/10.1016/j.wasman.2009.04.001. Elsevier Ltd.

Caballero-Guzman, A., Nowack, B., 2016. A critical review of engineered nanomaterial release data: are current data useful for material flow modeling? Environ. Pollut. 213, 502–517. Available from: https://doi.org/10.1016/j.envpol.2016.02.028. Elsevier Ltd.

Cakmak, I., Horst, W.J., 1991. Effect of aluminium on lipid peroxidation, superoxide dismutase, catalase, and peroxidase activities in root tips of soybean (Glycine max). Physiol. Plant 83 (3), 463–468. Available from: https://doi.org/10.1111/j.1399-3054.1991.tb00121.x.

Carroni, M., Saibil, H.R., 2016. Cryo electron microscopy to determine the structure of macromolecular complexes. Methods 95, 78–85. Available from: https://doi.org/10.1016/j.ymeth.2015.11.023. Elsevier Inc.

Carvalho, R., et al., 2019. Bactericidal activity of copper-zinc hybrid nanoparticles on copper-tolerant Xanthomonas perforans. Sci. Rep. 9 (1), 1–9. Available from: https://doi.org/10.1038/s41598-019-56419-6. Springer US.

Chen, X., et al., 2010. 2′,7′-Dichlorodihydrofluorescein as a fluorescent probe for reactive oxygen species measurement: forty years of application and controversy. Free Radic. Res. 44 (6), 587–604. Available from: https://doi.org/10.3109/10715761003709802.

Chen, J., et al., 2018. Phenotypic, transcriptional, physiological and metabolic responses to carbon nanodots exposure in Arabidopsis thaliana (L.). Environ. Sci. Nano . Available from: https://doi.org/10.1039/C8EN00674A.

Cota-Ruiz, K., et al., 2018. Current findings on terrestrial plants—engineered nanomaterial interactions: are plants capable of phytoremediating nanomaterials from soil? Curr. Opin.

Environ. Sci. Health 6, 9–15. Available from: https://doi.org/10.1016/j.coesh.2018.06.005. Elsevier Ltd.

Cox, A., et al., 2016. Silver and titanium dioxide nanoparticle toxicity in plants: a review of current research. Plant Physiol. Biochem. 107, 147–163. Available from: https://doi.org/10.1016/j.plaphy.2016.05.022. Elsevier Masson SAS.

Dan, Y., et al., 2016. Single particle ICP-MS method development for the determination of plant uptake and accumulation of CeO_2 nanoparticles. Anal. Bioanal. Chem. 408 (19), 5157–5167. Available from: https://doi.org/10.1007/s00216-016-9565-1.

Deng, Y., et al., 2017. Multiple method analysis of TiO_2 nanoparticle uptake in rice (Oryza sativa L.) plants. Environ. Sci. Technol. 51 (18), 10615–10623. Available from: https://doi.org/10.1021/acs.est.7b01364.

Dhindsa, R.S., Plumb-dhindsa, P., Thorpe, T.A., 1981. Leaf senescence: correlated with increased levels of membrane permeability and lipid peroxidation, and decreased levels of superoxide dismutase and catalase. J. Exp. Bot. 32 (1), 93–101. Available from: https://doi.org/10.1093/jxb/32.1.93.

Dimkpa, C.O., et al., 2012. CuO and ZnO nanoparticles: phytotoxicity, metal speciation, and induction of oxidative stress in sand-grown wheat. J. Nanopart. Res. 14 (9). Available from: https://doi.org/10.1007/s11051-012-1125-9.

Dimkpa, C.O., et al., 2013. Antifungal activity of ZnO nanoparticles and their interactive effect with a biocontrol bacterium on growth antagonism of the plant pathogen *Fusarium graminearum*. BioMetals 26 (6), 913–924. Available from: https://doi.org/10.1007/s10534-013-9667-6.

Elmer, W.H., White, J.C., 2016. The use of metallic oxide nanoparticles to enhance growth of tomatoes and eggplants in disease infested soil or soilless medium. Environ. Sci. Nano 3 (5). Available from: https://doi.org/10.1039/C6EN00146G.

Elmer, W.H., Ma, C., White, J.C., 2018. Nanoparticles for plant disease management. Curr. Opin. Environ. Sci. Health. Available from: https://doi.org/10.1016/J.COESH.2018.08.002. Elsevier.

Elmer, W.H., et al., 2021. Influence of single and combined mixtures of metal oxide nanoparticles on eggplant growth, yield, and Verticillium wilt severity. Plant. Dis. Am Phytopath Society, p. PDIS-07.

Fan, X., et al., 2018. Multiwall carbon nanotubes modulate paraquat toxicity in Arabidopsis thaliana. Environ. Pollut. 233, 633–641. Available from: https://doi.org/10.1016/j.envpol.2017.10.116. Elsevier Ltd.

Fang, T., et al., 2013. Does doping with aluminum alter the effects of ZnO nanoparticles on the metabolism of soil pseudomonads? Microbiol. Res. 168 (2), 91–98. Available from: https://doi.org/10.1016/j.micres.2012.09.001. Elsevier GmbH.

Flohé, L., Günzler, W.A., 1984. [12] Assays of glutathione peroxidase. Health (San. Francisco) 114–120. Available from: https://doi.org/10.1016/S0076-6879(84)05015-1.

Gallego, S.M., Benavídes, M.P., Tomaro, M.L., 1996. Effect of heavy metal ion excess on sunflower leaves: evidence for involvement of oxidative stress. Plant. Sci. 121 (2), 151–159. Available from: https://doi.org/10.1016/S0168-9452(96)04528-1.

Gardea-Torresdey, J.L., Rico, C.M., White, J.C., 2014. Trophic transfer, transformation, and impact of engineered nanomaterials in terrestrial environments. Environ. Sci. Technol. 48 (5), 2526–2540. Available from: https://doi.org/10.1021/es4050665.

Garner, K.L., Suh, S., Keller, A.A., 2017. Assessing the risk of engineered nanomaterials in the environment: development and application of the nanofate model. Environ. Sci. Technol. 51 (10), 5541–5551. Available from: https://doi.org/10.1021/acs.est.6b05279.

Geisler-Lee, J., et al., 2014. Reproductive toxicity and life history study of silver nanoparticle effect, uptake and transport in Arabidopsis thaliana. Nanomaterials 4 (2), 301–318. Available from: https://doi.org/10.3390/nano4020301.

Giese, B., et al., 2018. Risks, release and concentrations of engineered nanomaterial in the environment. Sci. Rep 8 (1), 1565. Available from: https://doi.org/10.1038/s41598-018-19275-4. Springer US.

Gottschalk, F., Sun, T., Nowack, B., 2013. Environmental concentrations of engineered nanomaterials: review of modeling and analytical studies. Environ. Pollut 181, 287–300. Available from: https://doi.org/10.1016/j.envpol.2013.06.003. Elsevier Ltd.

Graham, U.M., et al., 2020. Tissue specific fate of nanomaterials by advanced analytical imaging techniques—a review. Chem. Res. Toxicol. 33 (5), 1145–1162. Available from: https://doi.org/10.1021/acs.chemrestox.0c00072.

Hao, Y., et al., 2018. Engineered nanomaterials suppress Turnip mosaic virus infection in tobacco (Nicotiana benthamiana). Environ. Sci. Nano 5 (7), 1685–1693. Available from: https://doi.org/10.1039/c8en00014j. Royal Society of Chemistry.

Hao, Y., et al., 2019. Engineered nanomaterials inhibit Podosphaera pannosa infection on rose leaves by regulating phytohormones. Environ. Res. 170 (December 2018), 1–6. Available from: https://doi.org/10.1016/j.envres.2018.12.008. Elsevier Inc.

Hayashi, M., Nishimura, M., 2006. Arabidopsis thaliana-A model organism to study plant peroxisomes. Biochim. Biophys. Acta - Mol. Cell Res. 1763 (12), 1382–1391. Available from: https://doi.org/10.1016/j.bbamcr.2006.08.014.

He, L., et al., 2011. Antifungal activity of zinc oxide nanoparticles against Botrytis cinerea and Penicillium expansum. Microbiol. Res. 166 (3), 207–215. Available from: https://doi.org/10.1016/j.micres.2010.03.003. Elsevier GmbH.

He, X., et al., 2017. Xylem and phloem based transport of CeO_2 nanoparticles in hydroponic cucumber plants. Environ. Sci. Technol. 2–8. Available from: https://doi.org/10.1021/acs.est.6b05998.

Hu, P., et al., 2020. Nanoparticle charge and size control foliar delivery efficiency to plant cells and organelles. ACS Nano 14 (7), 7970–7986. Available from: https://doi.org/10.1021/acsnano.9b09178.

Huang, L., et al., 2005. Controllable preparation of Nano-MgO and investigation of its bactericidal properties. J. Inorg. Biochem. 99 (5), 986–993. Available from: https://doi.org/10.1016/j.jinorgbio.2004.12.022.

Jalali, M., et al., 2017. Physiological effects of repeated foliar application of magnetite nanoparticles on maize plants. J. Agron. Crop. Sci. 203 (6), 593–602. Available from: https://doi.org/10.1111/jac.12208.

Juárez-Maldonado, A., et al., 2016. Cu Nanoparticles absorbed on chitosan hydrogels positively alter morphological, production, and quality characteristics of tomato. J. Appl. Bot. Food Qual. 89, 183–189. Available from: https://doi.org/10.5073/JABFQ.2016.089.023.

Kachroo, A., Kachroo, P., 2007. Salicylic acid-, jasmonic acid- and ethylene-mediated regulation of plant defense signaling. Genet. Eng. 28, 55–83. Available from: https://doi.org/10.1007/978-0-387-34504-8_4.

Karny, A., et al., 2018. Therapeutic nanoparticles penetrate leaves and deliver nutrients to agricultural crops. Sci. Rep. 8 (1), 1–10. Available from: https://doi.org/10.1038/s41598-018-25197-y.

Khodakovskaya, M.V., et al., 2011. Complex genetic, photothermal, and photoacoustic analysis of nanoparticle-plant interactions. Proc. Natl Acad. Sci. U S A. 108 (3), 1028–1033. Available from: https://doi.org/10.1073/pnas.1008856108.

Kolackova, M., et al., 2019. Antioxidant, gene expression and metabolomics fingerprint analysis of *Arabidopsis thaliana* treated by foliar spraying of ZnSe quantum dots and their growth inhibition of *Agrobacterium tumefaciens*. J. Hazard. Mater. 365 (March 2018), 932–941. Available from: https://doi.org/10.1016/j.jhazmat.2018.11.065. Elsevier.

Koornneef, M., Meinke, D., 2010. The development of Arabidopsis as a model plant. Plant J. 61 (6), 909–921. Available from: https://doi.org/10.1111/j.1365-313X.2009.04086.x.

Kranjc, E., et al., 2018. Foliar surface free energy affects platinum nanoparticle adhesion, uptake, and translocation from leaves to roots in arugula and escarole. Environ. Sci. Nano 5 (2), 520–532. Available from: https://doi.org/10.1039/c7en00887b.

Kumar, A., et al., 2018. Carbon nanoparticles influence photomorphogenesis and flowering time in Arabidopsis thaliana. Plant. Cell Rep. 37 (6), 901–912. Available from: https://doi.org/10.1007/s00299-018-2277-6. Springer Berlin Heidelberg.

Leonardo, T., et al., 2016. Silver accumulation in the green microalga Coccomyxa actinabiotis: toxicity, in situ speciation, and localization investigated using synchrotron XAS, XRD, and TEM. Environ. Sci. Technol. 50 (1), 359–367. Available from: https://doi.org/10.1021/acs.est.5b03306.

Li, W., et al., 2020a. Evolutionary history of the heat shock protein 90 (Hsp90) family of 43 plants and characterization of Hsp90s in Solanum tuberosum. Mol. Biol. Rep. 47 (9), 6679–6691. Available from: https://doi.org/10.1007/s11033-020-05722-x. Springer Netherlands.

Li, Y., et al., 2020b. Multifunctional molybdenum disulfide-copper nanocomposite that enhances the antibacterial activity, promotes rice growth and induces rice resistance. J. Hazard. Mater. 394 (December 2019), 122551. Available from: https://doi.org/10.1016/j.jhazmat.2020.122551. Elsevier.

Liao, Y.Y., Strayer-Scherer, A.L., et al., 2019a. Nano-magnesium oxide: a novel bactericide against copper-tolerant xanthomonas perforans causing tomato bacterial spot. Phytopathology 109 (1), 52–62. Available from: https://doi.org/10.1094/PHYTO-05-18-0152-R.

Liao, Y.Y., Strayer-Scherer, A., et al., 2019b. Particle-size dependent bactericidal activity of magnesium oxide against Xanthomonas perforans and bacterial spot of tomato. Sci. Rep. 9 (1), 1–10. Available from: https://doi.org/10.1038/s41598-019-54717-7.

Lin, C., et al., 2009. Studies on toxicity of multi-walled carbon nanotubes on Arabidopsis T87 suspension cells. J. Hazard. Mater. 170 (2–3), 578–583. Available from: https://doi.org/10.1016/j.jhazmat.2009.05.025.

Lopez-Lima, D., et al., 2021. The bifunctional role of copper nanoparticles in tomato: effective treatment for Fusarium wilt and plant growth promoter. Sci. Hortic. 277 (October 2020). Available from: https://doi.org/10.1016/j.scienta.2020.109810.

López-Vargas, E.R., et al., 2018. Foliar application of copper nanoparticles increases the fruit quality and the content of bioactive compounds in tomatoes. Appl. Sci. (Switz.) 8 (7). Available from: https://doi.org/10.3390/app8071020.

Ma, C., et al., 2015. Metal-based nanotoxicity and detoxification pathways in higher plants. Environ. Sci. Technol. 49 (12), 7109–7122. Available from: https://doi.org/10.1021/acs.est.5b00685.

Ma, C., et al., 2019. Time-dependent transcriptional response of tomato (Solanum lycopersicum L.) to Cu nanoparticle exposure upon infection with Fusarium oxysporum f. sp. lycopersici. ACS Sustain. Chem. Eng. 7 (11), 10064–10074. Available from: https://doi.org/10.1021/acssuschemeng.9b01433.

Ma, C., et al., 2020. Advanced material modulation of nutritional and phytohormone status alleviates damage from soybean sudden death syndrome. Nature Nanotechnol. 15 (12), 1033–1042. Available from: https://doi.org/10.1038/s41565-020-00776-1. Springer US.

Maurel, C., 1997. Aquaporins and water permeability of plant membranes. Annu. Rev. Plant. Biol. 48, 399–429. Available from: https://doi.org/10.1146/annurev.arplant.48.1.399.

Montes, A., et al., 2017. Uptake and transformations of engineered nanomaterials: critical responses observed in terrestrial plants and the model plant *Arabidopsis thaliana*. Sci. Total Environ. 607–608, 1497–1516. Available from: https://doi.org/10.1016/j.scitotenv.2017.06.190. Elsevier B.V.

Navarro, D.A., Bisson, M.A., Aga, D.S., 2012. Investigating uptake of water-dispersible CdSe/ZnS quantum dot nanoparticles by Arabidopsis thaliana plants. J. Hazard. Mater. 211–212, 427–435. Available from: https://doi.org/10.1016/j.jhazmat.2011.12.012. Elsevier B.V.

Navarro, E., et al., 2008. Environmental behavior and ecotoxicity of engineered nanoparticles to algae, plants, and fungi. Ecotoxicology 17 (5), 372–386. Available from: https://doi.org/10.1007/s10646-008-0214-0.

Ochoa, L., et al., 2017. Modulation of CuO nanoparticles toxicity to green pea (Pisum sativum Fabaceae) by the phytohormone indole-3-acetic acid. Sci. Total Environ. 598, 513–524. Available from: https://doi.org/10.1016/j.scitotenv.2017.04.063. Elsevier B.V.

Pace, H.E., et al., 2011. Determining transport efficiency for the purpose of counting and sizing nanoparticles via single particle inductively coupled plasma mass spectrometry. Anal. Chem. 83 (24), 9361–9369. Available from: https://doi.org/10.1021/ac201952t.

Peralta-Videa, J.R., et al., 2011. Nanomaterials and the environment: a review for the biennium 2008-2010. J. Hazard. Mater. 186 (1), 1–15. Available from: https://doi.org/10.1016/j.jhazmat.2010.11.020. Elsevier B.V.

Piccinno, F., et al., 2012. Industrial production quantities and uses of ten engineered nanomaterials in Europe and the world. J. Nanopart. Res. 14 (9). Available from: https://doi.org/10.1007/s11051-012-1109-9.

Pradhan, S., Mailapalli, D.R., 2017. Interaction of engineered nanoparticles with the agri-environment. J. Agric. Food Chem. 65 (38), 8279–8294. Available from: https://doi.org/10.1021/acs.jafc.7b02528.

Sawai, J., et al., 2000. Antibacterial characteristics of magnesium oxide powder. World J. Microbiol. Biotechnol. 16 (2), 187–194. Available from: https://doi.org/10.1023/A:1008916209784.

Scandalios, J.G., 2005. Oxidative stress: molecular perception and transduction of signals triggering antioxidant gene defenses. Braz. J. Med. Biol. Res. 38 (7), 995–1014. Available from: https://doi.org/10.1590/S0100-879X2005000700003.

Servin, A., et al., 2015. A review of the use of engineered nanomaterials to suppress plant disease and enhance crop yield. J. Nanopart. Res. 17 (2), 1–21. Available from: https://doi.org/10.1007/s11051-015-2907-7.

Servin, A.D., et al., 2017. Weathering in soil increases nanoparticle CuO bioaccumulation within a terrestrial food chain. Nanotoxicology 11 (1), 98–111. Available from: https://doi.org/10.1080/17435390.2016.1277274.

Shah, J., 2003. The salicylic acid loop in plant defense. Curr. Opin. Plant Biol. 6 (4), 365–371. Available from: https://doi.org/10.1016/S1369-5266(03)00058-X.

Shang, H., et al., 2020. Copper sulfide nanoparticles suppressGibberella fujikuroiinfection in rice (Oryza sativa L.) by multiple mechanisms: contact-mortality, nutritional modulation and phytohormone regulation. Environ. Sci. Nano 7 (9), 2632–2643. Available from: https://doi.org/10.1039/d0en00535e. Royal Society of Chemistry.

Shen, C.X., et al., 2010. Induction of programmed cell death in Arabidopsis and rice by single-wall carbon nanotubes. Am. J. Bot. 97 (10), 1602–1609. Available from: https://doi.org/10.3732/ajb.1000073.

Stintzi, A., et al., 2001. Plant defense in the absence of jasmonic acid. PNAS 98 (22), 12837–12842. Available from: https://doi.org/10.1016/s1360-1385(01)02218-x.

Strayer-Scherer, A., et al., 2018. Advanced copper composites against copper-tolerant xanthomonas perforans and tomato bacterial spot. Phytopathology 108 (2), 196–205. Available from: https://doi.org/10.1094/PHYTO-06-17-0221-R.

Tamez, C., et al., 2019. Uptake, transport, and effects of nano-copper exposure in zucchini (Cucurbita pepo). Sci. Total Environ. 665, 100–106. Available from: https://doi.org/10.1016/J.SCITOTENV.2019.02.029. Elsevier.

Tombuloglu, H., et al., 2019. Uptake and translocation of magnetite (Fe_3O_4) nanoparticles and its impact on photosynthetic genes in barley (Hordeum vulgare L.). Chemosphere 226, 110–122. Available from: https://doi.org/10.1016/j.chemosphere.2019.03.075. Elsevier Ltd.

Valdes, C., et al., 2020. Ecotoxicology and Environmental Safety Antioxidant and defense genetic expressions in corn at early-developmental stage are differentially modulated by copper form exposure (nano, bulk, ionic): nutrient and physiological effects. Ecotoxicol. Environ. Saf. 206 (September), 111197. Available from: https://doi.org/10.1016/j.ecoenv.2020.111197. Elsevier Inc.

Wang, Y., et al., 2020. Improvement of nutrient elements and allicin content in green onion (*Allium fistulosum*) plants exposed to CuO nanoparticles. Sci. Total Environ. 725, 138387. Available from: https://doi.org/10.1016/j.scitotenv.2020.138387. Elsevier B.V.

White, J.C., Gardea-Torresdey, J., 2018. Achieving food security through the very small. Nat. Nanotechnol. 13 (8), 627–629. Available from: https://doi.org/10.1038/s41565-018-0223-y.

Xiong, T., et al., 2017. Copper oxide nanoparticle foliar uptake, phytotoxicity, and consequences for sustainable urban agriculture. Environ. Sci. Technol. 51 (9), 5242–5251. Available from: https://doi.org/10.1021/acs.est.6b05546.

Xu, T., et al., 2020. Enhancing agrichemical delivery and seedling development with biodegradable, tunable, biopolymer-based nanofiber seed coatings. ACS Sustain. Chem. Eng. 8 (25), 9537–9548. Available from: https://doi.org/10.1021/acssuschemeng.0c02696.

Xue, T., Hartikainen, H., Piironen, V., 2001. Antioxidative and growth-promoting effect of selenium on senescing lettuce. Plant Soil 237 (1), 55–61. Available from: https://doi.org/10.1023/A:1013369804867.

Yang, A., et al., 2018. Genotoxicity of zinc oxide nanoparticles in plants demonstrated using transgenic Arabidopsis thaliana. Bull. Environ. Contam. Toxicol. 0 (0), 0. Available from: https://doi.org/10.1007/s00128-018-2420-7. Springer US.

Ye, Y., et al., 2020. Manganese nanoparticles control salinity-modulated molecular responses in Capsicum annuum L. through priming: a sustainable approach for agriculture. ACS Sustain. Chem. Eng. 8 (3), 1427–1436. Available from: https://doi.org/10.1021/acssuschemeng.9b05615.

Yuan, J., et al., 2016. Internalization and phytotoxic effects of CuO nanoparticles in *Arabidopsis thaliana* as revealed by fatty acid profiles. Environ. Sci. Technol. 50 (19), 10437–10447. Available from: https://doi.org/10.1021/acs.est.6b02613.

Zhang, L., Chen, L., Dong, H., 2019. Plant aquaporins in infection by and immunity against pathogens—a critical review. Front. Plant Sci. 10 (May), 1–16. Available from: https://doi.org/10.3389/fpls.2019.00632.

Zhang, T., et al., 2020. Metallic fingerprints of carbon: label-free tracking and imaging of graphene in plants. Anal. Chem. 92 (2), 1948–1955. Available from: https://doi.org/10.1021/acs.analchem.9b04262.

Zhao, L., et al., 2012. Transport of Zn in a sandy loam soil treated with ZnO NPs and uptake by corn plants: electron microprobe and confocal microscopy studies. Chem. Eng. J. 184, 1–8. Available from: https://doi.org/10.1016/j.cej.2012.01.041. Elsevier B.V.

Zhao, S., et al., 2015. Toxicity and translocation of graphene oxide in Arabidopsis thaliana. Environ. Toxicol. Pharmacol. 39 (1), 145–156. Available from: https://doi.org/10.1016/j.etap.2014.11.014. Elsevier B.V.

Zhao, J., et al., 2020. Engineered nanomaterials in the environment: are they safe? Crit. Rev. Environ. Sci. Technol. 0 (0), 1–36. Available from: https://doi.org/10.1080/10643389.2020.1764279. Taylor & Francis.

Chapter 7

Multigenerational exposure of plants to engineered nanoparticles (ENPs)

Xingmao Ma and Xiaoxuan Wang
Department of Civil and Environmental Engineering, Texas A&M University, College Station, TX, United States

Chapter Outline

7.1 Introduction

The unique optical, electrical, and chemical properties of engineered nanoparticles (ENPs), derived primarily from their very large specific surface area and the quantum nature of the energy state at nanometer scale, lend numerous potentials for the applications of ENPs in consumer, industrial, agricultural, and biomedical products. The most frequently used ENPs are carbonaceous ENPs (primarily carbon nanotubes and fullerenes) and metallic ENPs (including both elemental ENPs and metal oxide ENPs). While most carbonaceous ENPs are incorporated into durable products and are expected to be bound in the matrix throughout the life cycle of these products, metallic ENPs are mostly used in dispersive applications and intentional and incidental releases are often expected. Consequently, rapid build-up of these metallic ENPs in the environment is highly anticipated. Many previous studies have reported the detection of metallic ENPs in wastewater streams (Limbach et al., 2008).

Plant Exposure to Engineered Nanoparticles. DOI: https://doi.org/10.1016/B978-0-323-85032-2.00002-6

Recently, metallic ENPs are increasingly detected in natural water bodies (Espinasse et al., 2018; Fréchette-Viens et al., 2019). These detections underscore the prevalence of ENPs in the environment as a result of the exponential growth of nanotechnology.

There had been some doubts with regard to the urgency or necessity to investigate the potential impact of ENPs on plants due to the perceived low concentrations of ENPs in the environment. However, it should be noted that the majority of ENPs in wastewater are typically concentrated in biosolids, and many of which are then land-applied to agricultural soils in the US and Europe. A previous study showed that over 90% of cerium oxide nanoparticles (CeO_2-NPs) in wastewater was retained in biosolids and half of which end up in agricultural soils (Limbach et al., 2008). According to the USEPA (Environmental Protection Agency), about 4.8 million tons of dry biosolids were generated in the United States in 2019 alone and about 2.5 million tons of them were directly land applied. Therefore, concentrations of ENPs in agricultural soils where plants grow might be much higher than what the mass flow models predicted. More importantly, nanotechnology is increasingly considered as a promising technology to significantly enhance the agricultural productivity without compromising environmental quality (Hofmann et al., 2020). Direct applications of ENP-incorporated agrichemicals markedly increase the chances of ENP accumulation in agricultural soils. Furthermore, the introduction of ENPs into agricultural soils will likely recur for many years, regardless of whether ENPs enter into agricultural soils through biosolid land applications, or through direct agrichemical application. Therefore some plants are expected to grow in soils containing ENPs for many generations. While short-term phytotoxic effect of ENPs on plants at high concentrations has been extensively investigated (Ma et al., 2010), studies on the chronic impact of ENPs on plants over several generations at environmentally and agriculturally relevant concentrations are relatively rare. This chapter will summarize the current understanding of the multigenerational impact of several common ENPs, identify critical knowledge gaps and highlight urgent research needs for a better understanding of the long-term consequences of plant exposure to ENPs over several generations. This chapter will be organized based on the nanoparticle types as detailed below, focusing primarily on metallic oxide nanoparticles.

7.2 Cerium oxide nanoparticles (CeO_2-NPs)

As the most abundant rare earth element, cerium is unique in the lanthanide group because it is stable in both trivalent (Ce^{3+}) and tetravalent (Ce^{4+}) oxidation states while all other elements in that group are exclusively stable in the trivalent state (Collin et al., 2014). In the lattice structure of pure CeO_2, each cerium cation is coordinated with eight oxygen anions at the corner of a cube while each oxygen anion is tetrahedrally-coordinated with four cerium cations. Depending on the temperature and pressure, oxygen anions (O^{2-}) are removed from the lattice structure to form oxygen vacancies (Gunawan et al., 2019). The electrostatic balance caused by the loss of O^{2-} is maintained by the reduction of two Ce^{4+} to Ce^{3+}. The concentration of oxygen vacancy and the percentage of Ce^{3+} on CeO_2-NPs surface increase with the decrease of particle size. A previous study showed that the percentage of Ce^{3+} on the surface of bare CeO_2-NPs increased from 9.7% to 22.9% when the nanoparticle size was reduced from 64 to 9 nm (Schwabe et al., 2014). An increase in Ce^{3+} strains the lattice and results in lattice expansion due to the relatively larger size of Ce^{3+} compared with Ce^{4+} (1.14 Å vs 0.97 Å) and therefore enhances the reactivity of CeO_2-NPs. The particle size dependent variation of Ce^{3+} partially explains the size dependent toxicity of CeO_2-NPs because our previous studies showed that Ce^{3+} is more toxic to plants than equivalent concentrations of CeO_2-NPs (Zhang et al., 2015), and the bulk CeO_2 particles exhibited lower phytotoxicity than CeO_2-NPs (Ma et al., 2015). However, some previous studies also reported the protective role of CeO_2-NPs against various abiotic stresses on plants. For example, treatment of plants growing under salt stress with CeO_2-NPs resulted in high biomass and greater net photosynthetic efficiency (Rossi et al., 2016). A subsequent study from the same researchers demonstrated that the alleviation of salt stress by CeO_2-NPs was due to the modification of the root anatomical structure of Brassica by CeO_2-NPs (Rossi et al., 2017a). The protective role of CeO_2-NPs may also derive from its unique redox chemistry because of the copresence of two different valences on the nanoparticle surface. At higher Ce^{3+}/Ce^{4+} ratio, CeO_2-NPs display the superoxide dismutase (SOD) mimetic property which converts superoxide ($O_2^{\cdot-}$) into hydrogen peroxide (H_2O_2) (Korsvik et al., 2007), likely through the following pathway:

$$O_2^{\cdot-} + Ce^{4+} \rightarrow O_2 + Ce^{3+}$$

$$O_2^{\cdot-} + Ce^{3+} + 2H^+ \rightarrow H_2O_2 + Ce^{4+}$$

At relatively lower Ce^{3+}/Ce^{4+} ratios, however, CeO_2-NPs were found to exhibit catalase (CAT) mimetic property which converts H_2O_2 to H_2O (Dhall et al., 2017), possibly through the following reactions:

$$H_2O_2 + 2Ce^{4+} \rightarrow O_2 + 2Ce^{3+} + 2H^+$$

$$H_2O_2 + 2Ce^{3+} + 2H^+ \rightarrow 2H_2O + 2Ce^{4+}$$

Importantly, the protective role of CeO_2-NPs was only reported in relatively short-term studies. The long-term benefit of CeO_2-NPs over several generations of plants appears minimal. For example, $10\,mg\,L^{-1}$ of CeO_2-NPs slightly enhanced the growth and yield of tomato plants through its life cycle (Wang et al., 2012). However, after the seeds were collected from the treated plants and exposed to the same concentration of CeO_2-NPs again, the second-generation seedlings developed from seeds of previously exposed mother plants had markedly smaller biomass than the control seedlings or seedlings developed from unexposed seeds (Wang et al., 2013). The roots from the second-generation seedlings whose mother plants were also exposed to CeO_2-NPs had greater root leakage. Interestingly, these roots also developed excessive root hairs, likely as a protective mechanism of plants to acquire more nutrients under stress. That study was the first report on the potentially negative long-term effect of ENPs on plants. A more detailed study with Brassica extended the generational studies to three generations and plants were exposed to three different concentrations of CeO_2-NPs (Ma et al., 2016). The results showed that plants in the second and third generation displayed slower growth and smaller biomass. After two generations of irrigation with water containing 10, 100, and 1000 mg L^{-1} of CeO_2-NPs, Brassica plants bore 39%, 59%, and 61% less of siliques, and the number of seeds per silique was reduced by about 50%, suggesting that repeated multigenerational exposure to CeO_2-NPs substantially lowered Brassica yield, Fig. 7.1.

The multigenerational study of CeO_2-NPs was extended to wheat recently, and the results showed that exposure to CeO_2-NPs to mother plants lowered the nutrient levels of the second-generation wheat grains such as Fe and Mn even though the second-generation plants were not exposed to CeO_2-NPs (Rico et al., 2017). When plants were exposed to CeO_2-NPs in both generations, more nutrient minerals in wheat grains were affected. The second-generation grains had lower Mn, Ca, K, Mg, and P compared with plants that were treated only once in the second generation. The results are consistent with the Brassica study that multigenerational exposure could potentially lower seed quality and nutritional values of grains from offspring

FIGURE 7.1 (A) Images of three generations of Brassica at the flowering stage after treating with 0, 10, 100 and 1000 mg L^{-1} of CeO_2-NPs; (B) Siliques collected from the third generation of plants treated with four different concentrations of CeO_2-NPs.

plants. A follow-up study from the authors found that CeO_2-NPs mainly affected the composition of seed fatty acid after two generations of exposure (Rico et al., 2020a). Intriguingly, the multigenerational impact of CeO_2-NPs on wheat was significantly affected by soil N levels, with low N favored the nutrient uptake by wheat when the plants were also exposed to CeO_2-NPs, suggesting that the long-term impact of CeO_2-NPs on plant health depends on the plant growing environment. Further metabolomic analysis showed that continuous exposure to CeO_2-NPs for three generations altered the metabolites of wheat grains. For example, nicotianamine, a metabolite closely associated with Fe storage was significantly reduced by 33% at high N soil, consistent with the 13%–16% lower Fe concentration in wheat grains from the same treatment (Rico et al., 2020b).

7.3 Silver nanoparticles (Ag-NPs)

Ag-NPs is one of the most extensively explored ENPs due to its strong antimicrobial property and potential uses in various industries. Ag-NPs display notable effects on plants in both hydroponic and soil systems (Geisler-Lee et al., 2012). Owing to its high dissolution rate in aerobic environment and the strong antimicrobial properties of Ag^+, there have been intense debates concerning the toxicity of Ag-NPs. For over a decade, many studies were conducted to answer the question that whether the observed impact of Ag-NPs on plants is due to Ag-NPs itself or the released Ag ions. Up to today, such debate is still not fully settled. However, the general consensus is that while dissolved Ag^+ contributes significantly to the toxicity of Ag-NPs, Ag-NPs themselves exert unique toxicity to plants, different from Ag ions. As a highly soluble ENP, Ag-NPs undergo extensive transformation in the environment after initial oxidation. Sulfidation to form silver sulfide (AgS_2) is likely the most common transformation product of Ag-NPs, but some Ag will exist in the form of elemental Ag (Impellitteri et al., 2013). The transformation of Ag-NPs in the environment will undoubtedly alter its long-term toxicity because the transformed product, Ag_2S-NPs, is more readily available for plant uptake, yet displays lower toxicity to plants. In a multigenerational study with Ag-NPs and plants, exposure to 12.5 $mg\ kg^{-1}$ of Ag-NPs to parent *Arabidopsis* significantly weakened the offspring plant, indicated by delayed flowering and significant downregulation of the genes regulating plant flowering (Ke et al., 2020). Importantly, even though plants were only exposed in the first generation, the second-generation plant seedlings were still not as robust as the control plants which had never been exposed to Ag-NPs. Plants exposed to Ag-NPs in two generations in a row exhibited the longest delay in flowering, suggesting that Ag-NPs affected the seed quality of *Arabidopsis*. The results are consistent with an earlier study which showed that continuous irrigation of Arabidopsis with Ag-NPs as low as 75 $\mu g\ L^{-1}$ prolonged the vegetative stage of *Arabidopsis* by 2–3 days (Geisler-Lee et al., 2014). The study also found that seeds from the offspring generations had lower germination rate than wild type seeds when they are treated similarly with Ag-NPs. Poor nutrient uptake caused by Ag-NPs might be one of the reasons for the negative effect of Ag-NPs.

7.4 Zinc oxide nanoparticles (ZnO-NPs)

ZnO-NPs are popular ingredients in cosmetics and sunscreens due to its UV adsorption property (Goh et al., 2014). The global production of ZnO-NPs is estimated to be 31,500 to 34,000 ton per year. Recently, interest in ZnO-NPs as an efficient slow release nanofertilizer has been growing because of the widespread Zn deficiency in agricultural soils globally. At higher concentrations, ZnO-NPs can function as an antimicrobial agent. This dual functionality of ZnO-NPs makes it an appealing candidate for broad applications in agriculture. Similar to Ag-NPs, ZnO-NPs undergo significant dissolution and transformation in the environment. Sulfidation is the most notable transformation for ZnO-NPs. In the presence of phosphate, ZnO-NPs also react with phosphate anions to form zinc phosphate $Zn_3(PO_4)_2$ hydrate or zinc phytate particles, depending on the relative ratios of Zn and P, and the pH (Lv et al., 2014; Ma et al., 2014).

Previous studies have shown that high concentrations of ZnO-NPs (800 mg kg^{-1}) significantly inhibited plant photosynthesis by 12% and corn yield by 38% in one generation (Zhao et al., 2015). However, the impact of ZnO-NPs on plants over multiple generations is still poorly understood. One study showed that ZnO-NPs significantly reduced the seed weight of radish (*Raphanus raphanistrum*) in the first generation, and the second generation seedlings developed with seeds from previously exposed mother plants had short roots and smaller biomass compared with seedlings from unexposed seeds (Singh and Kumar, 2018). However, in a different study with bean (*Phaseolus vulgaris*) plants, up to 500 mg kg^{-1} of ZnO-NPs showed minimal effect on the seed quality after two generations of exposure in a row (Medina-Velo et al., 2018). The only element that was negatively affected was nickel which was reduced by 41%–74% by ZnO-NPs. The surface property of ZnO-NPs seemed to play a role in their multigenerational impact because coated ZnO-NPs showed greater impact on Ni content in bean seeds than uncoated ZnO-NPs, however, detailed information on the coating material was not provided. The inconsistency of these two studies could arise from the different properties of ZnO-NPs used in these studies, and different soil properties and plant species.

7.5 Copper oxide nanoparticles (CuO-NPs)

CuO-NPs are incorporated into more than 200 commercial pesticide products (Chhipa, 2017). In addition, CuO-NPs was explored as a slow releasing source of Cu to meet the long-term crop needs because Cu is an essential micronutrient for plants (Printz et al., 2016). However, CuO-NPs are likely one of the most toxic ENPs among common metal oxide nanoparticles as evidenced by many previous studies involving comparative studies of different metallic nanoparticles. For example, when wheat plants were exposed to the same concentrations of CuO-NPs, titanium oxide nanoparticles (TiO_2-NPs) and iron oxide nanoparticles (Fe_2O_3-NPs), only CuO-NPs markedly lowered the essential nutrient of Zn and Fe in wheat grains (Wang et al., 2019). CuO-NPs also negatively affected the concentration of several essential amino acids such as isoleucine and histidine in wheat grains. Similar to Ag-NPs and ZnO-NPs, CuO-NPs also experience substantial transformation in the environment. The sulfidation of CuO-NPs generally leads to an intermediate amorphous Cu_xS first, but these amorphous particles eventually recrystallize into CuS (Gogos et al., 2017). However, how the transformation of CuO-NPs, or the primitive CuO-NPs affect plants over several generations is not clear because no multigenerational studies of CuO-NPs have been conducted. Considering the popularity of CuO-NPs and other Cu containing ENPs in agriculture and their potential impact on plants, it is surprising the multigenerational effect of CuO-NPs on plants has never been investigated.

7.6 Titanium oxide nanoparticles (TiO_2-NPs)

TiO_2-NPs is commonly detected in wastewater streams. TiO_2 has several crystalline structures, with the rutile and anatase structures being the most popular and stable. In each structure, a Ti^{4+} is surrounded by an octahedron of six O^{2-} ions. What differs between these two crystal structures is the distortion of the octahedron and assembly pattern, with anatase showing larger distortion (Cassaignon et al., 2007). Because of the different atomic arrangements, TiO_2-NPs with different crystal structures display different thermodynamic, Raman vibrational and optical properties, and thermal stability. Both crystal structures are commonly used in industry.

The literature is contradictory on the toxicity of TiO_2-NPs to plants. Positive effects of TiO_2-NPs on plants that have been reported include stimulation of germination and/or growth (Andersen et al., 2016), and enhancement of photosynthesis and nitrogen fixation (Li et al., 2020), which might derive from the semiconducting property of TiO_2-NPs. Negative effects of TiO_2-NPs exposure to a range of crop plants such as corn (*Zea mays*), wheat (*Triticum aestivum*), rice (*Oryza sativa*), and *Arabidopsis thaliana* have also been reported. However, the majority of studies were performed in short duration and with relatively high concentrations. There is a dearth of information on the long-term multigenerational impact of TiO_2-NPs on plants. The only study which investigated the multigenerational impact of TiO_2-NPs over two generations of basil (*Ocimum basilicum*) plants showed that TiO_2-NPs affected the nutrient accumulation in basil tissues (Tan et al., 2018). Importantly, the surface properties of TiO_2-NPs appeared to play a significant role in the accumulation of Ti and other nutrients.

For example, continuous treatment of basil plant with either hydrophobic or hydrophilic TiO_2-NPs over two generations markedly lowered the Ti concentration in plant roots compared with plants which were only exposed to the same TiO_2-NPs once; however, plants treated with pristine TiO_2-NPs had similar concentration of Ti in their roots, regardless whether they were exposed to TiO_2-NPs only in one generation, or two generations in a row. The nutrients in plant roots that were affected by TiO_2-NPs over two generations included Cu and Zn and the specific effects differed for TiO_2-NPs with different surface properties. Continuous exposure to TiO_2-NPs over two generations also significantly increased the carbohydrates in plant tissues, irrespective of the surface properties. However, only hydrophilic TiO_2-NPs significantly increased the reducing sugar contents compared with plants which were only exposed TiO_2-NPs once with the same surface coating or unexposed plants. TiO_2-NPs could modify seed quality as many other metal oxide nanoparticles. However, the impact was relatively mild and after wheat plant was exposed to 500 mg kg^{-1} of TiO_2-NPs, only the overall amino acids was increased (Wang et al., 2019).

7.7 Impact of engineered nanoparticles on soil health

In a plant-soil-ENP system, soil plays a fundamental role in plant growth because soil regulates water and nutrient availability, and disturbance of soil properties and functions by ENPs could greatly affect crop growth. Evidences have been reported that ENPs can significantly impact soil properties, which affect soil ecosystem services and agricultural productivity consequently (Sun et al., 2020; Wang et al., 2021a). For example, ENPs can modify the surface properties of soil minerals and their capability to adsorb and release various cations and anions (Guo et al., 2019). Several previous studies suggested that ENPs at agriculturally relevant concentrations are unlikely to significantly modify soil physical properties (Ma et al., 2019). However, they can significantly alter chemical and biological properties of soil. In addition, chemical bioavailability in soils can be modified by ENPs, causing different bioavailability of essential macro-and micro-nutrients and contaminants to plants (Conway and Keller, 2016). Previous studies also reported that metallic nanoparticles such as $Cu(OH)_2$-NPs increased ion release, particularly Mg^{2+}, roughly in proportion to the cation exchange capacity (CEC) of soil (Conway and Keller, 2016).

ENPs can also modify the biological properties of soil. For example, ZnO-NPs enhanced the concentration of phosphorous in wheat by 10.8% because Zn enhanced the activity of several phosphorous-solubilizing enzymes, leading to greater bioavailability of native phosphorous sources (Raliya et al., 2016). The critical role of soil biota in ecological and agricultural processes has long been recognized and most previous studies evaluating the impact of ENPs on soil health were on its biological indicators. In general, ENPs introduced to soil showed a strong impact on soil enzymatic activities associated with carbon, nitrogen, phosphorous, and sulfur cycling. The magnitude of such impact depends on the composition and concentration of ENPs and the duration of exposure. For example, Ag-NPs at the concentration of 1.6 mg kg^{-1} dry soil or higher significantly reduced the enzymatic activity of β-glucosaminidase after one hour of exposure in a silt loam soil but significantly increased its activity after one week (Eivazi et al., 2018). ENPs generally displayed stronger effects on enzymatic activities in soils with a coarse texture and low soil organic matter content.

Plant growth in a soil system depends on both its physiological health and the growing environment, which are both subject to the impact of ENPs. How to distinguish the direct and indirect impact of ENPs over generations of ENP applications is critical to understand the long-term effect of nanotechnology on plant health. Current advancement in artificial intelligence may allow the distinction of differential impacts of ENPs on various aspects of plant growth and help predict the long-term ENP effects on plant health (Wang et al., 2021b).

7.8 Future research needs

Even though the consequences of nanotechnology on plant health have been investigated for over a decade, many unknowns remain. Based on plant growth and other physiological parameters, most ENPs will exert greater and more negative effects on plants if the offspring generations of plants are continuously exposed to the same type and concentration of ENPs over several generations. It will be important to understand the molecular mechanisms leading to the susceptibility of offspring generations of plants to ENPs using different "-omics" techniques. While there have been some attempts to associate the unique physicochemical properties of ENPs with their phytotoxicity, such studies are essentially empty in long-term multigenerational studies due to the extensive requirements of resources and time. Importantly, many agriculturally popular ENPs undergo remarkable transformation, and the transformed products often display different toxicity to plants from the pristine ENPs applied to plants. Therefore unraveling the long-term fate and transformation of ENPs with different properties, and their contribution to the overall impact of multigenerational impact of ENPs will shed light on the sustainable applications of ENPs in agriculture.

Short-term studies on the phyto effect of ENPs have repeatedly shown the significant role of soil properties including the clay mineral contents and soil organic matter (Eivazi et al., 2018). Understandably, they will play an equally important role in the long-term effect of ENPs. However, how the long-term effect of ENPs to plants is affected by these factors has not been studied. Significantly, soils contain a range of other chemicals such as heavy metals, and previous research has shown that ENPs applied to soil interact closely with such heavy metals and alter their accumulation in plant tissues

(Wang et al., 2021b; Rossi et al., 2017b, 2018). These interactions have strong implications for food safety. Therefore evaluating the synergistic effect of ENPs with other co-existing contaminants in soil over several generations of plants is needed.

Finally, all multigenerational studies on ENPs effects on plants were conducted in systems where plants are root exposed to ENPs. With growing interests in foliar applications of ENPs which are more cost-effective and convenient to use, it is urgent to understand how ENPs from foliar application affect plant growth over multiple generations. This is important because ENPs exposed through foliar application have lower impact on soil health; therefore the long-term impact of ENPs from foliar application will shed more light on the direct impact of ENPs on plant health (Hong et al., 2021).

Acknowledgment

Xingmao Ma acknowledges the partial financial support from the National Science Foundation through grant #1900022.

References

Andersen, C.P., King, G., Plocher, M., Storm, M., Pokhrel, L.R., Johnson, M.G., et al., 2016. Germination and early plant development of ten plant species exposed to titanium dioxide and cerium oxide nanoparticles. Environ. Toxicol. Chem. 35 (9), 2223–2229.

Cassaignon, S., Koelsch, M., Jolivet, J.-P., 2007. Selective synthesis of brookite, anatase and rutile nanoparticles: thermolysis of $TiCl_4$ in aqueous nitric acid. J. Mater. Sci. 42 (16), 6689–6695.

Chhipa, H., 2017. Nanofertilizers and nanopesticides for agriculture. Environ. Chem. Lett. 15 (1), 15–22.

Collin, B., Auffan, M., Johnson, A.C., Kaur, I., Keller, A.A., Lazareva, A., et al., 2014. Environmental release, fate and ecotoxicological effects of manufactured ceria nanomaterials. Environ.-Sci. Nano 1 (6), 533–548.

Conway, J.R., Keller, A.A., 2016. Gravity-driven transport of three engineered nanomaterials in unsaturated soils and their effects on soil pH and nutrient release. Water Res. 98, 250–260.

Dhall, A., Burns, A., Dowding, J., Das, S., Seal, S., Self, W., 2017. Characterizing the phosphatase mimetic activity of cerium oxide nanoparticles and distinguishing its active site from that for catalase mimetic activity using anionic inhibitors. Environ. Sci. Nano 4 (8), 1742–1749.

Eivazi, F., Afrasiabi, Z., Elizabeth, J., 2018. Effects of silver nanoparticles on the activities of soil enzymes involved in carbon and nutrient cycling. Pedosphere 28 (2), 209–214.

Espinasse, B.P., Geitner, N.K., Schierz, A., Therezien, M., Richardson, C.J., Lowry, G.V., et al., 2018. Comparative persistence of engineered nanoparticles in a complex aquatic ecosystem. Environ. Sci. Technol. 52 (7), 4072–4078.

Fréchette-Viens, L., Hadioui, M., Wilkinson, K.J., 2019. Quantification of ZnO nanoparticles and other Zn containing colloids in natural waters using a high sensitivity single particle ICP-MS. Talanta 200, 156–162.

Geisler-Lee, J., Wang, Q., Yao, Y., Zhang, W., Geisler, M., Li, K., et al., 2012. Phytotoxicity, accumulation and transport of silver nanoparticles by *Arabidopsis thaliana*. Nanotoxicology 7 (3), 323–337.

Geisler-Lee, J., Brooks, M., Gerfen, J.R., Wang, Q., Fotis, C., Sparer, A., et al., 2014. Reproductive toxicity and life history study of silver nanoparticle effect, uptake and transport in *Arabidopsis thaliana*. Nanomaterials 4 (2), 301–318.

Gogos, A., Thalmann, B., Voegelin, A., Kaegi, R., 2017. Sulfidation kinetics of copper oxide nanoparticles. Environ. Sci. Nano 4 (8), 1733–1741.

Goh, E., Xu, X., McCormick, P., 2014. Effect of particle size on the UV absorbance of zinc oxide nanoparticles. Scr. Mater. 78, 49–52.

Gunawan, C., Lord, M.S., Lovell, E., Wong, R.J., Jung, M.S., Oscar, D., et al., 2019. Oxygen-vacancy engineering of cerium-oxide nanoparticles for antioxidant activity. ACS Omega 4 (5), 9473–9479.

Guo, B., Jiang, J., Serem, W., Sharma, V.K., Ma, X., 2019. Attachment of cerium oxide nanoparticles of different surface charges to kaolinite: molecular and atomic mechanisms. Environ. Res. 177, 108645.

Hofmann, T., Lowry, G.V., Ghoshal, S., Tufenkji, N., Brambilla, D., Dutcher, J.R., et al., 2020. Technology readiness and overcoming barriers to sustainably implement nanotechnology-enabled plant agriculture. Nat. Food 1 (7), 416–425.

Hong, J., Wang, C., Wagner, D.C., Gardea-Torresdey, J.L., He, F., Rico, C.M., 2021. Foliar application of nanoparticles: mechanisms of absorption, transfer, and multiple impacts. Environ. Sci. Nano 8 (5), 1196–1210.

Impellitteri, C.A., Harmon, S., Silva, R.G., Miller, B.W., Scheckel, K.G., Luxton, T.P., et al., 2013. Transformation of silver nanoparticles in fresh, aged, and incinerated biosolids. Water Res. 47 (12), 3878–3886.

Ke, M., Li, Y., Qu, Q., Ye, Y., Peijnenburg, W., Zhang, Z., et al., 2020. Offspring toxicity of silver nanoparticles to *Arabidopsis thaliana* flowering and floral development. J. Hazard. Mater. 386, 121975.

Korsvik, C., Patil, S., Seal, S., Self, W.T., 2007. Superoxide dismutase mimetic properties exhibited by vacancy engineered ceria nanoparticles. Chem. Commun. 10, 1056–1058.

Li, Z., Juneau, P., Lian, Y., Zhang, W., Wang, S., Wang, C., et al., 2020. Effects of titanium dioxide nanoparticles on photosynthetic and antioxidative processes of scenedesmus obliquus. Plants 9 (12), 1748.

Limbach, L.K., Bereiter, R., Müller, E., Krebs, R., Gälli, R., Stark, W.J., 2008. Removal of oxide nanoparticles in a model wastewater treatment plant: influence of agglomeration and surfactants on clearing efficiency. Environ. Sci. Technol. 42 (15), 5828–5833.

Lv, J., Zhang, S., Wang, S., Luo, L., Huang, H., Zhang, J., 2014. Chemical transformation of zinc oxide nanoparticles as a result of interaction with hydroxyapatite. Colloids Surf. A Physicochem. Eng. Asp. 461, 126–132.

Ma, X., Geiser-Lee, J., Deng, Y., Kolmakov, A., 2010. Interactions between engineered nanoparticles (ENPs) and plants: phytotoxicity, uptake and accumulation. Sci. Total Environ. 408 (16), 3053–3061.

Ma, R., Levard, C., Judy, J.D., Unrine, J.M., Durenkamp, M., Martin, B., et al., 2014. Fate of zinc oxide and silver nanoparticles in a pilot wastewater treatment plant and in processed biosolids. Environ. Sci. Technol. 48 (1), 104–112.

Ma, X., Wang, Q., Rossi, L., Zhang, W., 2015. Cerium oxide nanoparticles and bulk cerium oxide leading to different physiological and biochemical responses in Brassica rapa. Environ. Sci. Technol. 50 (13), 6793–6802.

Ma, X., Wang, Q., Rossi, L., Ebbs, S.D., White, J.C., 2016. Multigenerational exposure to cerium oxide nanoparticles: physiological and biochemical analysis reveals transmissible changes in rapid cycling *Brassica rapa*. NanoImpact 1, 46–54.

Ma, X., Sharifan, H., Dou, F., Sun, W., 2019. Simultaneous reduction of arsenic (As) and cadmium (Cd) accumulation in rice by zinc oxide nanoparticles. Chem. Eng. J. 123802.

Medina-Velo, I.A., Zuverza-Mena, N., Tamez, C., Ye, Y., Hernandez-Viezcas, J.A., White, J.C., et al., 2018. Minimal transgenerational effect of ZnO nanomaterials on the physiology and nutrient profile of *Phaseolus vulgaris*. ACS Sustain. Chem. Eng. 6 (6), 7924–7930.

Printz, B.;, Lutts, S.;, Hausman, J.-F.;, Sergeant, K., 2016. Copper trafficking in plants and its implication on cell wall dynamics. Front. Plant. Sci. 7, 601.

Raliya, R., Tarafdar, J.C., Biswas, P., 2016. Enhancing the mobilization of native phosphorus in the mung bean rhizosphere using ZnO nanoparticles synthesized by soil fungi. J. Agric. Food Chem. 64 (16), 3111–3118.

Rico, C.M., Johnson, M.G., Marcus, M.A., Andersen, C.P., 2017. Intergenerational responses of wheat (*Triticum aestivum* L.) to cerium oxide nanoparticles exposure. Environ. Sci. Nano 4 (3), 700–711.

Rico, C.M., Abolade, O.M., Wagner, D., Lottes, B., Rodriguez, J., Biagioni, R., et al., 2020a. Wheat exposure to cerium oxide nanoparticles over three generations reveals transmissible changes in nutrition, biochemical pools, and response to soil N. J. Hazard. Mater. 384, 121364.

Rico, C.M., Wagner, D., Abolade, O., Lottes, B., Coates, K., 2020b. Metabolomics of wheat grains generationally-exposed to cerium oxide nanoparticles. Sci. Total Environ. 712, 136487.

Rossi, L., Zhang, W., Lombardini, L., Ma, X., 2016. The impact of cerium oxide nanoparticles on the salt stress responses of *Brassica napus* L. Environ. Pollut. 219, 28–36.

Rossi, L., Zhang, W., Ma, X., 2017a. Cerium oxide nanoparticles alter the salt stress tolerance of *Brassica napus* L. by modifying the formation of root apoplastic barriers. Environ. Pollut. 229, 132–138.

Rossi, L., Zhang, W., Schwab, A.P., Ma, X., 2017b. Uptake, accumulation, and in planta distribution of coexisting cerium oxide nanoparticles and cadmium in Glycine max (L.) Merr. Environ. Sci. Technol. 51 (21), 12815–12824.

Rossi, L., Sharifan, H., Zhang, W., Schwab, A.P., Ma, X., 2018. Mutual effects and in planta accumulation of co-existing cerium oxide nanoparticles and cadmium in hydroponically grown soybean (Glycine max (L.) Merr.). Environ. Sci. Nano.

Schwabe, F., Schulin, R., Rupper, P., Rotzetter, A., Stark, W., Nowack, B., 2014. Dissolution and transformation of cerium oxide nanoparticles in plant growth media. J. Nanopart. Res. 16 (10), 1–11.

Singh, D., Kumar, A., 2018. Investigating long-term effect of nanoparticles on growth of *Raphanus sativus* plants: a trans-generational study. Ecotoxicology 27 (1), 23–31.

Sun, W., Dou, F., Li, C., Ma, X., Ma, L.Q., 2020. Impacts of metallic nanoparticles and transformed products on soil health. Crit. Rev. Environ. Sci. Technol. 973–1002.

Tan, W., Du, W., Darrouzet-Nardi, A.J., Hernandez-Viezcas, J.A., Ye, Y., Peralta-Videa, J.R., et al., 2018. Effects of the exposure of TiO_2 nanoparticles on basil (*Ocimum basilicum*) for two generations. Sci. Total Environ. 636, 240–248.

Wang, Q., Ma, X., Zhang, W., Pei, H., Chen, Y., 2012. The impact of cerium oxide nanoparticles on tomato (*Solanum lycopersicum* L.) and its implications for food safety. Metallomics 4 (10), 1105–1112.

Wang, Q., Ebbs, S.D., Chen, Y., Ma, X., 2013. Trans-generational impact of cerium oxide nanoparticles on tomato plants. Metallomics 5 (6), 753–759.

Wang, Y., Jiang, F., Ma, C., Rui, Y., Tsang, D.C., Xing, B., 2019. Effect of metal oxide nanoparticles on amino acids in wheat grains (*Triticum aestivum*) in a life cycle study. J. Environ. Manag. 241, 319–327.

Wang, X., Li, X., Dou, F., Sun, W., Chen, K., Wen, Y., et al., 2021a. Elucidating the impact of three metallic nanoagrichemicals and their bulk and ionic counterparts on the chemical properties of bulk and rhizosphere soils in rice paddies. Environ. Pollut. 290, 118005.

Wang, X., Liu, L., Zhang, W., Ma, X., 2021b. Prediction of plant uptake and translocation of engineered metallic nanoparticles by machine learning. Environ. Sci. Technol. 55 (11), 7491–7500.

Zhang, W.L., Ebbs, S.D., Musante, C., White, J.C., Gao, C.M., Ma, X.M., 2015. Uptake and accumulation of bulk and nanosized cerium oxide particles and ionic cerium by radish (Raphanus sativus L.). J. Agric. Food Chem. 63 (2), 382–390.

Zhao, L., Sun, Y., Hernandez-Viezcas, J.A., Hong, J., Majumdar, S., Niu, G., et al., 2015. Monitoring the environmental effects of CeO_2 and ZnO nanoparticles through the life cycle of corn (Zea mays) plants and in situ μ-XRF mapping of nutrients in kernels. Environ. Sci. Technol. 49 (5), 2921–2928.

Chapter 8

Nanomaterial transformation in root–soil interface: a function of root exudate or microbial activity?*

Peng Zhang[1,2], Wanqin Dai[1,3], Chaonan Dong[1] and Zhiyong Zhang[1,3]

[1]*Key Laboratory for Biological Effects of Nanomaterials and Nanosafety, Institute of High Energy Physics, Chinese Academy of Sciences, Beijing, China,* [2]*School of Geography, Earth and Environmental Science, University of Birmingham, Birmingham, United Kingdom,* [3]*School of Nuclear Science and Technology, University of the Chinese Academy of Sciences, Beijing, China*

Chapter Outline

* This chapter is modified from a review (Zhang P, Guo ZL, Zhang ZY, Fu HL, White JC, Lynch I. Small, 2020, 16: e2000705)

Plant Exposure to Engineered Nanoparticles. DOI: https://doi.org/10.1016/B978-0-323-85032-2.00004-X

8.1 Introduction

Nanomaterials (NMs) are highly dynamic and reactive; various physical, chemical, or biological transformations may occur in the environment (Valsami-Jones and Lynch, 2015). These processes change the properties of NMs, thereby affecting the subsequent behavior, including transport in soil, uptake and translocation in the plant, and their toxicity to organisms. For example, some metal-based NMs such as nano-Ag (Das et al., 2018), nano-Cu (Conway et al., 2015), and nano-ZnO (David et al., 2012) may dissolve quickly in the environment. Acquisition of an evolving coating of environmental or biological molecules (called "corona") on the surface of these NMs could occur, which either accelerates or slows the dissolution (Collin et al., 2016). The released metal ions can be accumulated by the plant directly or as complexes with other components from the environment. For other NMs such as nano-Au and nano-TiO_2, their chemical structures are persistent in environment, but physical transformation such as aggregation, agglomeration, or corona formation may occur, which can modify their surface charge and chemistry and influence subsequent behavior and bioavailability (Lowry et al., 2012). In terms of the agricultural application of NMs, the initially designed analyte function may be compromised due to these transformation processes that may occur both in soil before the material interacts with the plant or at the nanoparticle–plant interface. Therefore a full understanding of the transformation mechanisms, processes, and drivers in the soil–plant system is essential for accurate NM risk assessment, as well as safe and effective use of these materials in agriculture.

The transformations of NMs in the soil–plant system were discussed in a few previous review papers (Lv et al., 2019; Shukla et al., 2016; Zhang et al., 2015). The present review does not intend to be a comprehensive literature review. Rather, we will focus on the behavior of NMs at the root–soil interface (rhizosphere), such as aggregation, antiaggregation, redox reaction and transformation (Wang et al., 2020b). Root–soil interface is the most active area in soil-plant ecosystems with intensive interactions among soil particles, organic compounds, plant roots and microorganisms. Key questions to be discussed here include (1) the locations where NM transformation occurs; (2) rhizosphere processes on the transformation of NMs; (3) how NM transformation affects their uptake, translocation, and phytotoxicity in plants. We also highlight the knowledge gaps in this area and provide our opinions on the directions for future studies.

8.2 Types of nanomaterial transformation in the environment

Physical transformation occurs for all types of NMs in the environment. In soil, NMs tend to aggregate, including both homoaggregation and heteroaggregation (Hotze et al., 2010). Homoaggregation is the aggregation of particles of the same nature triggered by the compression of the electrical double layers on the particle surfaces in high ionic strength conditions. As the aggregation process continues, larger sizes of particles and clusters form. In soil, heteroaggregation occurs between NMs and larger soil particles, effectively rendering the NMs as part of the soil particle. NMs can also adsorb and acquire an evolving coating (called a "corona") of environmental or biological macromolecules such as humic substances, proteins, nucleic acids, and polysaccharides, which can substantially change the surface properties of the material (Lundqvist et al., 2011).

Chemical transformation implies compositional or structural changes of the NMs. Dissolution is a common chemical transformation of metal-based NMs, and this transformation process is routinely accompanied by other chemical reactions such as reduction, oxidation, sulfidation, phosphorylation, or chelation. For example, oxidation is the first step for the release of Ag^+ from Ag NMs (Stabryla et al., 2018). During oxidation, an oxide layer may wholly or partially coat the NM surface, which may then reduce further oxidation and dissolution processes, although those effects may be reversed as the oxidation layer is further transformed or lost. For example, nanoscale CeO_2 can be reduced and release Ce^{3+} ions (Zhang et al., 2012b,c). Further transformation processes, following redox reaction and dissolution, such as precipitation, may occur for metal based NMs, for example, sulfidation of Fe (Kumar et al., 2018), Cu (Gogos et al., 2018), Zn (Banerjee and Jain, 2018), and Ag (Kampe et al., 2018), chlorination of Ag (Singh et al., 2019), or phosphorylation of Ce (Zhang et al., 2017b) and Zn (Xiao et al., 2018). Organic ligands may bind to the released metal ions. For example, organic acids can chelate Ce^{3+}, forming Ce-carboyxlates (Zhang et al., 2019a). In addition to metal-based NMs, carbon based NM can also undergo redox reactions. Carbon nanotubes or graphene can be oxidized by hydroxyl radicals which are normally present in plant leaves at low levels (Allen et al., 2009; Huang et al., 2018). The oxidation may generate aromatic hydrocarbons upon partial degradation or release CO_2 upon complete oxidation. Graphene oxide can be reduced in soil via the respiration activities of bacteria (Salas et al., 2010).

Biotransformation is inevitable as the environment is replete with biota such as plants, fungi, protozoa, and bacteria/archae. Biotransformation may involve the physical and chemical transformation processes mentioned above, but is mediated by biological processes. NMs may aggregate in soil, root–soil interface, or in biological fluids (e.g., xylem and phloem sap).

Dissolution, degradation, redox reactions, or chelation may take place in the cell wall/membrane, cytoplasma, vacuole, or extracellularly via enzyme-assisted reactions or through interaction with reactive oxygen species.

8.3 Locations of nanomaterial transformation in soil–plant system

The transformation of NMs is driven by various factors in both soil and the plant. The major entry pathway of NMs into plants is through soil exposure; therefore transformation of NMs in the terrestrial ecosystem is primarily driven and regulated by the physical and chemical environment of the soil. Soil is a highly complex mixture generally composed of 50% solids (minerals, organic matter) and 50% voids, which are approximately equal parts pore water and air (Speight, 2019). The soil texture (mineral composition), soil pH, ionic strength, redox potential, humidity, and organic matter content (quality and quantity) may all affect the transformation of NMs in soil matrices. Conversely, soil is also teeming with biota, including bacteria, fungi, algae, protozoa, and larger soil fauna such as earthworms and nematodes. The presence of this biotic community can cause tremendous spatial heterogeneity of soil; the physical and chemical properties of soil at different microenvironments in areas such as the rhizosphere can vary dramatically (Frey, 2015). Bacteria represent the largest population of organisms in soil. Bacteria are an essential community involved in many critical geochemical and biochemical processes, including carbon and nitrogen cycling (Ren et al., 2018). Their biological activity (e.g., respiration, excretion of organic acids) can trigger various chemical transformations of NMs. Nano-plant interfaces such as nano-root and nano-leaf interfaces are also important locations of NM transformation before their entry into plants. Plants can excrete metabolites at the root and leaf surface which may react with NMs and change the physicochemical properties of these materials (Zhang et al., 2012b,c). There is also a large community of bacteria and other microorganisms colonized at the root–soil interface, where plant exudates are abundant, forming narrow but critical zone called the "rhizosphere" (Philippot et al., 2013). The microbial community and physicochemical properties at this region are very different from other parts of the plant and soil and this region will play an important role in the transformation of NMs. NM transformation inside plant tissues or fluids may be also possible, though little has been done in this area. This section will discuss the transformation of NMs at these locations, with the emphasis on root–soil interface.

8.3.1 Drivers of nanomaterial transformation in soil

Nanomaterial transformation can occur in soil prior to contact with the plant surface (Fig. 8.1A and Table 8.1). A key driver in soil inducing physical

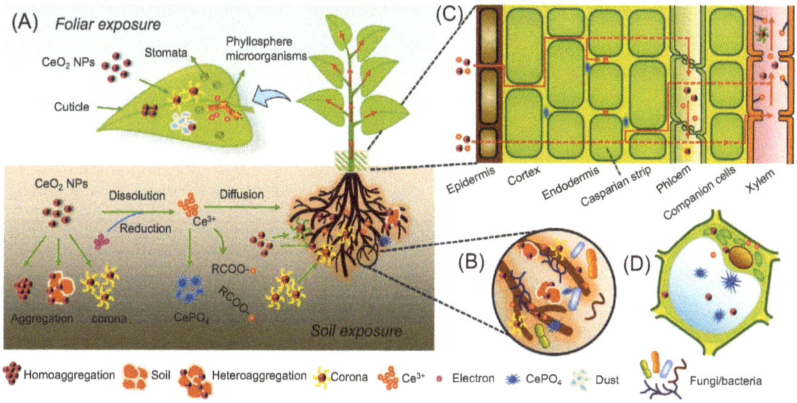

FIGURE 8.1 Transformation of nanomaterials (NMs) in soil and plants using CeO_2 NMs as an example. (A) Transformation of CeO_2 NMs at root and leaf interfaces. (B) Magnification of root showing the transformation of CeO_2 NMs at the rhizosphere. (C) Transport and transformation of CeO_2 NMs in the plant vascular structure. (D) Transformation of CeO_2 NMs in a plant cell.

transformation of NMs is the presence of various inorganic (minerals, ions) and organic components (natural organic matter, NOM) (Baalousha, 2017; Duan et al., 2017). Differences in soil texture may greatly affect the aggregation state of NMs in soil. Soil texture indicates the relative content of mineral particles of different sizes, including gravel, sand, silt, and clay. Silty and clayey soils contain large amounts of negative sites for analyte adsorption and binding (Manrique et al., 1991). Therefore these sites can strongly bind positively charged NMs and cause stronger heteroaggregation of NMs than a sandier soil. Soil texture also affects the humidity and redox potential of soil by indirectly impacting the water and air content. Soils with high clay content have higher water holding capacity and less pore space for air than sandy soil, which creates a reducing environment that favors the reduction of NMs (Li et al., 2017). Sandy soils have higher oxygen content which could favor oxidative transformations of NMs (e.g., Ag and Cu NMs).

Natural organic matters (NOMs) account for about 1%–5% of agricultural soil composition. They are naturally occurring macromolecules containing large amounts of protein, humic substances, polysaccharides, etc., which are derived from the breakdown of plant and animal tissues at various stages. They can drive the transformation of NMs by forming a NOM coating or "corona" on the particle surface, thus changing the surface of NMs and affecting their subsequent behavior and fate. The relative amounts of NOMs vs NMs in the soil environment are very large, and the modification of NMs properties by NOM could be both substantial and dynamic. However, the effects of this transformation on the properties of NMs are largely dependent on the composition of the NOMs and other soil conditions, such as pH and ionic strength (Lin et al., 2017). NOMs can be divided into two distinct

TABLE 8.1 Drivers of engineered nanomaterial transformation in soil.

Drivers	Transformation	References
Natural organic matter	Aggregation	Duan et al. (2017), Lin et al. (2017), Luo et al. (2018), and Degenkolb et al. (2019)
Ionic strength	Aggregation	El Badawy et al. (2010), Zhang et al. (2012a), and Tang and Cheng (2018)
Soil pH	Aggregation, dissolution	Tang et al. (2002) and Tang and Cheng (2018)
Redox potential	Reduction, oxidation	Wang et al. (2015b) and Li et al. (2017)
Inorganic ligands	Sulfidation, chlorination, phosphorylation	Ma et al. (2013), Sekine et al. (2015), Wang et al. (2015b), Li et al. (2017), Zhang et al. (2017b), Banerjee and Jain (2018), Gogos et al. (2018), Kampe et al. (2018), Kumar et al. (2018), Xiao et al. (2018), and Singh et al. (2019)
Microorganisms	All physical and chemical	Allen et al. (2009) and Xie et al. (2019a)

compositions, hydrophilic and hydrophobic. The hydrophilic fraction contains humic substances, which represent the largest portion (60%–80%) of NOMs. Attachment of hydrophilic and hydrophobic fractions to NMs surfaces will cause opposite consequences in terms of surface charge. NOMs with larger molecular weight may form coherent and thick coatings that fully or nearly fully cover the NM surface. Depending on the properties of the NOMs, this coating may cause electrostatic stabilization of NMs or may cause flocculation and aggregation due to bridging effects between NOM molecules.

Ionic strength in soil will greatly affect the stability of NMs. Based on classic Derjaguin, Landau, Verwey, and Overbeek (DLVO) theory, the stability of colloidal particles is determined by the relative balance between the van der Waals attraction and the electrostatic repulsion. NMs are stable when there is sufficient charge within the electrostatic double layer (EDL) that contributes to the electrostatic repulsion. The thickness of the EDL is directly related to the surrounding conditions, such as ionic strength, pH, and ion valences. In general, high ionic strength compresses the EDL and drives NM aggregation. The effects vary depending on solution pH, valency, and type of ions. High valent ions have stronger effects than low valent ions. For example, the addition of Ca^{2+} electrolyte causes greater aggregation of negatively charged citrate functionalized Ag NMs than does Na^{+} at the same ionic strength (El Badawy et al., 2010). Ca^{2+} can bind with the carboxyl groups on citrate, which reduces the charge of the Ag NMs and induces subsequent aggregation. By contrast, Ca^{2+} or Na^{+} did not change the zeta potential and induce aggregation of uncharged PVP-coated Ag NMs regardless of pH range, suggesting that the presence of ions does not necessarily affect the stability of all NMs, with sterically stabilized NMs being more resistant to aggregation.

Soil pH is another important driver of NM transformation. For example, it affects aggregation as different NMs have different isoelectric points (IEP), which is the pH at which NMs have neutral surface charge and are subject to aggregation. The IEP varies greatly among different NMs, ranging from extremely acidic to basic; therefore the effect of pH on NMs stability is quite different. Soil pH also affects the chemical transformation of NMs, especially metal-based particles. The dissolution of metal-based NMs can be enhanced significantly in acidic soil. Soil pH can also determine the chemical speciation of NMs. For example, ZnO NMs at a pH lower than 6 will tend to release Zn^{2+}, but at more basic pH, a $Zn(OH)_2$ layer may form at the particle surface, and further transformation to zincate ions, $Zn(OH)_n^{(n-2)-}$, could occur under highly basic conditions (pH > 12) (Tang et al., 2002).

Reduction and oxidation reactions occur widely in soil and are involved in many essential biogeochemical processes. These reactions also play important roles in the transformation of NMs that are sensitive to the redox potential (E_h). The E_h is used to determine the overall reducing or oxidizing capacity of an environment. The E_h is related to the relative amounts of oxidants and reductants in soil. In soil, common oxidants include oxygen, nitrate, nitrite, iron, sulfate, CO_2, etc. For example, ferrous ions (Fe^{2+}) can reduce graphene oxide (Wang et al., 2015b). In an aerated soil with high oxygen content, oxidation of NMs is more favorable. This may result in the formation of an oxide layer on the surface of some metallic NMs which can act as a protective coating for the particle core. However, for other NMs, for example, Ag NMs, formation of an Ag_2O layer triggers the dissolution and release of Ag^+. In unaerated (anoxic) conditions such as paddy soil, wetlands, and organic-rich soils, the environment is predominantly reducing, which drives the reduction of NMs. Li et al. (2017) compared the Ag NM transformation in paddy soil under different conditions. Under aerobic conditions, 90% of Ag remained as Ag NMs after 2 days incubation in soil, whereas under anaerobic conditions, 95% was oxidized and transformed to Ag_2S. However, again, the microenvironment of soil may be highly heterogeneous. For example, even a paddy soil may have highly localized oxygen-rich areas, which can impact the transformation of NMs. Such complexity of soil need to be considered as a reality in all of these situations mentioned above.

Inorganic ligands such as sulfide, chloride, and phosphate are key drivers involved in the chemical transformation of metal-based NMs. Sulfide is ubiquitous in the environment, and sulfidation is a very common chemical processes for many types of metal-based NMs (e.g., Ag, Cu, ZnO, PbO), based on the Pearson acid-base concept (Pearson, 1963). Metal sulfides are usually less soluble than their oxide counterparts. Despitethis, a low degree of sulfidation has only a minor effect on the release of Zn^{2+} from ZnO; only a high extent of sulfidation ($>50\%$) will significantly reduce the dissolution (Ma et al., 2013). The ZnS may further transform to $Zn_3(PO_4)_2$ because Zn^{2+} has a higher affinity to phosphate ions ($Ksp = 9.0 \times 10^{-33}$) than to sulfide ($Ksp = 2.0 \times 10^{-25}$). For Ag NMs, the dominant species depends on the soil pH. Acidic conditions favor AgCl formation while neutral or alkaline conditions are favorable for sulfidation of Ag (Sekine et al., 2015). Ag_2S is chemically stable and has low Ag lability over extended periods. Depending on the chemical composition of the soil, the formation or reformation of NMs may also occur. For example, CeO_2 NMs could be reduced by bacteria in soil, release Ce^{3+} and possibly further form $CePO_4$. $CePO_4$ is a representative phosphate mineral (rhabdophane) in soil that forms during the weathering of apatite. Redissolution of $CePO_4$ in the presence of biogenic organic acids that are ubiquitous in soil (e.g., oxalic acid, citric acids, ascorbic acid) has been reported (Cervini-Silva et al., 2005); the released Ce^{3+} may be oxidized to Ce^{4+} and precipitated as CeO_2 (nanoscale) again on the surface of $CePO_4$. Other examples are the transformation of metal ions to metallic nanoparticles (e.g., Au, Ag) directly in plants or by plant extracts (Gardea-Torresdey et al., 2002; Roy et al., 2017).

Microorganisms are ubiquitous in soil, representing the greatest contributor of biomass. The microbial activity is a key driver for NM transformation in the soil environment. Bacteria-mediated transformation may include all the previously mentioned physical and chemical transformation processes. He et al. (2015) illustrated that the dissociation and dissolution of rare earth (RE) oxide NMs on the surface of bacteria were faster than in the culture medium. This implies that the interaction between nanoparticles and biological surfaces promotes transformation of NMs. A more recent investigation of the same group showed that CeO_2 NMs were reduced at the surface of *Bacillus subtilis* and the released Ce^{3+} ions were chelated by the carboxyl functional groups in the bacteria cell wall (Xie et al., 2019a). Species of the genus *Shewanella* are widespread in nature in various habitats. Some of them can utilize Fe minerals as terminal electron acceptors for anaerobic respiration and affect the fate and transport of Fe minerals in the subsurface. Luo et al. (2017) proved the formation of hematite nanoparticles after the biotransformation of hematite. Wang et al. (2020a) reported that the biotransformation of four carbon NM including graphene oxide, reduced graphene oxide, single walled carbon nanotubes, and oxidized (carboxylated) single walled carbon nanotubes by *Labrys* sp. WJW resulted in morphology changes, defect increases and functional group changes. The underlying mechanism of carbon NM biodegradation was mediated by extracellular Fenton-like reaction in which the $\cdot OH$ production was regulated by reduction of H_2O_2 in a continuous cycle of Fe^{3+}/Fe^{2+}. Carbon nanotubes can be oxidized by bacteria-generated enzymes or radicals, causing oxidative degradation (Allen et al., 2009). Facultative anaerobes such as *Shewanella* can cause reduction of NMs (e.g., graphene oxide) by donating electrons mediated by cytochromes during their respiration process (Salas et al., 2010). These reactions may take place at the cell membrane, as well as in the cell cytoplasm or even outside of the cell. Microorganisms release large amounts of enzymes into soil, which mediate and catalyze many biogeochemical reactions, such as the decomposition of plant/animal/microbial debris and nutrient (e.g., N, C, P, S) cycling, thereby facilitating the release of inorganic nutrients from minerals for plant growth (Lewis et al., 2019; Salas et al., 2010; Visser and Parkinson, 1992). These enzymes, including oxidases, reductases, hydrolases, and lyases (Kandeler and Dick, 2006), may also catalyze the transformation of NMs.

As discussed, the transformation of NMs in soil is a highly complex process due to the diverse and heterogenous nature of the soil environment. Many processes may occur simultaneously or sequentially. The dominant and final species of NMs will depend on the residence time in the soil, the relative amount of absorbed ligands and soil environmental factors such as pH and oxygen content.

8.3.2 Drivers of nanomaterial transformation in plants

8.3.2.1 *Transformation of nanomaterials at the rhizosphere and phyllosphere*

Rhizosphere is one of the most important locations for transformation of NMs (Fig. 8.1B). This narrow region around roots (extending 1–3 mm from the root surface) is governed by complex interactions between the plant and other associated organisms, including microorganisms and invertebrates (e.g., nematode). Studies have shown that NM transform to a greater extent at the rhizosphere region than in the soil away from the roots (Kuzyakov and Razavi, 2019; Rico et al., 2018; Zhang et al., 2012b,c). The transformation of NMs in this region is driven by the chemistry of the rhizosphere. Plant root exudates are secondary metabolites and can be generally divided into two classes based on molecular weight. High molecular weight compounds contain mucilage and cellulose, while small molecules include organic acids, amino acids, proteins, sugars, and phenols which can be utilized by the associated microbes. Mucilage is primarily polysaccharides which can protect root cells during the penetration into the soil. It may form a coating on NMs that significantly changes their surface properties. Small organic molecules appear to play more significant roles in the chemical transformation of NMs in the rhizosphere. Excretion of organic acids into the rhizosphere causes lower local soil pH around the root compared to the bulk soil, thereby enhancing the dissolution of metal based NMs (Huang et al., 2017). In addition to dissolution, significant redox reactions may also occur at the rhizosphere. Plant-secreted reducing substances such as phenols and ascorbic acid can reduce CeO_2 NMs and release Ce^{3+}, which can further transform to $CePO_4$ or Ce carboxylates on the root surface (Zhang et al., 2017a). For Cu NMs, a more complex oxidation and reduction can occur, which produces Cu(II), Cu(I), and Cu(0) species in the presence of root exudates (McManus et al., 2018). The exact composition of root exudates is plant species dependent. For example, mugineic acids, which can promote the dissolution of NMs, are specifically excreted by gramineous species. A larger diversity of organic acids are found in the exudates of cabbage (*Brassica oleracea var. gongylodes L.*) than in that of cucumber (*Cucumis sativus L.*) (Vančura and Hovadík, 1965). Therefore the type and extent of the transformation of a given NMs may vary with plant species.

As mentioned in 8.3.1, rhizosphere microorganisms may transform NMs through microbially mediated redox reactions and secreting secondary compounds to mediate the dissolution and antiaggregation of NMs. Both plant roots and microbes contribute to the transformation of NM in the rhizosphere. The "rhizosphere effect" among plant species is governed by the type and quantity of root exudates, leading to the differential microbial abundance and composition. However, to what extent the association of NMs with rhizosphere microbial cells can enhance bioreactivity of NMs needs further confirmation, more in planta researches should be conducted (Wang et al., 2020b).

Transformation processes such as aggregation, redox reactions, and dissolution may also occur at the leaf interface (Fig. 8.1A). Aggregation of NMs could occur in the air due to the interaction with aerosol particles or interaction with dust or microorganisms (phyllosphere) on the leaf surface. The aggregation state may change due to the interaction with water excreted from the leaf or derived from rain or irrigation. Logically, chemical transformation of NMs at leaf surfaces may also occur by interacting with the epiphytes, which are a population of microorganisms including bacteria, fungi, and yeast that are colonized on the phyllosphere, the above-ground part of the plant that is colonized by microorganisms (Kim et al., 2010). The extracellular polymeric substances (EPS) excreted by the microorganisms may change the surface chemistry of NMs by forming a corona on the particle surface or by causing dissolution of the particles. The role of the phyllosphere in the transformation of NMs is almost completely unknown as of yet.

8.3.2.2 *Transformation of nanomaterials inside plants*

Although many studies have reported an alteration of NM chemical species in plant tissues, few have differentiated transformation processes that occur outside (in soil or at the nano-root interface) or inside plants (e.g., in plant cells, sap, etc.). Ma et al. (2017) demonstrated the main site for CeO_2 NM transformation in cucumber plants by using split-root and petiole exposure methods. In split-root experiments, the authors separated the root manually into two parts and placed them into two separate containers, one of which was filled with CeO_2 NM suspensions. The exposure lasted for 3 h to allow the uptake of CeO_2 NMs into the plant. Then they cut off the roots exposed to CeO_2 NMs and continued the growth of the remaining seedlings for 7 days and examined the chemical species of Ce. They found no transformation of CeO_2 NMs after exposure for 3 h and 7 days, suggesting that short term root exposure did not cause any transformation and those CeO_2 particles that entered the plant cannot be readily transformed. Unlike CeO_2 NMs, transformation of CuO inside maize plants was reported by Wang et al. using a similar split-root method (Wang et al., 2012). The authors reported the transformation of CuO NMs to Cu_2O and Cu_2S inside the plants during xylem and phloem translocation. Not surprisingly, the difference between CeO_2 and CuO fate suggests that the occurrence of transformation inside the plant depends on the NMs type. The plant sap components (e.g., reducing sugars, enzymes) may also be involved in the reduction and transformation of NMs (Fig. 8.1C). Xylem sap, which is largely responsible for moving water- and soil-acquired nutrients from roots to shoots, consists primarily of water and inorganic ions, as well lesser amounts of hormones, organic acids, and other biomolecules (Rodríguez-Celma et al., 2016). One possible NM transformation process in the xylem is aggregation, largely due to the presence of inorganic ions. Phloem sap contains much higher concentrations of inorganic ions and sugars, amino acids, proteins, organic acids, and hormones. Therefore both physical and chemical transformation may readily occur. The organic acids may drive the dissolution of NMs, whereas the reducing sugars such as glucose and fructose can induce NM formation from ions (Wang et al., 2012). Although there is no direct evidence showing whether and how sap transforms NMs, based on the sap composition, the chance of transformation of NMs in phloem is higher than in xylem. The transformation may be also affected by sap flow rate, which will essentially control residence time. The flow rate of xylem sap is 0.47–4.8 mm s^{-1}, which is much higher than that of phloem sap (0.07–0.58 mm s^{-1}) (Su et al., 2019).

Once the NMs reach the xylem via the root, certain particles will travel upward very quickly and distribute into other tissues, allowing little time for the interaction of the NMs with the sap components. In the phloem, the opposite appears to be true. As suggested by Ma et al. (2017), even in root rhizosphere, short term exposure cannot transform CeO_2 NMs, suggesting that a critical reaction time is required for transformation of particles (e.g., CeO_2) that are intrinsically stable.

In addition to sap, transformation in the plant leaf is also possible. Experience gained from the green synthesis of NMs has shown that plant leaves can serve as a "factory" for synthesizing various kinds of NMs such as Ag, Au, ZnO, and CuO NMs (Ahmed et al., 2016; Qu et al., 2011; Sharma et al., 2015; Su et al., 2019). A common method to synthesize metallic NMs is to use reactive phytochemicals in plant leaf extracts (e.g., flavonoids, glucose, and fructose) to reduce the metal salt pre-cursor (Singh et al., 2018). It is reasonable to conclude that NMs may also be reduced by those plant-generated substances. For example, CuO may be reduced to Cu(I) or Cu(0) species, undergoing similar processes in plant vascular fluids. Alternatively, oxidation of NMs in the leaf is also possible, driven by the free radicals which are generated by various physiological processes in plant. Oxidative transformation of graphene to CO_2 in rice leaves was recently reported, and hydroxyl radicals ($OH\cdot$) were considered to be responsible for the transformation (Huang et al., 2018). The exact type of NMs transformation in plant tissues may greatly depend on the location of the NMs within the organism. Although less studied, the presence of NMs and their transformed species in intercellular spaces cytoplasm, vacuoles, and chloroplasts has been reported (Fig. 8.1D); notably, the mechanisms driving those transformations remain largely unexplored.

8.4 Effect of nanomaterial transformation on uptake and in planta translocation

The uptake of NMs by plants, with particle accumulation in plant tissues especially in the edible parts (e.g., carrot roots, lettuce leaves, tomato fruit), is a major concern in terms of the risk of NMs in agricultural ecosystems. Trophic transfer along the food chain and biomagnification of NMs within agricultural ecosystems have been reported, and there is a likelihood of human exposure through dietary uptake. NM accumulation in plants may also affect the nutritional quality of plant-derived food. The uptake and translocation of NMs are controlled not only by plant species, which vary in terms of both morphology and physiology, but is also highly dependent on the physicochemical properties of NMs which may be altered during transformation processes. For example, aggregation decreases both the mobility of NMs in soil and their reactivity, due to the decrease of available surface area. Larger aggregates may not be able to transport across size-restrictive biological barriers such as the cell wall/or membrane; therefore NM bioavailability in the plant will be reduced. Chemical transformations change properties of NMs such as structure and composition, the behavior of which may differ significantly from the original form.

8.4.1 Effect of transformation on foliar uptake of nanomaterials

Although plants have many physiological barriers to prevent the entry of substances that are not required for growth, there are many pathways that permit NM entry. For example, the leaf epidermis is covered by a waxy hydrophobic cuticle that is perforated by pores with sizes of ≈ 5 nm; this may permit entry of small NMs. The plant leaf also has large stomata (e.g., 13 μm for spinach and cucumber, 7 μm for tomato and water-melon, etc.), which may allow larger particles or aggregates to enter when the pores are open. The uptake of NMs in plants via the foliar pathway, with subsequent translocation and accumulation, has been explored only recently. Foliar-applied CeO_2 NMs either as airborne powder or as suspension can accumulate in leaves and translocate to the root (Hong et al., 2016). The stomatal pathway has been demonstrated as a pathway for NM uptake in leaves. The uptake efficiency is highly dependent on the size of NMs and of the stomata. For example, polystyrene (PS) particles with a size of 43 nm could penetrate into leaves via the stomata while larger sized (1.1 μm) PS particles could not. Physical transformation such as aggregation could result in large clusters that prevent NM entry into leaves via the stomata (Fig. 8.2A).

Chemical transformation of NMs in the air and at the nano-leaf interface is also possible. For instance, redox sensitive NMs such as Ag and Cu can interact with oxidants in the air (e.g., radicals, oxygen), resulting in an oxide coating on the particle surface. Graphene oxide exposed to the air could be reduced upon exposure to sunlight (Shams et al., 2019). These processes can greatly impact the uptake of NMs into the leaf. For example, degradation of graphene oxide reduces the amount of hydrophilic functional groups and increases hydrophobicity, which may reduce the binding of this material to the hydrophobic leaf surface but also increase entry through the stomata. The dissolution of metal-based NMs releases metal ions, the uptake and translocation mechanisms of which are quite different from the particulate forms (Xie et al., 2019b). Corona formation on NM surfaces may increase particle stability and prevent aggregation, thereby increasing the likelihood of entry into leaf. Importantly, our current knowledge in this area is very limited. NM transformation on the plant leaf surface is an area in need of significant investigation.

8.4.2 Effects of transformation on root uptake and translocation of nanomaterials

Under root exposure, NMs must traverse several physiological barriers in order to reach the vascular structure for translocation to the shoots. In general, this process will follow three steps: (1) NM arrival at or deposition on the root surface. (2) NM movement across physiological barriers including the cuticle, epidermis, cortex, endodermis, and Casparian strip; and (3) NM translocation within the xylem or phloem to the shoot system (Su et al., 2019). The apoplastic pathway is one pathway for NM uptake and transport

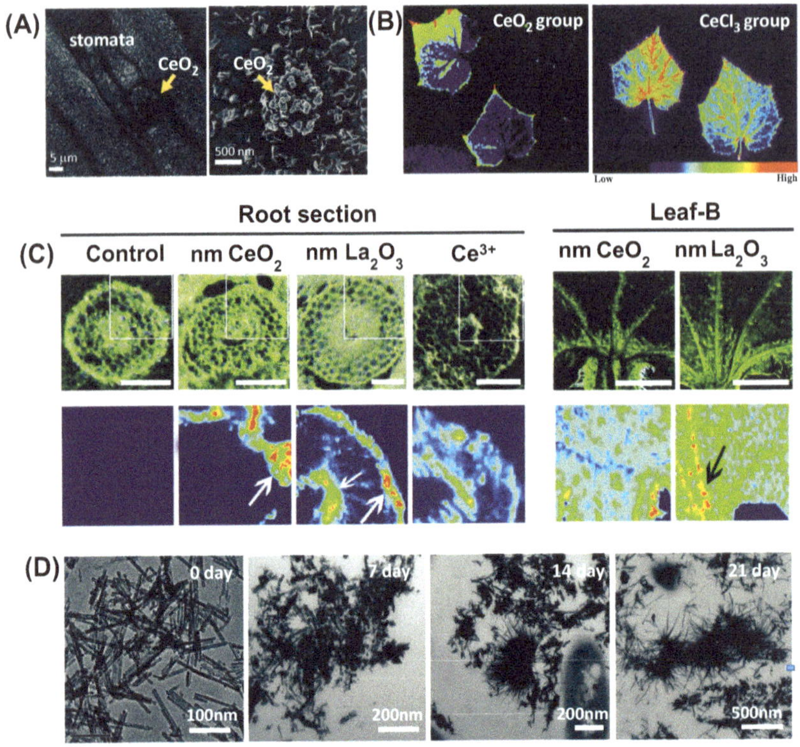

FIGURE 8.2 Effect of engineered nanomaterials transformation on the uptake and translocation of nanomaterials (NMs) in plants. (A) CeO_2 NMs aggregate near the closed stomata (left) and integrated into the leaf wax (right). (B) Different distribution patterns of CeO_2 NMs (left) and $CeCl_3$ (right) in cucumber leaves. (C) Comparison of the distribution pattern of CeO_2 and La_2O_3 NMs and $CeCl_3$ in cucumber root sections and basal leaves. (D) Gradual transformation of CeO_2 nanorods on the root surface over 21 days. *(A) Adapted with permission from Birbaum et al. (2010). Copyright 2010, American Chemical Society. (B) Adapted with permission from Zhang et al. (2011). Copyright 2011, Royal Society of Chemistry. (C) Adapted with permission from Ma et al. (2015). Copyright 2015, Taylor & Francis. (D) Adapted with permission from Zhang et al. (2017b). Copyright 2017, American Chemical Society.*

via the plant root (Schwab et al., 2016). Here, the NM enters the epidermal tissue and diffuses through the intercellular space. The symplastic pathway involves cell to cell transport via the plasmodesmata. In either case, the NMs need to cross the cell wall, which typically has 5–20 nm pores. Therefore particle size is one of the most important properties controlling entry into the roots. Theoretically, NMs larger than 20 nm would not be able to cross the cell wall and cell membrane. However, if the structures are damaged, the penetration of larger NMs is possible. After entering the xylem, NMs can be translocated to the shoot system. The particles must also pass through pores (called "pits") of the connecting cells in xylem; the size of these structures is relatively large (43–340 nm), although significant NM aggregation could restrict movement during translocation.

Given the presence of these barriers, aggregation in soil or at the root surface can effectively prevent the uptake of NMs into the roots. For example, the accumulation of NMs in plants cultured in soil is typically lower than that in plants cultured in hydroponic solution due to homo- or hetero-aggregation of NMs in soils. Supplementing NMs with NOM such as humic substances can induce acquisition of a hydrophilic organic corona that increases particle mobility and uptake (Zhao et al., 2012). The corona can also change the NM surface charge, which can significantly alter uptake and translocation in the plant (Mensch et al., 2017). A positively charged corona increases the adsorption of NMs onto negatively charged root surfaces, whereas negatively charged NMs are more readily translocated once in the plant (Spielman-Sun et al., 2019).

As discussed, the chemical transformation of NMs may occur in soil prior to reaching the root surface or at the root interface. It is known that NM dissolution can significantly enhance uptake by roots and also that translocation and distribution patterns of particles and ions within plants are different. For example, CeO_2 NMs tend to accumulate in leaf margins while Ce^{3+} ions accumulate mostly in vascular bundles (Fig. 8.2B) (Zhang et al., 2011). Nanoscale La_2O_3 have similar distribution patterns as Ce^{3+}, largely due to significant dissolution of the metal oxide after interaction with plants (Fig. 8.2C) (Ma et al., 2015). Dissolution is typically accompanied by other concurrent or subsequent chemical processes, such as precipitation or complexation. In soil, Zn^{2+} ions released from ZnO NMs can further transform to ZnS, $Zn_3(PO_4)_2$ or be incorporated into soil minerals such as γ-Al_2O_3 and goethite (Lv et al., 2019); this process immobilizes the Zn in soil and reduces availability to plants. Phosphorylation has also been reported for RE oxide NMs such as La_2O_3 (Ma et al., 2011), Yb_2O_3 (Zhang et al., 2012b,c), and CeO_2 (Zhang et al., 2017a), all of which can release RE^{3+} ions to transform to $REPO_4$. Similar processes also occur at the root–soil interfaces, where these processes may be accelerated by root-and microorganism-generated exudates. Although transformation of NMs after interaction with plants has been reported, direct investigation of the effect of this transformation on the uptake and translocation of NMs in plant is scarce. Wang et al. (2017) compared the uptake of $AgNO_3$, Ag NPs, and Ag_2S NMs in cucumber (*C. sativus*) and wheat (*Triticum aestivum*) and reported similar amounts of Ag accumulation in shoot and root while suggesting different distribution patterns of Ag in tissues. Ag_2S mainly accumulated in the veins while $AgNO_3$ accumulated mostly in leaf tip and margins. Another study found that the uptake of $Cu(OH)_2$ NMs with higher solubility in roots of wheat (*T. aestivum*) is greater than those with lower solubility (e.g., CuO and CuS NMs) after 1 h hydroponic exposure (Spielman-Sun et al., 2018). In a follow-up depuration study, the authors removed the exposure solutions and found that both CuO and CuS NMs persisted in or on the roots and continued to release Cu for translocation to plant leaves over 48 h. There is also some indirect evidence from studies such as that by Zhang and coworkers; here, the authors reported that changing the plant culture medium (Rui et al., 2015), plant species (Zhang et al., 2019a), or morphology (Zhang et al., 2017b) of the CeO_2 NMs affected analyte transformation and subsequent translocation. Importantly, removal of phosphates from the culture media dramatically enhanced Ce accumulation in the leaves (Zhang et al., 2019a). Rod-shaped CeO_2 NMs have greater reactivity than other morphologies, including octahedral and cubical, and thus have greater dissolution and accumulation in plant leaves (Zhang et al., 2017b). Significantly higher amounts of $CePO_4$ needle-like clusters have also been observed on root surfaces (Fig. 8.2D). Transformation of CeO_2 NMs also depends on the plant species (Fig. 8.3); monocot and dicot plants have different capacities to transform CeO_2 NMs into Ce(III) species at the root surface and thus have different Ce translocation patterns within the plant (Zhang et al., 2019a).

8.5 Implication of engineered nanomaterial transformation and future research

NMs are being discharged into agricultural environments, both intentionally (e.g., NM application in agriculture) and incidentally (e.g., biosolid application to soil as fertilizer), and will have impacts on agricultural ecosystems that may be either positive or negative. To evaluate those impacts, one must understand the behavior and fate of NMs in this highly complex system. Given the high reactivity and dynamic nature of NMs, their transformation

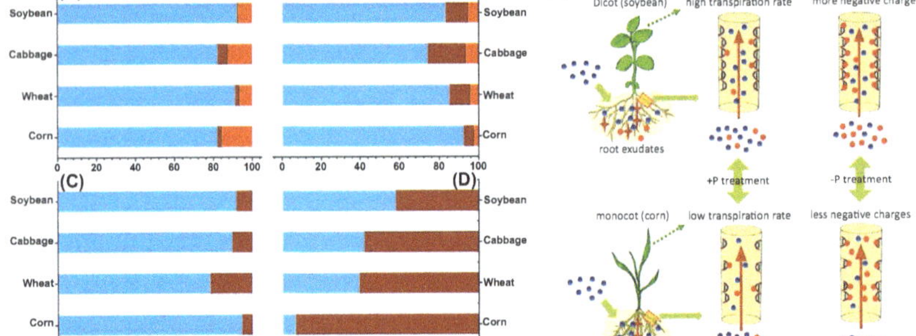

FIGURE 8.3 Plant-species-dependent transformation of CeO_2 NMs in cucumber plants. (A–D) Fraction of the Ce species in the four types of plants treated with 2000 mg L^{-1} CeO_2 NMs for 2 weeks. (A) Root, +P; (B) root, −P; (C) shoot, +P; (D) shoot, −P. +P and −P indicate whether the nutrient solution is supplemented with or without phosphates, respectively. (E) Schematic illustration of the transformation of CeO_2 NMs on the root surface and the translocation of CeO_2 and Ce^{3+} ions in the root xylem of a dicot (soybean) and monocot (corn) with and without the presence of phosphates. *(A)-(D) Adapted with permission Zhang, P., et al., 2019a. Plant species-dependent transformation and translocation of ceria nanoparticles. Environ. Sci.-Nano 6 (1), 60–67. Copyright 2019, Royal Society of Chemistry.*

in agricultural systems is inevitable, and will be affected by many factors, including NM physicochemical properties, the soil environment and plant species. These dynamic transformation processes complicate our understanding of NM fate and impact because of the presence of both the "acquired" and "original" identity. For example, a long-standing question in the nanotoxicology of metal-based NMs is whether the source of toxicity is from the particle or from the ionic form that is transformed/ released from the prisitine NMs (Lubick, 2008). Indeed, the toxicity of many NMs such as ZnO, Ag, CeO_2, La_2O_3, Yb_2O_3, and CdS quantum dots are all reported to be positively correlated with their dissolution, although the particle effect cannot be entirely excluded (Wang et al., 2016). Metal ions can move more freely than the particles in plants and as such, cause greater toxicity. The presence of inorganic ligands which can precipitate the released metal ions may partially reduce toxicity. For example, S^{2-} can reduce the toxicity of Ag by forming Ag_2S (Devi et al., 2015); PO_4^{3-} can alleviate the toxicity of RE oxide NMs by forming phosphate precipitates (Zhang et al., 2019a). Some processes, for example, precipitation with inorganic ligands and complexation with organic ligands, may occur competitively. The formation of nanoscale $CePO_4$ immobilizes the Ce^{3+} and reduces bioavailability, while complexes of Ce^{3+} with small molecule organic acid may improve or maintain the transportability of Ce^{3+} in plants (Zhang et al., 2012b,c; Bao et al., 2020). The final fate and impact depends on the net outcome of these reactions. NM transformation also affects their efficacy for agricultural applications. For example, some nanofertilizers are designed to release nutrients slowly or in a controllable fashion (Kottegoda et al., 2017); this function may be confounded by transformation processes. Surface change induced by corona formation may result in targeted NMs being delivered to a nontarget location, as has been reported in targeted drug delivery in nanomedicine due to overcoating of the targeting ligands (Corbo et al., 2016). Transformation also complicates the prediction of NM behavior in the environment, since the primary nanoscale descriptors (e.g., size, surface charge, hydrophobicity, porosity, etc.) (Lynch et al., 2014) measured for the original NMs may change dramatically and dynamically during transformation.

Despite the clear importance of these processes, our understanding of NM transformation and the consequences for plant uptake, biodistribution, and toxicity is quite limited. Most current toxicity and fate studies continue to use pristine NMs that may well have little relevance to the particles present in complex environments, especially those applied via biosolids. More soil- and field-based experiments are needed to obtain environmentally relevant results with regard to the impacts and transformation products of agriculturally relevant NMs. Subsequent subsections would identify several key knowledge gaps and highlight critical future steps to address the current lack of understanding.

8.5.1 Elucidating the role of plant metabolites in transformation of nanomaterials

Plant metabolites, such as root exudate components, are primary drivers of NM transformation at the root–soil interface. When studying the interaction of NMs with root exudates, knowledge could be leveraged from the study of mineral-root exudate interactions. Plants acquire a large quantity of nutrients from soil minerals by secreting exudates to directly chelate mineral nutrients or to reduce the rhizosphere pH to increase nutrient availability. For example, as a response to iron deficiency, plants can secrete organic acids to increase the dissolution of Fe-minerals and riboflavin/phenolic compounds to facilitate Fe acquisition (Jin et al., 2007). Published in vitro experiments have shown that some key components, for example, small organic acids (Zhang et al., 2012b,c), reducing sugars and phenols, (Wang et al., 2012) are responsible for the dissolution and reduction of NMs at the root surface. However, the compositions of root exudates are complex and differ between plant species. Identification and quantification of the critical exudate components is therefore necessary for studying plant species-specific NM transformations.

The metabolites in xylem and phloem fluids are also important for in planta transformation of NMs, but their role in this process is largely unexplored. The vascular fluids contain abundant substances such as inorganic ions and organic compounds which may mediate NM transformation. For example, organic acids can cause dissolution while other organic constituents may cause corona formation or evolution during their translocation within the xylem and phloem. As a result, NM behavior and fate may change during transport. The transformation of NMs in plant leaves is also an important aspect but is largely unknown. For example, CNTs have been shown to enter the plant cell and chloroplast to enhance the photosynthetic efficiency (Giraldo et al., 2019). However, if transformation occurs as a function of longer residence time of CNTs in plant leaves, functionality may be lost and the safety of transformation products may become a concern. Unlike transformation occurring in soil or at the root surface, transformation inside plants may act in a mode known as the "Trojan effect," where initially innocuous materials become toxic over time (Hsiao et al., 2015).

8.5.2 Exploring the role of microorganisms at the rhizosphere and phyllosphere in transformation of nanomaterials

A robust microbiome exists in both the root (rhizosphere) and shoot (phyllosphere). These communities play vital roles in plant physiology, with the vast majority being beneficial (e.g., legume rhizobia) to plant growth and a very small minority being potentially pathogenic. Similar to the plant root, the microbiome also excretes a large amount of metabolites known as EPS

(Costa et al., 2018). These substances and their metabolic activity (e.g., respiration) contribute to the transformation of NMs. Considering the size and activity of the microbiome, the effects on transformation of NMs are expected to be significant. A microbiome also colonizes the plant leaf and stem surface, forming the phyllosphere. Transformation of NMs at/by the phyllosphere has not been studied, but their activity is expected to be significant for NMs designed for foliar application. The in planta microbiome, also known as endophytes (Santoyo et al., 2016), may also cause transformation of NMs in plant tissues, affecting the translocation and impact of NMs in plants. Thus the role of the various microbiomes in transformation of NMs is wide open for exploration and may provide important insights for the design of functional nanoagrochemicals.

8.5.3 Determination of the kinetics of transformation of nanomaterials

Nanomaterial transformation is a dynamic process. For example, NMs may dissolve at the root surface and be transported to plant tissues where reducing environments may induce the reformation of nanoscale materials in intracellular spaces or inside plant cells. Some transformation processes (e.g., dissolution of ZnO and CuO NMs) may occur quickly at the root surface, resulting in the immediate uptake of ions and subsequent acute toxicity to the plant. Such rapid transformation can cause the loss of the desired functions of NMs for agricultural application. A corona can form on the surface of NMs and alter material translocation temporarily but then later breakdown inside the plant. The corona will likely only partially cover the particle surface and will be highly dynamic, being replaced by high affinity biomolecules and/or degraded by enzymes in the local environment. A biomolecular corona may also serve as a platform for biofilm growth; the initial abiotic corona will degrade whereas the biofilm will exude EPS to sustain the biofilm growth. The kinetics of each step of these transformation will be critical for understanding the subsequent behavior and final fate of NMs, and may lead to redesign of NMs for precision agriculture and sustained or targeted release that account for these transformation at the design stage (safe and benign by design approaches).

8.5.4 Unraveling the long-term effects of transformation of nanomaterials on agricultural ecosystems

In reviewing the existing literature, many NMs show low to moderate phytotoxicity. However, most of the current studies are carried out in simple or model systems (e.g., hydroponic), with short term exposures to relatively high and often environmentally unrealistic doses. Long term retention of NMs in soil is possible with continuous application of biosolids to agricultural land

and possibly repeated application of nanotechnology in agricultural and soil remediation scenarios. Therefore the ultimate ecosystem effects of NMs will be largely determined by the transformed products rather than the pristine materials. Long term impacts that need to be evaluated include the effects of NMs and their transformed products on the soil and plant microbiome, plant growth and nutritional content, trophic transfer of NMs along food chain, and possible transgenerational effects. Impacts on ecosystem services such a nutrient cycling, water purification and food production, as well as accumulation and persistence of NM residues in edible plant tissues, also need investigation.

8.5.5 Obtaining a systems-level view of the transformation of nanomaterials in ecosystems

The complexity of transformation is not only due to the various physicochemical properties of NMs, but also because of the dynamics of the soil and biological environment. NMs can be divided into many different types in terms of size, shape, surface charge/chemistry, crystal structure, bandgaps, etc., each of which may affect the transformation process. Transformation of NMs is also greatly affected by soil chemistry, the soil microbiome and the plant rhizosphere chemistry/microbiome; all of which are interconnected and further impacted by the environment and climate conditions. We therefore argue that these factors cannot be considered separately but rather must be considered and interpreted at a systems level. The interconnections between the plant, microbiome and NMs are shown in Fig. 8.4. Transformation of NMs at the root interface is largely determined by the rhizosphere, with its robust root exudates and microbiome. However, NM exposure may also alter

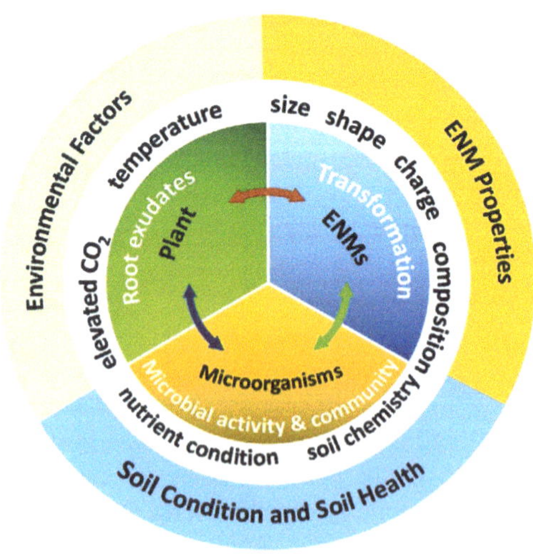

FIGURE 8.4 Interdependence of nanomaterials transformation, plants, and microorganisms.

the plant root exudate composition, for example, enhancing the excretion of organic acids and metal chelating substances, which can in turn further alter transformation of NMs. NMs can also interact with the rhizosphere microbiome, causing alteration of the microbiome community structure and activity (e.g., respiration, composition of EPS), which in turn affects NM transformation. The plant and soil microbiomes are also interconnected. Any change in the root exudate profile can have significant impact on microbiome activity and structure in the rhizosphere, and any alteration of microbiome (e.g., inhibition of legume rhizobia) can affect plant growth and root exudate secretion. Moreover, environmental factors such as climate change-induced elevation of CO_2 and changing soil moisture and temperature, can change microbial diversity and plant root exudate secretion, which will further affect NM transformations. Clearly, only a systems-level approach will enable understanding of complex NM transformation processes in the ecosystem, including application of nanoinfomatics approaches.

8.5.6 Taking advantage of the emergence of state-of-the-art analytical techniques

Given the complexity of NM transformation processes, especially rhizosphere processes, advanced analytical techniques are required to investigate these phenomena. For example, identification and quantification of transformation products in soil and plant tissues are challenging due to the presence of high background interferences in such samples. Advanced techniques such as synchrotron radiation based techniques (e.g., X-ray absorption fine structure spectroscopy) combined with imaging techniques including scanning transmission X-ray microscopy (STXM) and micro X-ray fluorescence (μ-XRF) (Chen et al., 2013), is one of sophisticated techniques that can be brought to bare for studying transformation of NMs.

Synchrotron based X-ray imaging techniques, because of their high spatial resolutions, sensitivity and abilities to identify elements, and their low matrix effects and nondestructiveness, have attracted great attention in the process of NM transformation. The application of synchrotron radiation based XRD, XPS, SAXS, Raman spectroscopy, FTIR, micro/nano-CT and so on provides better resolution and sensitivity, less acquisition time and in situ measurement compared to instruments using conventionally incident photons in physicochemical characterization of NM. For example, in complex biological samples such as plant tissues, the FTIR spectrum is the sum of contributions obtained from all the biomolecules. SR-μFTIR takes advantage of the characteristics of SR: brightness and energy tenability to provide high-contrast 2D images in live cells (Marine Cotte et al., 2009). And because SR-μFTIR employs non-ionizing radiation, it does not induce any detectable side-effects (Holman et al., 2002). SR-μFTIR achieves contrast from molecular vibration modes, but is not element-specific. Furthermore, synchrotron-based

CD spectroscopy possesses the capacity to obtain structural information at vacuum ultraviolet wavelength with only small sample amounts; in this way, the structural stability and changes of biomolecules bound to ENMs can be analyzed (Wang et al., 2015a).

Because of its unique chemical sensitivity, XAS can render precise fingerprint measurements of the ENM structure, which sheds light on the molecular mechanism of biophysicochemical interactions, such as adsorption, dissolution, phase transformation, and oxidative–reductive reactions (Bertsch and Hunter, 2001). XRF allows nondestructive measurement of the elemental composition of materials and could offer an excitation spot size of less than $0.15 \times 0.15\ \mu m^2$, and detection limits range from 5×10^{-20} to $3.9 \times 10^{-19}\ mol\ \mu m^{-2}$ (McRae et al., 2009; Twining et al., 2003). Synchrotron-based soft X-ray microscopy or scanning transmission X-ray microscopy (STXM) can study the distribution of certain chemical elements at high spatial resolution (Maser et al., 2000). It can also provide spectroscopic signals that reflect the chemical bonding state of major low Z constituents, without the disordered inelastic scattering background that is present in electron energy loss spectroscopy. It is thus possible to image whole, hydrated cells up to several micrometers in thickness with nanometer-scale resolution (Chao et al., 2009).The combined application of STXM coupled with XAFs can analyze thin samples in situ with a spatial resolution of better than 30 nm, with no need for prior chemical extraction or staining (Ma et al., 2011; Zhang et al., 2012b,c). The combination of XRF with XAFS can provide information about both the spatial distribution and chemical species of metals of interest (Li et al., 2015). Hernandez-Viezcas et al. used XRF and XANES to study ZnO NPs uptake and transformation in mesquite tissues (Hernandez-Viezcas et al., 2011). In addition, synchrotron tomography (SR-CT) has been a current focus in characterizing root system architecture and associated plant–soil interactions on biologically relevant (sub) mm scales in 3D, due to its superiority to μCT in several aspects such as low scan time, high spatial resolution and without gross intra-scan movement (Lombi and Susini, 2009; Mooney et al., 2012). For example, Keyes et al. (2013) used SR-CT to image the root hairs of wheat plants grown in soil at a resolution of *c*.1 μm.

Isotope labeling approaches (Zhang et al., 2012a, 2019b) coupled with in situ detection techniques such as nano secondary ion mass spectrometry (Nano-SIMS) (Aubert et al., 2012), autoradiography (Zhang et al., 2011), laser ablation inductively coupled plasma mass spectrometry (LA-ICP-MS) (Koelmel et al., 2013) and single particle/cell ICP-MS is another advanced technique in studying the transformation of NMs. The combination of isotope labeling approaches and Nano-SIMS allows next-level imaging of isotope distributions on resolutions beyond individual cells (50–100 nm) and identifies processes in situ that occur on a biologically relevant scale (Clode et al., 2009; Hoppe et al., 2013). Despite its great potential, there are considerably fewer Nano-SIMS facilities (35 ± 5) in the world than there are synchrotrons

(Oburger and Schmidt, 2016). As a result, accessing a Nano-SIMS device for an experiment can potentially be difficult for many researchers. Compared with some other approaches, such as LA-ICP-MS or Nano-SIMS, access to autoradiography facilities is likely not too difficult. This approach also has a range of other advantages, including being able to examine hydrated plant tissues, large samples or even entire plants. Another major advantage is the ability to separate background isotopes of an element (i.e., those natively present in the plant tissues) from the radioisotope of the same element added exogenously (Zhang et al., 2008). The resolution achievable with autoradiography varies between approximately 25 and 1000 μm (Kopittke et al., 2020). When compared to the other techniques, the main advantages of isotope labeling approaches coupled LA-ICP-MS is the higher sensitivity, lower matrix effect and simplicity (Becker, 2010; Becker et al., 2010; Pozebon et al., 2014).

Metabolomics (Schauer and Fernie, 2006), proteomics (Tan et al., 2017), and secretomics (Tanveer et al., 2014) are emerging powerful tools to identify the components of root exudates, EPS, and plant/microbial metabolite response to NM exposure. The data sets obtained could be combined with machine learning approaches (e.g., nanoinformatics) (Barnard et al., 2019) to identify key drivers and critical steps of NM transformation, and to predict NM transformations thereby facilitating safe-by-design of NMs for sustainable application in agriculture. Metabolomics can be used to acquire comprehensive information on the composition of a metabolite pool to provide a functional screen of the cellular state (Jorge et al., 2016). For example, Zhang et al. (2020) combined the 16S rRNA gene sequencing with metabolomics to investigate the response of soil microbial communities to engineered nanomaterials in presence of maize (*Zea mays L.*) plants. The corona composed of proteins (protein corona) is spontaneously and stably formed, when NMs is exposed to intracellular milieu (Matysiak et al., 2016). Proteomics is suited for corona composition analysis for better recognition of the mechanisms of protein corona formation and turnover in different environments and understanding the responses of cells and organisms to NMs (Monopoli et al., 2011; Treuel et al., 2014). Furthermore, plants are well known to actively secrete a large array of different molecules from polymers (e.g., extracellular RNA and DNA) to small compounds (e.g., ATP, phytochemicals, secondary metabolites, phytohormones). The plant secretome is usually considered in the frame of proteomics, for characterizing extracellular proteins, their biological roles, and the mechanisms accounting for their secretion in the extracellular space (Vincent et al., 2020).

Acknowledgments

This work is supported by Marie Skłodowska-Curie Individual Fellowships (NanoLabels Grant Agreement No. 750455) and National Natural Science Foundation of China (Grant Nos. 11875267 and 12075262).

References

Ahmed, S., et al., 2016. A review on plants extract mediated synthesis of silver nanoparticles for antimicrobial applications: a green expertise. J. Adv. Res. 7 (1), 17–28.

Allen, B.L., et al., 2009. Mechanistic investigations of horseradish peroxidase-catalyzed degradation of single-walled carbon nanotubes. J. Am. Chem. Soc. 131 (47), 17194–17205.

Aubert, T., et al., 2012. Root uptake and phytotoxicity of nanosized molybdenum octahedral clusters. J. Hazard. Mater. 219–220, 111–118.

Baalousha, M., 2017. Effect of nanomaterial and media physicochemical properties on nanomaterial aggregation kinetics. NanoImpact 6, 55–68.

Banerjee, P., Jain, P.K., 2018. Mechanism of sulfidation of small zinc oxide nanoparticles. RSC Adv. 8 (60), 34476–34482.

Bao, Y., et al., 2020. Citric acid enhances Ce uptake and accumulation in rice seedlings exposed to CeO_2 nanoparticles and iron plaque attenuates the enhancement. Chemosphere 240.

Barnard, A.S., et al., 2019. Nanoinformatics, and the big challenges for the science of small things. Nanoscale 11 (41), 19190–19201.

Becker, J.S., 2010. Bioimaging of metals in brain tissue from micrometre to nanometre scale by laser ablation inductively coupled plasma mass spectrometry: state of the art and perspectives. Int. J. Mass Spectrom. 289 (2), 65–75.

Becker, J.S., et al., 2010. Bioimaging of metals by laser ablation inductively coupled plasma mass spectrometry (LA-ICP-MS). Mass. Spectrom. Rev. 29 (1), 156–175.

Bertsch, P.M., Hunter, D.B., 2001. Applications of synchrotron-based X-ray microprobes. Chem. Rev. 101 (6), 1809–1842.

Birbaum, K., et al., 2010. No evidence for cerium dioxide nanoparticle translocation in maize plants. Environ. Sci. Technol. 44 (22), 8718–8723.

Cervini-Silva, J., et al., 2005. Biogenic dissolution of a soil cerium-phosphate mineral. Am. J. Sci. 305 (6–8), 711–726.

Chao, W., et al., 2009. Demonstration of 12 nm resolution fresnel zone plate lens based soft X-ray microscopy. Opt. Express 17 (20), 17669–17677.

Chen, C., et al., 2013. Advanced nuclear analytical and related techniques for the growing challenges in nanotoxicology. Chem. Soc. Rev. 42 (21), 8266–8303.

Clode, P.L., et al., 2009. In situ mapping of nutrient uptake in the rhizosphere using nanoscale secondary ion mass spectrometry. Plant. Physiol. 151 (4), 1751–1757.

Collin, B., et al., 2016. Effect of natural organic matter on dissolution and toxicity of sulfidized silver nanoparticles to: Caenorhabditis elegans. Environ. Sci. Nano 3 (4), 728–736.

Conway, J.R., et al., 2015. Aggregation, dissolution, and transformation of copper nanoparticles in natural waters. Environ. Sci. Technol. 49 (5), 2749–2756.

Corbo, C., et al., 2016. The impact of nanoparticle protein corona on cytotoxicity, immunotoxicity and target drug delivery. Nanomedicine 11 (1), 81–100.

Costa, O.Y.A., et al., 2018. Microbial extracellular polymeric substances: ecological function and impact on soil aggregation. Front. Microbiol. 9 (JUL).

Das, P., et al., 2018. Mechanism of toxicity and transformation of silver nanoparticles: inclusive assessment in earthworm-microbe-soil-plant system. Geoderma 314, 73–84.

David, C.A., et al., 2012. Dissolution kinetics and solubility of ZnO nanoparticles followed by AGNES. J. Phys. Chem. C 116 (21), 11758–11767.

Degenkolb, L., et al., 2019. The variable fate of Ag and TiO_2 nanoparticles in natural soil solutions—sorption of organic matter and nanoparticle stability. Water Air Soil Pollut. 230 (3).

Devi, G.P., et al., 2015. Sulfidation of silver nanoparticle reduces its toxicity in zebrafish. Aquat. Toxicol. 158, 149–156.

Duan, L., et al., 2017. Removal of graphene oxide nanomaterials from aqueous media via coagulation: effects of water chemistry and natural organic matter. Chemosphere 168, 1051–1057.

El Badawy, A.M., et al., 2010. Impact of environmental conditions (pH, ionic strength, and electrolyte type) on the surface charge and aggregation of silver nanoparticles suspensions. Environ. Sci. Technol. 44 (4), 1260–1266.

Frey, S. D., 2015. Soil Microbiology, Ecology and Biochemistry, p. 223.

Gardea-Torresdey, J.L., et al., 2002. Formation and growth of Au nanoparticles inside live alfalfa plants. Nano Lett. 2 (4), 397–401.

Giraldo, J.P., et al., 2019. Nanobiotechnology approaches for engineering smart plant sensors. Nat. Nanotechnol. 14 (6), 541–553.

Gogos, A., et al., 2018. Influence of organic compounds on the sulfidation of copper oxide nanoparticles. Environ. Sci. Nano 5 (11), 2560–2569.

He, X., et al., 2015. Quantifying the total ionic release from nanoparticles after particle-cell contact. Environ. Pollut. 196, 194–200.

Hernandez-Viezcas, J.A., et al., 2011. Spectroscopic verification of zinc absorption and distribution in the desert plant Prosopis juliflora-velutina (velvet mesquite) treated with ZnO nanoparticles. Chem. Eng. J. 170 (2), 346–352.

Holman, H.-Y., et al., 2002. Synchrotron infrared spectromicroscopy as a novel bioanalytical microprobe for individual living cells: cytotoxicity considerations." 7% J. J. Biomed. Opt. (3).

Hong, J., et al., 2016. Foliar applied nanoscale and microscale CeO_2 and CuO alter cucumber (*Cucumis sativus*) fruit quality. Sci. Total. Env. 563–564, 904–911.

Hoppe, P., et al., 2013. NanoSIMS: technical aspects and applications in cosmochemistry and biological geochemistry. Geostand. Geoanal. Res. 37 (2), 111–154.

Hotze, E.M., et al., 2010. Nanoparticle aggregation: challenges to understanding transport and reactivity in the environment. J. Environ. Qual. 39 (6), 1909–1924.

Hsiao, I.L., et al., 2015. Trojan-horse mechanism in the cellular uptake of silver nanoparticles verified by direct intra- and extracellular silver speciation analysis. Environ. Sci. Technol. 49 (6), 3813–3821.

Huang, Y., et al., 2017. Interactions, transformations, and bioavailability of nano-copper exposed to root exudates. Environ. Sci. Technol. 51 (17), 9774–9783.

Huang, C., et al., 2018. Transformation of 14 C-labeled graphene to 14 CO 2 in the shoots of a rice plant. Angew. Chem. 130 (31), 9907–9911.

Jin, C.W., et al., 2007. Iron deficiency-induced secretion of phenolics facilitates the reutilization of root apoplastic iron in red clover. Plant Physiol. 144 (1), 278–285.

Jorge, T.F., et al., 2016. Mass spectrometry-based plant metabolomics: metabolite responses to abiotic stress. Mass Spectrom. Rev. 35 (5), 620–649.

Kampe, S., et al., 2018. Silver nanoparticles in sewage sludge: bioavailability of sulfidized silver to the terrestrial isopod Porcellio scaber. Environ. Toxicol. Chem. 37 (6), 1606–1613.

Kandeler, E., Dick, R.P., 2006. Soil enzymes: spatial distribution and function in agroecosystems. Biodivers. Agric. Prod. Syst. 263–286.

Keyes, S.D., et al., 2013. High resolution synchrotron imaging of wheat root hairs growing in soil and image based modelling of phosphate uptake. N. Phytol. 198 (4), 1023–1029.

Kim, H.S., et al., 2010. Atmospherically stable nanoscale zero-valent iron particles formed under controlled air contact: characteristics and reactivity. Environ. Sci. Technol. 44 (5), 1760–1766.

Koelmel, J., et al., 2013. Investigation of gold nanoparticles uptake and their tissue level distribution in rice plants by laser ablation-inductively coupled-mass spectrometry. Environ. Pollut. 174, 222–228.

Kopittke, P.M., et al., 2020. Methods to visualize elements in plants. Plant Physiol. 182 (4), 1869.

Kottegoda, N., et al., 2017. Urea-hydroxyapatite nanohybrids for slow release of nitrogen. ACS Nano 11 (2), 1214–1221.

Kumar, N., et al., 2018. Sulfidation mechanisms of Fe(iii)-(oxyhydr)oxide nanoparticles: a spectroscopic study. Environ. Sci. Nano 5 (4), 1012–1026.

Kuzyakov, Y. & B. S. Razavi, 2019. Soil Biology and Biochemistry.

Lewis, R.W., et al., 2019. Nanotoxicity of engineered nanomaterials (ENMs) to environmentally relevant beneficial soil bacteria–a critical review. Nanotoxicology 13 (3), 392–428.

Li, Y.F., et al., 2015. Synchrotron radiation techniques for nanotoxicology. Nanomedicine 11 (6), 1531–1549.

Li, M., et al., 2017. The transformation and fate of silver nanoparticles in paddy soil: effects of soil organic matter and redox conditions. Environ. Sci. Nano 4 (4), 919–928.

Lin, D., et al., 2017. Role of pH and ionic strength in the aggregation of TiO2 nanoparticles in the presence of extracellular polymeric substances from Bacillus subtilis. Environ. Pollut. 228, 35–42.

Lombi, E., Susini, J., 2009. Synchrotron-based techniques for plant and soil science: opportunities, challenges and future perspectives. Plant Soil. 320 (1–2), 1–35.

Lowry, G.V., et al., 2012. Transformations of nanomaterials in the environment. Environ. Sci. Technol. 46 (13), 6893–6899.

Lubick, N., 2008. Nanosilver toxicity: ions, nanoparticles-or both?". Environ. Sci. Technol. 42 (23), 8617-8617.

Lundqvist, M., et al., 2011. The evolution of the protein corona around nanoparticles: a test study. ACS Nano 5 (9), 7503–7509.

Luo, H.-W., et al., 2017. Probing the biotransformation of hematite nanoparticles and magnetite formation mediated by Shewanella oneidensis MR-1 at the molecular scale. Environ. Sci. Nano 4 (12), 2395–2404.

Luo, M., et al., 2018. Properties of different natural organic matter influence the adsorption and aggregation behavior of TiO_2 nanoparticles. J. Saudi Chem. Soc. 22 (2), 146–154.

Lv, J., et al., 2019. Uptake, translocation, and transformation of metal-based nanoparticles in plants: recent advances and methodological challenges. Environ. Sci. Nano 6 (1), 41–59.

Lynch, I., et al., 2014. A strategy for grouping of nanomaterials based on key physico-chemical descriptors as a basis for safer-by-design NMs. Nano Today 9 (3), 266–270.

Ma, Y., et al., 2011. Phytotoxicity and biotransformation of La_2O_3 nanoparticles in a terrestrial plant cucumber (Cucumis sativus). Nanotoxicology 5 (4), 743–753.

Ma, R., et al., 2013. Sulfidation mechanism for zinc oxide nanoparticles and the effect of sulfidation on their solubility. Environ. Sci. Technol. 47 (6), 2527–2534.

Ma, Y.H., et al., 2015. Origin of the different phytotoxicity and biotransformation of cerium and lanthanum oxide nanoparticles in cucumber. Nanotoxicology 9 (2), 262–270.

Ma, Y., et al., 2017. Xylem and phloem based transport of CeO_2 Nanoparticles in hydroponic cucumber plants. Environ. Sci. Technol. 51 (9), 5215–5221.

Manrique, L.A., et al., 1991. Predicting cation-exchange capacity from soil physical and chemical-properties. Soil. Sci. Soc. Am. J. 55 (3), 787–794.

Marine Cotte, E. C., V. Mazel, V. Armando Solé, P. Richardin, Y. Taniguchi, P. Walter, et al., 2009. Combination of FTIR and X-Rays Synchrotron-Based Micro-Imaging Techniques for the Study of Ancient Paintings. A Practical Point of View.

Maser, J., et al., 2000. Soft X-ray microscopy with a cryo scanning transmission X-ray microscope: I. Instrumentation, imaging and spectroscopy. J. Microscopy-Oxford 197, 68–79.
Matysiak, M., et al., 2016. Proteomic approach to nanotoxicity. J. Proteom. 137, 35–44.
McManus, P., et al., 2018. Rhizosphere interactions between copper oxide nanoparticles and wheat root exudates in a sand matrix: influences on copper bioavailability and uptake. Environ. Toxicol. Chem. 37 (10), 2619–2632.
McRae, R., et al., 2009. In situ imaging of metals in cells and tissues. Chem. Rev. 109 (10), 4780–4827.
Mensch, A.C., et al., 2017. Natural organic matter concentration impacts the interaction of functionalized diamond nanoparticles with model and actual bacterial membranes. Environ. Sci. Technol. 51 (19), 11075–11084.
Monopoli, M.P., et al., 2011. Nanobiotechnology: nanoparticle coronas take shape. Nat. Nanotechnol. 6 (1), 11–12.
Mooney, S.J., et al., 2012. Developing X-ray computed tomography to non-invasively image 3-D root systems architecture in soil. Plant Soil 352 (1–2), 1–22.
Oburger, E., Schmidt, H., 2016. New methods to unravel rhizosphere processes. Trends Plant Sci. 21 (3), 243–255.
Pearson, R.G., 1963. Hard and soft acids and bases. J. Am. Chem. Soc. 85 (22), 3533–3539.
Philippot, L., et al., 2013. Going back to the roots: the microbial ecology of the rhizosphere. Nat. Rev. Microbiol. 11 (11), 789–799.
Pozebon, D., et al., 2014. Review of the applications of laser ablation inductively coupled plasma mass spectrometry (LA-ICP-MS) to the analysis of biological samples. J. Anal. At. Spectromet. 29 (12), 2204–2228.
Qu, J., et al., 2011. Synthesis of ZnO nanoparticles from Zn-hyperaccumulator (Sedum alfredii Hance) plants. Micro Nano Lett. 6 (3), 174–176.
Ren, M., et al., 2018. Diversity and contributions to nitrogen cycling and carbon fixation of soil salinity shaped microbial communities in Tarim Basin. Front. Microbiol. 9 (MAR).
Rico, C.M., et al., 2018. Cerium oxide nanoparticles transformation at the root-soil interface of barley (Hordeum vulgare L.). Environ. Sci. Nano 5 (8), 1807–1812.
Rodríguez-Celma, J., et al., 2016. Plant fluid proteomics: delving into the xylem sap, phloem sap and apoplastic fluid proteomes. Biochim. Biophys. Acta—Proteins Proteom. 1864 (8), 991–1002.
Roy, P., et al., 2017. Green synthesis of silver nanoparticles using azadirachta indica leaf extract and its antimicrobial study. Appl. Nanosci. (Switz.) 7 (8), 843–850.
Rui, Y., et al., 2015. Transformation of ceria nanoparticles in cucumber plants is influenced by phosphate. Environ. Pollut. 198, 8–14.
Salas, E.C., et al., 2010. Reduction of graphene oxide via bacterial respiration. ACS Nano 4 (8), 4852–4856.
Santoyo, G., et al., 2016. Plant growth-promoting bacterial endophytes. Microbiol. Res. 183, 92–99.
Schauer, N., Fernie, A.R., 2006. Plant metabolomics: towards biological function and mechanism. Trends Plant Sci. 11 (10), 508–516.
Schwab, F., et al., 2016. Barriers, pathways and processes for uptake, translocation and accumulation of nanomaterials in plants—critical review. Nanotoxicology 10 (3), 257–278.
Sekine, R., et al., 2015. Speciation and lability of Ag-, AgCl-, and Ag2S-nanoparticles in soil determined by X-ray absorption spectroscopy and diffusive gradients in thin films. Environ. Sci. Technol. 49 (2), 897–905.
Shams, M., et al., 2019. Influence of functional groups on the degradation of graphene oxide nanomaterials. Environ. Sci. Nano 6 (7), 2203–2214.

Sharma, J.K., et al., 2015. Green synthesis of CuO nanoparticles with leaf extract of Calotropis gigantea and its dye-sensitized solar cells applications. J. Alloy Compd. 632, 321–325.
Shukla, P. K., et al., 2016. Uptake, translocation, accumulation, transformation, and generational transmission of nanoparticles in plants. In:Kole, C., Kumar, D. S. Khodakovskaya, M. V. (Eds.), Plant Nanotechnology, pp. 183–218.
Singh, J., et al., 2018. 'Green' synthesis of metals and their oxide nanoparticles: applications for environmental remediation. J. Nanobiotechnol. 16 (1).
Singh, A., et al., 2019. Roles of silver-chloride complexations in sunlight-driven formation of silver nanoparticles. Environ. Sci. Technol.
Speight, J. G., 2019. Global Climate Change Demystified.
Spielman-Sun, E., et al., 2019. Nanoparticle surface charge influences translocation and leaf distribution in vascular plants with contrasting anatomy. Environ. Sci. Nano 6 (8), 2508–2519.
Spielman-Sun, E., et al., 2018. Temporal evolution of copper distribution and speciation in roots of Triticum aestivum exposed to CuO, Cu(OH)(2), and CuS nanoparticles. Environ. Sci. Technol. 52 (17), 9777–9784.
Stabryla, L.M., et al., 2018. Emerging investigator series: it's not all about the ion: support for particle-specific contributions to silver nanoparticle antimicrobial activity. Environ. Sci. Nano 5 (9), 2047–2068.
Su, Y., et al., 2019. Delivery, uptake, fate, and transport of engineered nanoparticles in plants: a critical review and data analysis. Environ. Sci. Nano 6 (8), 2311–2331.
Tan, B.C., et al., 2017. Proteomics in commercial crops: an overview. J. Proteom. 169, 176–188.
Tang, F., et al., 2002. Electrophoretic deposition behavior of aqueous nanosized zinc oxide suspensions. J. Am. Ceram. Soc. 85 (9), 2161–2165.
Tang, Z., Cheng, T., 2018. Stability and aggregation of nanoscale titanium dioxide particle ($nTiO_2$): effect of cation valence, humic acid, and clay colloids. Chemosphere 192, 51–58.
Tanveer, T., et al., 2014. Plant secretomics: identification, isolation, and biological significance under environmental stress. Plant Signal. Behav. 9 (JUN), e29426-29421-e29426-29412.
Treuel, L., et al., 2014. Physicochemical characterization of nanoparticles and their behavior in the biological environment. Phys. Chem. Chem. Phys. 16 (29), 15053–15067.
Twining, B.S., et al., 2003. Quantifying trace elements in individual aquatic protist cells with a synchrotron X-ray fluorescence microprobe. Anal. Chem. 75 (15), 3806–3816.
Valsami-Jones, E., Lynch, I., 2015. How safe are nanomaterials?". Science 350 (6259), 388–389.
Vančura, V., Hovadík, A., 1965. Root exudates of plants - II. Composition of root exudates of some vegetables. Plant Soil 22 (1), 21–32.
Vincent, D., et al., 2020. The Multiple Facets of Plant–Fungal Interactions Revealed Through Plant and Fungal Secretomics. 10 (1626).
Visser, S., Parkinson, D., 1992. Soil biological criteria as indicators of soil quality: soil microorganisms. Am. J. Altern. Agric. 7 (1–2), 33–37.
Wang, Z., et al., 2012. Xylem- and phloem-based transport of CuO nanoparticles in maize (Zea mays L.). Environ. Sci. Technol. 46 (8), 4434–4441.
Wang, B., et al., 2015a. Probing the interaction at nano-bio interface using synchrotron radiation-based analytical techniques. Sci. China Chem. 58 (5), 768–779.
Wang, F., et al., 2015b. Transformation of graphene oxide by ferrous iron: environmental implications. Environ. Toxicol. Chem. 34 (9), 1975–1982.
Wang, D., et al., 2016. Where does the toxicity of metal oxide nanoparticles come from: the nanoparticles, the ions, or a combination of both? J. Hazard. Mater. 308, 328–334.
Wang, P., et al., 2017. Characterizing the uptake, accumulation and toxicity of silver sulfide nanoparticles in plants. Environ. Sci.-Nano 4 (2), 448–460.

Wang, J., et al., 2020a. Bacteria mediated Fenton-like reaction drives the biotransformation of carbon nanomaterials. Sci. Total Env. 746, 141020.

Wang, Z., et al., 2020b. Nano-enabled improvements of growth and nutritional quality in food plants driven by rhizosphere processes. Environ. Int. 142, 105831.

Xiao, H., et al., 2018. Accelerated effects of nano-ZnO on phosphorus removal by Chlorella vulgaris: formation of zinc phosphate crystallites. Sci. Total Environ. 635, 559–566.

Xie, C., et al., 2019a. Bacillus subtilis causes dissolution of ceria nanoparticles at the nano-bio interface. Environ. Sci. Nano 6 (1), 216–223.

Xie, C.J., et al., 2019b. Effects of foliar applications of ceria nanoparticles and $CeCl_3$ on common bean (Phaseolus vulgaris). Environ. Pollut. 250, 530–536.

Zhang, M., et al., 2008. Image deconvolution in digital autoradiography: a preliminary study. Med. Phys. 35 (2), 522–530.

Zhang, Z.Y., et al., 2011. Uptake and distribution of ceria nanoparticles in cucumber plants. Metallomics 3 (8), 816–822.

Zhang, P., et al., 2012a. Distribution and bioavailability of ceria nanoparticles in an aquatic ecosystem model. Chemosphere 89 (5), 530–535.

Zhang, P., et al., 2012b. Comparative toxicity of nanoparticulate/bulk Yb_2O_3 and $YbCl_3$ to cucumber (Cucumis sativus). Environ. Sci. Technol. 46 (3), 1834–1841.

Zhang, P., et al., 2012c. Biotransformation of ceria nanoparticles in cucumber plants. ACS Nano 6 (11), 9943–9950.

Zhang, P., et al. (2015). Interactions between engineered nanomaterials and plants: phytotoxicity, uptake, translocation, and biotransformation. In: Nanotechnology Plant Sciences, Springer: 77–99.

Zhang, P., et al., 2017a. Phytotoxicity, uptake and transformation of nano-CeO_2 in sand cultured romaine lettuce. Environ. Pollut. 220, 1400–1408.

Zhang, P., et al., 2017b. Shape-dependent transformation and translocation of ceria nanoparticles in cucumber plants. Environ. Sci. Technol. Lett. 4 (9), 380–385.

Zhang, P., et al., 2019a. Plant species-dependent transformation and translocation of ceria nanoparticles. Environ. Sci. Nano 6 (1), 60–67.

Zhang, P., et al., 2019b. Stable isotope labeling of metal/metal oxide nanomaterials for environmental and biological tracing. Nat. Protoc. 14 (10), 2878–2899.

Zhang, W., et al., 2020. Response of soil microbial communities to engineered nanomaterials in presence of maize (Zea mays L.) plants. Environ. Pollut. 267, 115608.

Zhao, L.J., et al., 2012. Effect of surface coating and organic matter on the uptake of CeO_2 NPs by corn plants grown in soil: insight into the uptake mechanism. J. Hazard. Mater. 225, 131–138.

Chapter 9

Future questions and approaches in plant–nanoparticle research

Luca Pagano and Marta Marmiroli
Department of Chemistry, Life Sciences and Environmental Sustainability, University of Parma, Parco Area delle Scienze 11/A, Parma, Italy

Chapter Outline

9.1 The present state and the next challenges

The general interest on engineered nanoparticles (ENPs) increases with particular regard to the implications on environmental fate, physico–chemical properties, exposure effects on living organisms, and trophic transfer (Gardea-Torresdey et al., 2014; Keller et al., 2013; Ma et al., 2018; Majumdar et al., 2019). On one hand, Ma et al. (2018) described in detail the general molecular effects of ENPs in plant cells, tissues and organs from exposure to response to optimize the safe and sustainable application of nanotechnology in agriculture for enhancing yield and food quality. Majumdar et al. (2019) highlighted the functional influence of physico–chemical modification, such as chemistry and surface coating, on ENPs fate and effects. Gardea-Torresdey et al. (2014) focused the attention on trophic transfer studies to predict quantitative modeling of ENPs fate, the use of nano-enabled agricultural products, and the improvement of nanotoxicity assessment in plants in more realistic scenarios of exposure. For this purpose,

Plant Exposure to Engineered Nanoparticles. DOI: https://doi.org/10.1016/B978-0-323-85032-2.00009-9

it is necessary to consider the long-term effects as well as chronic and low dose scenarios, including the effects on the plants, and on their environment. Keller et al. (2013) has shown how the material flow of ENPs can be used to quantify emissions in terms of fate and transport models necessary to estimate ENPs concentrations in the environment. Nonetheless, basic research keeps on investigating the principia of the ENPs exposure in terms of molecular pathways involved, but also improving the methods to assess ENPs exposure and effects on plant tissues and organs. In this context, the physico−chemical properties of the ENPs are becoming central for the understanding of the ENPs bioactive forms, their modes of action, and their functional targets (Burkard et al., 2020; Marmiroli et al., 2020; Pagano et al., 2018).

Basic research fulfills a fundamental role in ENPs applicability by highlighting the importance of inherent nanomaterial hazard determination, assessing the actual risk posed to agricultural systems, and the need of more sensitive endpoints to mechanistically characterize the effects of ENPs (positive, neutral or negative) (Servin and White, 2016). Basic research plays also a pivotal role in terms of designing safer and sustainable formulations for agrifood production purposes such as plant nutrients delivery, growth-promoting compounds, and disease management (Kah et al., 2019; Lowry et al., 2019; Ma et al., 2020; Mali et al., 2020). Contextually, Mali et al. (2020) described the advances in the fields of nanobiotechnology by providing novel toolsets able to manipulate and enhance crop production including nanoparticles, nanoemulsions, nanofibres, or nanocapsules. ENPs provide functional platforms for agrochemicals delivery needed for plant growth enhancement and to increase resistance to stresses. Relevant examples can be found in Ma et al. (2020) who described a thermodynamic model on how morphology and matrix effects are implicated in metal release and how pathogens suppression and nutrient release of metal-based ENPs alleviate the effects of potential diseases in widely used crops. The findings showed the importance of tuning the physico−chemical properties of ENPs (i.e., morphology, composition, and dissolution) to develop safer and more sustainable nanoformulations for agriculture in terms of targeting and delivery, stress management, and efficient resources utilization (Lowry et al., 2019).

The present advances in ENPs basic and applied research pose new questions and major challenges that should be addressed in the future: (1) how to improve available tools on ENPs exposure and effects assessment; (2) how to characterize relevant physico−chemical forms implicated in the physiological and molecular response; (3) how to optimize ENPs chemistry in terms of safety and sustainability. The present chapter aims to illustrate the major questions and challenges in plant-ENMs interactions, from the implications at research level to the applicability in actual agrifood production (Fig. 9.1).

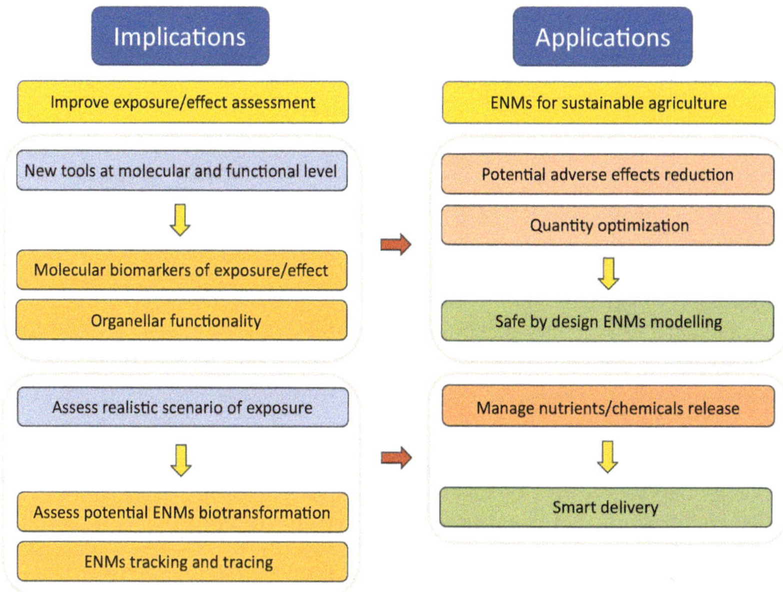

FIGURE 9.1 Schematic representation of the main challenges and potential solutions involved in exploring the plant- engineered nanoparticles interaction from basic research (implication) to potential application in agrifood production.

9.2 Assessment of exposure and effects to improve the available tools

Owing to the multifacet applications of ENPs, there are still no consensus on endpoints common to different ENP classes (e.g., physico–chemical properties, biological effects). Specific biomarkers for exposure, effect, and susceptibility to ENPs remain under investigation in plants: (1) biomarkers of exposure, which include secondary metabolites and physiological parameters that reflect the dose/effects relationships; (2) biomarkers of effects, which highlight modifications at cellular and molecular levels and reflect gene expression/protein abundance of specific targets by ENPs; (3) biomarkers of susceptibility, which indicate the constitutive responsiveness to a contaminant and whose manifestation can be related to tolerance/resistance pathways or phenotypes considered hypersensitive. Some of these types of biomarkers reflect the responses to ENP interaction at whole organism level for providing useful information to risk assessment protocols (Schudoma et al., 2012).

Examples of ENP biomarkers of exposure in plants are related to photosynthetic activity, cellular viability, and oxidative stress (Ma et al., 2015). In this context, the relevant involvement of sub-cellular targets such as mitochondria and chloroplasts appear to be crucial in ENPs exposure/effects. These organelles are primarily involved in metabolic functions and scavenging of reactive oxygen species, but there are more factors to consider, since mitochondria and chloroplast are not only the cellular energy power plant, but they also have important regulatory and developmental roles in cell function (Pagano et al., 2018). Specifically, mitochondria show a high level of preservation in response to ENPs in simpler to higher eukaryotic models, while evidence of chloroplasts conservation in ENP response across different plant species has also been reported (Pagano et al., 2018). This important involvement is testified by organelle targets such as the chloroplast gene *ORF31*, whose gene product (petL) is implicated in electron transport at the chloroplast inner membrane system as a part of the cytochrome b_6f complex. *ORF31* is highly responsive during abiotic stress conditions, particularly oxidative stress. The same gene has been found to be strongly modulated by stress due to single- or co-exposure to different ENPs (Pagano et al., 2017). This finding not only underlies the central role of the chloroplast in ENP cellular response, but also pinpoints the role of *ORF31* gene as a biomarker of exposure/effect of ENP treatment in plant.

By coupling physiological and molecular evidences, it is possible to provide mechanistic insights related to cellular components, molecular pathways, and biological processes involved in plant responses to ENP exposure. The application of biomarkers opens up new perspectives on monitoring, screening, and assessing ENPs effects in plants. This will lead the way to complete risk assessment in plants (from model organisms to species of agricultural crops) and its environmental implications. In a more applied context, the biomarker strategy can be instrumental to plant selection and variety improvement, biofortification, and heavy metal remediation purposes (Pagano et al., 2017).

9.3 Relationships between physico–chemical forms and biological effectiveness

In recent years, major advancements on determining ENPs physico–chemical state and associated risks have been achieved in controlled systems as well as in more realistic scenarios (Servin and White, 2016). Particular relevance has been posed to dosimetry, surface charge, particle aggregation (DeLoid et al., 2017), and subsequently on ENPs availability and potential effects (positive, neutral, or negative) on plant physiology and molecular integrity (Du et al., 2018; Schwab et al., 2016) to characterize and predict the pathways involved in the response. In this context, it increases the relevance of the knowledge regarding ENPs biological behavior in relation not only to the physico–chemical properties of the pristine particles, but even more with the modifications and functionalization to which the particle can be subjected when in contact with the plant systems. This interaction can be defined as biotransformation (Marmiroli et al., 2020). The study of the biotransformation can shed light on the fate of ENPs within the plant tissues and organs which provides mechanistic understanding of their behavior and supports the modeling of the ENPs exposure in its complexity (Milosevic et al., 2020).

To achieve this, several points have to be considered, starting from the dynamics of the transformation, which mainly depends on the particle physico–chemical properties (size, shape, surface charge, composition) that govern their stability and capacity to dissolve. In a second instance, particle interacts with cellular structures such as corona proteins, sugars, glutathione, and secondary metabolites. These may determine potential modifications or functionalization with consequences on particle transport, reactivity, and stability. These new particle forms, become in many cases, responsible on the physiological and molecular responses observed (Marmiroli et al., 2020). In studying such complex dynamics, the analytical methodologies of metal/isotope detection and quantification were coupled with synchrotron-based X-ray fluorescence mapping (μ-XRF) and X-ray absorption spectroscopy (XAS) techniques, which are able to localize the element presence and determine its speciation and atomic environment within the plant tissues (Castillo-Michel et al., 2017). A relevant example highlighting the differences between ENP physico–chemical stability and their relative fate during exposure is the report of Servin et al. (2017), which analyzed the effects of CuO-NPs at different trophic levels starting from lettuce (*Lactuca sativa*), a primary consumer (cricket, *Acheta domesticus*), and a secondary consumer (lizard, *Anolis carolinensis*). The experiments were focused not only on the trophic transfer but also on the effects of ENP aging (weathering) within the soil. Interestingly, μ-XRF analysis of roots exposed to weathered CuO-NPs showed a homogenous distribution of Cu within the tissues while μ-XANES analysis showed a near complete transformation of Cu(II) to Cu(I) bond with sulfur and oxide complexes in the tissues; however, analysis of non-weathered treatment revealed that Cu remained mostly as CuO-NPs. In another work, Servin et al. (2018) demonstrated the bioaccumulation of CeO_2-NPs within tissues of the earthworm *Eisenia fetida*. The study was coupled with the utilization of different types of biochar, with the aim of clarifying the concerted biochar/ENPs effects on internalization in and viability of earthworm. Both μ-XRF and μ-XANES were used for cerium localization, speciation, and persistence within earthworm tissues at CeO_2 and biochar/CeO_2 conditions of exposure. Analysis by μ-XANES showed that, within the earthworm tissues, Ce remained predominantly in the form of Ce (IV), while only a minimal quantity was reduced to Ce(III) form, confirming the high particle stability of CeO_2-NPs. On the other hand, Hernandez-Viezcas et al. (2013) highlighted the effects of CeO_2-NPs and ZnO-NPs in soybean (*Glycine max*) where using synchrotron μ-XRF and μ-XANES analyses were performed to determine the localization and forms of Ce and Zn present in the treated plants. This study highlighted no residues of ZnO-NPs within the tissues but Zn-citrate have been observed. μ-XANES analyses performed on CeO_2-NP-treated plant tissues showed that Ce remained mostly as nanoparticle within the plant. In this case, while a small percentage of Ce^{4+} ionic form was observed, Ce^{3+} was the most abundant ionic form. Another interesting case is represented by quantum dots (QDs) due to their peculiar physico−chemical properties, their relevant effects on living organisms, and also considering their important fields of application. Marmiroli et al. (2020), through a structural extended X-ray absorption fine structure (EXAFS) spectroscopic analysis of *Arabidopsis thaliana* tissues exposed to CdS QDs, highlighted a potential pathway of biotransformation of a particular type of quantum dots based on CdS. XAS data showed how CdS QDs are subject to biotransformation within the plant tissues and organs. The particle original structure was modified but not totally lost and Cd atoms were not simply released as Cd^{2+} ions.

In addition to this approach, several other techniques can be employed to detect particle presence and specific location within organs and tissues. More information can be obtained in tracking transport from source to accumulation in plant tissues (top-down approach) or tracing the source of exposure (bottom-up approach) of the ENPs both in vitro and/or in vivo (Lv et al., 2019). Contextual examples can be taken into account, with regard to relevant elements such as Cu, Fe, Zn, which can show potential toxic effects at specific concentrations or conditions, but that are also fundamental for plant homeostasis and metabolic functions (Yin et al., 2017). Davis et al. (2017) showed the possibility to determine, through a non-invasive quantitative tracking procedure, the location and accumulation sites of radiolabeled Cu-based ENPs, during in vitro exposure in *L. sativa*. Furthermore, the study investigated ENPs transport and distribution over time using autoradiography, positron emission tomography (PET)/computed tomography (CT), obtaining insights on the filtering effect of plant cell walls along the water transport pathway as well as the roots to shoots dynamics of the ENPs transfer.

The complexity can be even increased, when considering cases of more realistic *in vivo* conditions of exposure, or, in other words, considering the plant−soil interactions. The mechanisms implicated in ENPs dissolution and transformation at the level of rhizosphere surface can be not only affected by the molecular processes that occur during ENPs intake but also from the effects of soil pH, organic matter, mineral constituents, and microbial community of the soil (Coutris et al., 2012; Sillen et al., 2015). Laycock et al. (2017), for example, utilized a multi-isotope tracer approach based on differential ratios of zinc stable isotope (^{68}Zn, ^{64}Zn) during treatment with NPs or salts of ZnO. The study was performed on earthworm *Eisenia andrei* exposed to soil amended with Zn-based compounds. Analyses performed on soils, pore water, and earthworm tissues by inductively coupled plasma mass spectrometry (ICP-MS) showed the distribution and uptake of the different Zn forms, which underlined the particle dissolution in the root−soil system.

9.4 From reducing adverse effects to smart delivery

Model organisms and edible plants can be used to gain information about ENPs release in the environment with regard to environmental and agrifood protection, but still many questions remain unanswered (Nel et al., 2015; Reddy et al., 2016). This lack of understanding mainly depends on the complexity/heterogeneity of ENPs modes of action, but also on the lack of suitable benchmark and standard protocols for testing nanotoxicity (Hjorth et al., 2017; Petersen et al., 2014). From environmental perspective, some information are available on trophic transfer (Gardea-Torresdey et al., 2014), aging (Servin et al., 2017), and interaction with other contaminants (Deng et al., 2017) including other ENPs (Pagano et al., 2017). From the physico–chemical point of view, models of dosimetry (DeLoid et al., 2017) and quantitative structure activity relationship (QSAR) (Burello, 2017) have been provided. The continuous deployment of ENPs technologies raises the urgency to develop more accurate methods of assessing ENPs exposure/effects as well as strategies able to minimize their negative effects and optimize the quantity utilized. In these contexts, several parameters need to be taken into account, such as the presence of coatings which may facilitate tracing and detection, at the cost of invalidating measurements related to the biological effects (Majumdar et al., 2019; Petersen et al., 2014). Furthermore, physico–chemical parameters related to surface charge or particle size cannot be necessarily preserved when introduced into complex exposure environmental scenarios (DeLoid et al., 2017; Deng et al., 2017; Marmiroli et al., 2020). Reproducibility in environmental conditions can be subject to a certain level of bias ascribable to exposure-specific parameters that include interactions with the medium utilized (Deng et al., 2017), the interaction with cellular components or biomolecules (e.g., corona proteins) (Nierenberg et al., 2018), and the biotransformation derived from the interaction with/within the organisms (Marmiroli et al., 2020). These types of interaction can be considered key factors when the ENPs are applied to products of interest in agriculture and food production. The impact on the ecosystems, from the rhizosphere microbiome to the plant tissues and organs (including edible parts), became of primary interest when considering the potential negative effects of ENPs in soils (Servin and White, 2016). Plant root microbiome plays a key role in ENPs detoxification or stabilization prior to root uptake, soil fertility, plant production, and ENPs tolerance/detoxification by plants (Sillen et al., 2015). This effort should not only provide alternative formulations for ENPs, but also minimizes the inherent adverse effects and optimizes quantity utilized. The goal is to develop more sustainable ENPs specifically synthetized to be "safer by design" (Burello and Worth, 2015; Geraci et al., 2015).

Accordingly, new and more effective approaches applied to the design of nanomaterials find their completion in products able to optimize the delivery of nutrient or pesticide by taking into account a series of physico–chemical properties which include stability, solubility, effectiveness, time-controlled release, targeting activity, safety of delivery and disposal (Kah et al., 2018a; 2019). These parameters have to be assessed not only in controlled condition (e.g., laboratory, green house) but also in more realistic scenarios of exposure where there is high variability of conditions (e.g., soil type, temperature, humidity, plant evapotranspiration, and water regime). In addition, physiological barriers related to ENPs uptake and translocation must be considered (Schwab et al., 2016).

Differential classes of compounds can be taken into account: (1) particles or particles emulsions, to clarify if ENPs can be candidates for alternative strategies of fertilization/pest control (Elmer et al., 2018; Ma et al., 2020); (2) encapsulated nanostructures, with aim to control the metal/nutrient release (Marchiol et al., 2019); (3) nanocomposites incorporated in organic polymers as a carrier, in many cases derived from transformed biological residues (e.g., biochar) (Marmiroli et al., 2018). Few relevant examples can be reported. Elmer et al. (2018) observed that the foliar treatment of watermelon (*Citrullus lanatus*) with CuO-based ENPs suppressed infection from *Fusarium oxysporum* spp., with an increase in plant biomass and production yield in comparison with the untreated controls and infected control plants. This suggested the potential role of CuO-NPs in delivering nutrients and, moreover, serving as potential disease suppressor. Marchiol et al. (2019), on the other hand, showed the potential of nano-hydroxyapatite as nanofertilizer, in particular as phosphorous supplier and carrier of micronutrients and molecules. Germination of *Solanum lycopersicum* was not affected by high concentration of stabilized nano-hydroxyapatite, while root elongation was strongly stimulated. In hydroponic growth condition, no phytotoxic effects were observed. While considering potential systems of delivery for nanopesticides, Kah et al. (2018b) investigated photodegradation and sorption of the insecticide clothianidin in different nanoformulations to estimate key parameters (such as durability) for the exposure assessment of nanopesticides. Results suggested that the main functional parameter was how clothianidin nanoformulations were rapidly released from the nanocarrier systems, highlighting a limited durability in water and in soil.

Nanocomposites used in carrying agrochemicals offer smart solutions and open possibilities in the adoption of new and sustainable strategies in agricultural fertilization to improve soil fertility, conferring the benefits of biodegradability and biocompatibility. As an example, biochars, which are transformed residual biomasses of plant and animal through pyrolysis produce new materials with nanometric scale and properties, improve soil quality due to the increased porosity, water retention, and capacity for metal adsorption from soil and irrigation water (Marmiroli et al., 2018). On the other hand, polymeric nanoparticles represent one major class of nanocomposites that can be considered for potential sustainable solution in agriculture as they can be used to efficiently deliver pesticides and growth promoters for crops and antibiotics for livestock (Shakiba et al., 2020). Relevant examples of polymeric carriers include zein, chitosan, poly-caprolactone (PCL) or poly-lactic-co-glycolic acid (PLGA), all of which demonstrated high targeting or delivery efficiency and enhanced uptake compared to traditional formulations (De Oliveira et al., 2018; Makadia and Siegel, 2011; Shakiba et al., 2020).

Nanoformulations open new possibilities for sustainable agricultural practices and offer great opportunities on plant nutrition optimization and stress tolerance; however, limitations still need to be carefully considered before market implementation. The adoption of new procedures for proper characterization of the risk related to environment and human health is still crucial and certainly remains one of the main challenges in future research (Zulfiqar et al., 2019).

9.5 Conclusions

The study of ENP–plant interactions is various and complex ranging from physics and chemistry to biology and ecology. The complexity of the field and the new information gained from the multitude of studies published in recent years are pushing forward the research to the resolution of new questions and challenges. There appears to be circularity in experimental approach from basic research (implication), where the different aspects are explored, to agrifood production (application) in which the knowledge gained become functional to potential utilization. Conversely, the information related to the ENPs applicability opens new questions from technological and methodological points of view.

Several examples of physiological biomarkers of ENPs exposure and effect have been suggested, including those based on genes and proteins involved in photosynthetic activity, cellular viability, oxidative stress, and DNA damage (Ma et al., 2018). Biological and statistical models to identify molecular markers across different plant species have been developed, but still research is required. More efficient physiological measurements, coupled with information related to the reprogramming of specific biochemical and molecular pathways involved in the ENPs exposure, can provide also new mechanistic insight on plant response during ENPs–plant interaction. In this scenario, the discovery of ENP's tolerant and sensitive species represents a functional and informational approach for highlighting targets and pathways exploited during the ENPs treatments, from simple eukaryotes to higher plants (Gallo et al., 2021; Marmiroli et al., 2014, 2016). Furthermore, the investigation of the biotransformation of ENPs within plant tissues, which represent a key path to understand ENPs response, has gained more relevance in the scientific panorama on planning and approaching the study of plant–ENP interaction (Marmiroli et al., 2020; Servin et al., 2017).

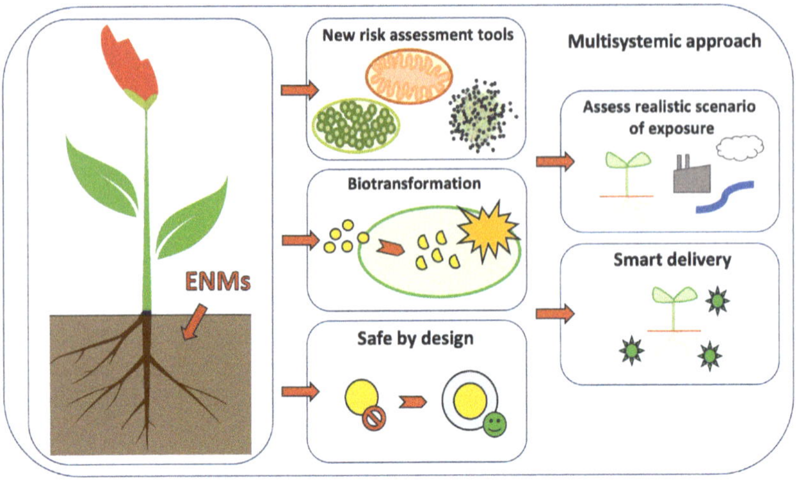

FIGURE 9.2 From research to relevant applications: an example of multisystems approach for the study of engineered nanoparticle-plant interactions.

Considering the expansion of nanotechnology, integrated approaches (Fig. 9.2) will aid in the accurate assessment of risk and ensure the sustainable use ENPs across a range of applications in important fields such as agriculture.

Acknowledgments

The authors wish to thank the support of FIL ("Fondi Locali per la Ricerca") 2018 and acknowledge Emeritus Prof. Nelson Marmiroli for inspiring us and revising the chapter.

References

Burello, E., 2017. Review of (Q)SAR models for regulatory assessment of nanomaterials risks. Nano Impact. 8, 48–58.

Burello, E., Worth, A.P., 2015. A rule for designing safer nanomaterials: do not interfere with the cellular redox equilibrium. Nanotoxicology 9, 116–117.

Burkard, M., Betz, A., Schirmer, K., Zupanic, A., 2020. Common gene expression patterns in environmental model organisms exposed to engineered nanomaterials: a *meta*-analysis. Environ. Sci. Technol. 54 (1), 335–344.

Castillo-Michel, H.A., Larue, C., Pradas del Real, A.E., Cotte, M., Sarret, G., 2017. Practical review on the use of synchrotron based micro- and nano- X-ray fluorescence mapping and X-ray absorption spectroscopy to investigate the interactions between plants and engineered nanomaterials. Plant Physiol. Biochem. 110, 13–32.

Coutris, C., Joner, E.J., Oughton, D.H., 2012. Aging and soil organic matter content affect the fate of silver nanoparticles in soil. Sci. Tot. Environ. 420, 327–333.

Davis, R.A., Rippner, D.A., Hausner, S.H., Parikh, S.J., McElrone, A.J., Sutcliffe, J.L., 2017. In vivo tracking of Copper-64 radiolabeled nanoparticles in Lactuca sativa. Environ. Sci. Technol. 51 (21), 12537–12546.

De Oliveira, J.L., Campos, E.V.R., Pereira, A.E.S., Pasquoto, T., Lima, R., Grillo, R., et al., 2018. Zein nanoparticles as eco-friendly carrier systems for botanical repellents aiming sustainable agriculture. J. Agric. Food Chem. 66, 1330–1340.

DeLoid, G.M., Cohen, J.M., Pyrgiotakis, G., Demokritou, P., 2017. Preparation, characterization, and in vitro dosimetry of dispersed, engineered nanomaterials. Nat. Protoc. 12, 355–371.

Deng, R., Lin, D., Zhu, L., Majumdar, S., White, J.C., Gardea-Torresdey, J.L., et al., 2017. Nanoparticle interactions with coexisting contaminants: joint toxicity, bioaccumulation and risk. Nanotoxicology 11, 591–612.

Du, W., Xu, Y., Yin, Y., Ji, R., Guo, H., 2018. Risk assessment of engineered nanoparticles and other contaminants in terrestrial plants. COESH 6, 21–28.

Elmer, W., De La Torre-Roche, R., Pagano, L., Majumdar, S., Zuverza-Mena, N., Dimkpa, C., et al., 2018. Effect of metalloid and metal oxide nanoparticles on fusarium wilt of watermelon. Plant Disease 102 (7), 1394–1401.

Gallo, V., Zappettini, A., Villani, M., Marmiroli, N., Marmiroli, M., 2021. Comparative analysis of proteins regulated during cadmium sulfide quantum dots response in arabidopsis thaliana wild type and tolerant mutants. Nanomaterials 11, 615.

Gardea-Torresdey, J.L., Rico, C.M., White, J.C., 2014. Trophic transfer, transformation, and impact of engineered nanomaterials in terrestrial environments. Environ. Sci. Technol. 48, 2526–2540.

Geraci, C., Heidel, D., Sayes, C., Hodson, L., Schulte, P., Eastlake, A., et al., 2015. Perspectives on the design of safer nanomaterials and manufacturing processes. J. Nanopart. Res. 17, 366.

Hernandez-Viezcas, J.A., Castillo-Michel, H., Andrews, J.C., Cotte, M., Rico, C., Peralta-Videa, J.R., et al., 2013. In situ synchrotron X-ray fluorescence mapping and speciation of CeO_2 and ZnO nanoparticles in soil cultivated soybean (*Glycine max*). ACS Nano 7 (2), 1415–1423.

Hjorth, R., Holden, P.A., Foss Hansen, S., Colman, B.P., Grieger, K., Ogilvie Hendren, C., 2017. The role of alternative testing strategies in environmental risk assessment of engineered nanomaterials. Environ. Sci. Nano 4, 292.

Kah, M., Kookana, R.S., Gogos, A., Bucheli, T.D., 2018a. A critical evaluation of nanopesticides and nanofertilizers against their conventional analogues. Nat. Nanotechol. 13, 677–684.

Kah, M., Walch, H., Hofmann, T., 2018b. Environmental fate of nanopesticides: durability, sorption and photodegradation of nanoformulated clothianidin. Environ. Sci. Nano 5, 882–889.

Kah, M., Tufenkji, N., White, J.C., 2019. Nano-enabled strategies to enhance crop nutrition and protection. Nat. Nanotechnol. 14, 532–540.

Keller, A.A., McFerran, S., Lazareva, A., Suh, S., 2013. Global life cycle releases of engineered nanomaterials. J. Nanopart. Res. 15, 1692–1709.

Laycock, A., Romero-Freire, A., Najorka, J., Svendsen, C., van Gestel, C.A.M., Rehkamper, M., 2017. Novel multi-isotope tracer approach to test ZnO nanoparticle and soluble Zn bioavailability in joint soil exposures. Environ. Sci. Technol. 51, 12756–12763.

Lowry, G.V., Avellan, A., Gilbertson, L.M., 2019. Opportunities and challenges for nanotechnology in the agri-tech revolution. Nat. Nanotechnol. 14, 517–522.

Lv, J., Christie, P., Zhang, S., 2019. Uptake, translocation, and transformation of metal-based nanoparticles in plants: recent advances and methodological challenges. Environ. Sci.: Nano 6, 41–59.

Ma, C., White, J.C., Dhankher, O.P., Xing, B., 2015. Metal-based nanotoxicity and detoxification pathways in higher plants. Environ. Sci. Technol. 49, 7109–7122.

Ma, C., White, J.C., Zhao, J., Zhao, Q., Xing, B., 2018. Uptake of engineered nanoparticles by food crops: characterization, mechanisms, and implications. Annu. Rev. Food Sci. Technol. 9, 129–153.

Ma, C., Borgatta, J., Hudson, B.G., Abbaspour-Tamijani, A., De La Torre-Roche, R., Zuverza-Mena, N., et al., 2020. Advanced material modulation of nutritional and phytohormone status alleviates damage from soybean sudden death syndrome. Nat. Nanotechnol. . Available from: https://doi.org/10.1038/s41565-020-00776-1.

Majumdar, S., Ma, C., Villani, M., Zuverza-Mena, N., Pagano, L., Huang, Y., et al., 2019. Surface coating determines the response of soybean plants to cadmium sulfide quantum dots. NanoImpact 14100151. Available from: https://doi.org/10.1016/j.impact.2019.100151.

Makadia, H.K., Siegel, S.J., 2011. Poly lactic-co-glycolic acid (PLGA) as biodegradable controlled drug delivery carrier. Polymers 3, 1377–1397.

Mali, S.C., Raj, S., Trivedi, R., 2020. Nanotechnology a novel approach to enhance crop productivity. Biochem. Biophys. Rep. 24100821.

Marchiol, L., Filippi, A., Adamiano, A., Degli Esposti, L., Iafisco, M., Mattiello, A., et al., 2019. Influence of hydroxyapatite nanoparticles on germination and plant metabolism of tomato (*Solanum lycopersicum* L.): preliminary evidence. Agronomy 9, 161.

Marmiroli, M., Pagano, L., Savo Sardaro, M.L., Villani, M., Marmiroli, N., 2014. Genome-wide approach in Arabidopsis thaliana to assess the toxicity of cadmium sulfide quantum dots. Environ. Sci. Technol. 48, 5902–5909.

Marmiroli, M., Pagano, L., Pasquali, F., Zappettini, A., Tosato, V., Bruschi, C.V., et al., 2016. A genome-wide nanotoxicology screen of Saccharomyces cerevisiae mutants reveals the basis for cadmium sulphide quantum dot tolerance and sensitivity. Nanotoxicology 10 (1), 84–93.

Marmiroli, M., Bonas, U., Imperiale, D., Lencioni, G., Mussi, F., Marmiroli, N., et al., 2018. Structural and functional features of chars from different biomasses as potential plant amendments. Front. Plant Sci. 9, 1119.

Marmiroli, M., Lepore, G.O., Pagano, L., d'Acapito, F., Gianoncelli, A., Villani, M., et al., 2020. The fate of CdS quantum dots in plants as revealed by extended X-ray absorption fine structure (EXAFS) analysis. Environ. Sci.: Nano. Available from: https://doi.org/10.1039/C9EN01433K.

Milosevic, A., Romeo, D., Wick, P., 2020. Understanding nanomaterial biotransformation: an unmet challenge to achieving predictive nanotoxicology. Small, 1907650.

Nel, A.E., Parak, W.J., Chan, W.C., Xia, T., Hersam, M.C., Brinker, C.J., et al., 2015. Where are we heading in nanotechnology environmental health and safety and materials characterization? ACS Nano 9, 5627–5630.

Nierenberg, D., Khaled, A.R., Flores, O., 2018. Formation of a protein corona influences the biological identity of nanomaterials. Rep. Pract. Oncol. Radiother 23 (4), 300–308.

Pagano, L., Pasquali, F., Majumdar, S., De La Torre-Roche, R., Zuverza-Mena, N., Villani, M., et al., 2017. Exposure of Cucurbita pepo to binary combinations of engineered nanomaterials: physiological and molecular response. Environ. Sci. Nano 4 (7), 1579–1590. Available from: https://doi.org/10.1039/c7en00219j.

Pagano, L., Maestri, E., Caldara, M., White, J.C., Marmiroli, N., Marmiroli, M., 2018. Engineered nanomaterial activity at the organelle level: impacts on the chloroplasts and mitochondria. ACS Sustain. Chem. Eng. 6, 12562–12579.

Petersen, E.J., Henry, T.B., Zhao, J., MacCuspie, R.I., Kirschling, T.L., Dobrovolskaia, M.A., et al., 2014. Identification and avoidance of potential artifacts and misinterpretations in nanomaterial ecotoxicity measurements. Environ. Sci. Technol. 48, 4226–4246.

Reddy, P.V.L., Hernandez-Viezcas, J.A., Peralta-Videa, J.R., Gardea-Torresdey, J.L., 2016. Lessons learned: are engineered nanomaterials toxic to terrestrial plants? Sci. Total Environ. 568, 470–479.

Schudoma, C., Steinfath, M., Sprenger, H., van Dongen, J.T., Hincha, D., Zuther, E., et al., 2012. Conducting molecular biomarker discovery studies in plants. In: Normanly, J. (Ed.), High-throughput phenotyping in plants: methods and protocols, 918. pp. 127–150.

Schwab, F., Zhai, G., Kern, M., Turner, A., Schnoor, J.L., Wiesner, M., 2016. Barriers, pathways and processes for uptake, translocation and accumulation of nanomaterials in plants—critical review. Nanotoxicology 10 (3), 257–278.

Servin, A.D., White, J.C., 2016. Nanotechnology in agriculture: next steps for understanding engineered nanoparticle exposure and risk. NanoImpact. 1, 9–12.

Servin, A.D., Pagano, L., Castillo-Michel, H., De la Torre-Roche, R., Hawthorne, J., Hernandez-Viezcas, J.A., et al., 2017. Weathering in soil increases nanoparticle CuO bioaccumulation within a terrestrial food chain. Nanotoxicology 11, 98–111.

Servin, A.D., Castillo-Michel, H., Hernandez-Viezcas, J.A., De Nolf, W., De La Torre-Roche, R., Pagano, L., et al., 2018. Bioaccumulation of CeO_2 nanoparticles by earthworms in biochar-amended soil: a synchrotron microspectroscopy study. J. Agri. Food Chem. 66 (26), 6609–6618.

Shakiba, S., Astete, C.E., Paudel, S., Sabliov, C.M., Rodrigues, D.F., Louie, S.M., 2020. Emerging investigator series: polymeric nanocarriers for agricultural applications: synthesis, characterization, and environmental and biological interactions. Environ. Sci.: Nano 7, 37–67.

Sillen, W.M.A., Thijs, S., Abbamondi, G.R., Janssen, J., Weyens, N., White, J.C., et al., 2015. Effects of silver nanoparticles on soil microorganisms and maize biomass are linked in the rhizosphere. Soil Biol. Biochem. 91, 14–22.

Yin, Y., Tan, Z., Hu, L., Yu, S., Liu, J., Jiang, G., 2017. Isotope tracers to study the environmental fate and bioaccumulation of metal-containing engineered nanoparticles: techniques and applications. Chem. Rev. 117 (5), 4462–4487.

Zulfiqar, F., Navarro, M., Ashraf, M., Akram, N.A., Munné-Bosch, S., 2019. Nanofertilizer use for sustainable agriculture: advantages and limitations. Plant Sci. 289110270.

Index

Note: Page numbers followed by "*f*" and "*t*" refer to figures and tables, respectively.

D

E

F

G

Q

R

Kolibri 360 GmbH
Ferdinand-Jühlke-Straße 7
99095 Erfurt, Deutschland
produktsicherheit@kolibri360.de